Data Analysis with Microsoft Excel 5.0 for Windows™

Data Analysis with Microsoft® Excel 5.0 for Windows™

Kenneth N. Berk
Illinois State University

Patrick Carey
University of Wisconsin at Madison

Course
TECHNOLOGY

Course Technology, Inc. One Main Street, Cambridge, MA 02142

An International Thomson Publishing Company

I⟨T⟩P

Albany ▪ Bonn ▪ Boston ▪ Cincinnati ▪ London ▪ Madrid ▪ Melbourne ▪ Mexico City
New York ▪ Paris ▪ San Francisco ▪ Singapore ▪ Tokyo ▪ Toronto ▪ Washington

Data Analysis with Microsoft Excel 5.0 for Windows is published by Course Technology, Inc.

Managing Editor: Mac Mendelsohn
Product Manager: Joan Carey
Production Editor: Catherine D. Griffin
Text Designer: Jean Hammond
Cover Designer: John Gamache

©1995 Course Technology, Inc.
A Division of International Thomson Publishing, Inc.

For more information contact:
Course Technology, Inc.
One Main Street
Cambridge, MA 02142

International Thomson Publishing Europe
Berkshire House 168-173
High Holborn
London WCIV 7AA
England

International Thomson Publishing GmbH
Königswinterer Strasse 418
53227 Bonn
Germany

Thomson Nelson Australia
102 Dodds Street
South Melbourne, 3205
Victoria, Australia

International Thomson Publishing Asia
211 Henderson Road
#05-10 Henderson Building
Singapore 0315

Nelson Canada
1120 Birchmount Road
Scarborough, Ontario
Canada M1K 5G4

International Thomson Publishing Japan
Hirakawacho Kyowa Building, 3F
2-2-1 Hirakawacho
Chiyoda-ku, Tokyo 102
Japan

International Thomson Editores
Campos Eliseos 385, Piso 7
Col. Polanco
11560 Mexico D.F. Mexico

Trademarks

Course Technology and the open book logo are registered trademarks of Course Technology, Inc.

I(T)P The ITP logo is a trademark under license.

Excel 5.0 for Windows is a registered trademark of Microsoft Corporation and Windows is a trademark of Microsoft Corporation.

Some of the product names and company names used in this book have been used for identification purposes only and may be trademarks or registered trademarks of their respective manufacturers and sellers.

Disclaimer

Course Technology, Inc. reserves the right to revise this publication and make changes from time to time in its content without notice.

ISBN 1-56527-525-X
Printed in the United States of America
10 9 8 7 6 5 4 3 2

From the Publisher

At Course Technology, Inc., we believe that technology will transform the way that people teach and learn. We are very excited about bringing you, college professors and students, the most practical and affordable technology-related products available.

The Course Technology Development Process

Our development process is unparalleled in the higher education publishing industry. Every product we create goes through an exacting process of design, development, review, and testing.

Reviewers give us direction and insight that shape our manuscripts and bring them up to the latest standards. Every manuscript is quality tested. Students whose backgrounds match the intended audience work through every keystroke, carefully checking for clarity, and point out errors in logic and sequence. Together with our own technical reviewers, these testers help us ensure that everything that carries our name is error-free and easy to use.

Course Technology Products

We show both *how* and *why* technology is critical to solving problems in college and in whatever field you choose to teach or pursue. Our time-tested, step-by-step instructions provide unparalleled clarity. Examples and applications are chosen and crafted to motivate students.

The Course Technology Team

This book will suit your needs because it was delivered quickly, efficiently, and affordably. In every aspect of our business, we rely on a commitment to quality and the use of technology. Every employee contributes to this process. The names of all of our employees are listed below:

Tim Ashe, David Backer, Stephen M. Bayle, Josh Bernoff, Michelle Brown, AnnMarie Buconjic, Jody Buttafoco, Kerry Cannell, Jim Chrysikos, Barbara Clemens, Susan Collins, John M. Connolly, Kim Crowley, Myrna D'Addario, Lisa D'Alessandro, Jodi Davis, Howard S. Diamond, Kathryn Dinovo, Joseph B. Dougherty, MaryJane Dwyer, Chris Elkhill, Don Fabricant, Jeff Goding, Laurie Gomes, Eileen Gorham, Andrea Greitzer, Catherine Griffin, Tim Hale, Jamie Harper, Roslyn Hooley, John Hope, Matt Kenslea, Susannah Lean, Laurie Lindgren, Kim Mai, Margaret Makowski, Elizabeth Martinez, Debbie Masi, Don Maynard, Dan Mayo, Kathleen McCann, Jay McNamara, Mac Mendelsohn, Kim Munsell, Amy Oliver, Michael Ormsby, Kristine Otto, Debbie Parlee, Kristin Patrick, Charlie Patsios, Darren Perl, Kevin Phaneuf, George J. Pilla, Nicole Jones Pinard, Cathy Prindle, Nancy Ray, Marjorie Schlaikjer, Christine Spillett, Michelle Tucker, David Upton, Mark Valentine, Karen Wadsworth, Ann Marie Walker, Renee Walkup, Tracy Wells, Donna Whiting, Janet Wilson, Lisa Yameen.

Preface

Purpose

Spreadsheets have become one of the most popular forms of computer software, second only to word processors. Spreadsheet software allows the user to combine data, mathematical formulas, text, and graphics together in a single report or workbook. For this reason, spreadsheets have been indispensable tools for business, although they have also become popular in scientific research. Excel in particular has won a great deal of acclaim for its ease of use and power.

The *Data Analysis with Microsoft Excel 5.0 for Windows* text and accompanying software harness the power and accessibility of Excel and transform it into a tool for teaching basic statistical analysis. This text can serve as the core text for an introductory statistics course or as a supplemental text. The students learn statistics in the context of analyzing data. For example, nonparametric tests are considered as they naturally come up, as alternatives to corresponding parametric tests.

We feel that it is important for students to work with real data, analyzing real world problems so that they understand the subtleties and complexities of analysis that make statistics such an integral part of understanding our world. The data set topics range from business examples to physiological studies on NASA astronauts. Because students work with real data, they can appreciate that in statistics no answers are completely final and that intuition and creativity are as much a part of data analysis as is plugging numbers into a software package.

The student who uses this book need not have any experience with Excel, although previous experience would be helpful. The first two chapters of the book cover basic concepts of mouse and Windows operation, data entry, formulas and functions, charts, and editing and saving workbooks. The interested student who wants to learn more about Excel's basic spreadsheet features can work through any of Course Technology's three texts on Excel: *Excel Illustrated*, *An Introduction to Microsoft Excel 5.0 for Windows*, or *Comprehensive Microsoft Excel 5.0 for Windows*. After Chapter 2, the emphasis of the book is on teaching statistics with Excel as the instrument.

Using Excel in a Statistics Course

As spreadsheets have expanded in power and ease-of-use, there has been increased interest in using them in the classroom. There are many advantages to using Excel in an introductory statistics course. An important advantage is that students, particularly business students, are more likely to be familiar with spreadsheets and are more comfortable working with data entered into a spreadsheet. An additional advantage is the ability of spreadsheets to handle data as well as the formulas, text, and graphics that are used to create reports, allowing students to complete homework assignments using a single software package. Finally, since spreadsheet software is very common at colleges and universities, a statistics instructor can teach a course without requiring students to purchase an additional software package.

Having identified the strengths of Excel for teaching basic statistics, it would be unfair not to include a few warnings. Spreadsheets are not statistics packages and there are limits to what they can do in replacing a full-featured statistics package. Using Excel for anything other than an introductory statistics course would probably not be appropriate due to its limitations. For example, Excel can easily perform balanced two-way analysis of variance, but not unbalanced two-way analysis of variance. Spreadsheets are also limited in handling data with missing values. While we recommend Excel for a basic statistics course, we feel it is not appropriate for more advanced analysis.

Excel's Statistical Capabilities

Basic Excel comes with 81 statistical functions and 59 mathematical functions. There are also functions devoted to business and engineering problems. The statistical functions that basic Excel provides include descriptive statistics such as means, standard deviations and rank statistics. There are also cumulative distribution and probability density functions for a variety of distributions, both continuous and discrete.

The Analysis ToolPak

The Analysis ToolPak is an *add-in* that is bundled with Excel. Add-ins expand Excel's capabilities. Chapter 3 shows you how to check whether you have already loaded the Analysis ToolPak Add-In and shows you how to activate it. If you have not loaded the Analysis ToolPak, you will have to install it from your original Excel installation disks, following the instructions in your Excel *User's Guide*.

The Analysis ToolPak adds the following capabilities to Excel:

- Analysis of variance, including one-way, two-way without replication, and two-way balanced with replication
- Correlation and covariance matrices
- Tables of descriptive statistics
- One-parameter exponential smoothing
- Histograms with user-defined bin values
- Moving averages
- Random number generation for a variety of distributions
- Rank and percentile scores
- Multiple linear regression
- Random sampling
- *t*-tests, including paired and two sample, assuming equal and unequal variances
- Z-tests

In this book we make extensive use of the Analysis ToolPak for multiple linear regression problems and analysis of variance.

The CTI Statistical Add-Ins

Since the Analysis ToolPak does not do everything that an introductory statistics course requires, this textbook comes with an additional add-in called the *CTI Statistical Add-Ins* that fills in some of the gaps left by basic Excel and the Analysis ToolPak.

Additional commands provided by the CTI Statistical Add-Ins are:

- Stack columns
- Unstack columns
- Make indicator variables
- Make two way table, to format the data properly so you can use the Analysis ToolPak's ANOVA commands
- 1-Sample Wilcoxon tests
- Means matrix for one way analysis of variance
- Correlation matrices, for Pearson and Spearman correlation statistics (including *p*-values)
- Boxplots
- Scatterplot matrices

- Histograms, including a superimposed normal curve
- Multiple histograms
- Normal probability plot
- ACF plots
- Seasonal adjustment of time series data
- Exponential smoothing using Holt-Winters models
- XBAR charts
- Range charts
- C-charts
- P-charts
- Inserting ID labels on scatterplots

A full description of these commands is included in the Reference section and through on-line help.

The philosophy of this book is to use basic Excel in appropriate situations that do not require too many operations on the part of the user. If basic Excel cannot handle a task, we look to the Analysis ToolPak. If neither basic Excel nor the Analysis ToolPak can perform the task, we have provided a new command with the CTI Statistical Add-Ins. In this way we hope to minimize the amount of time a student will spend typing commands and functions, and increase the amount of time spent on interpreting data.

Instructional Templates

Excel 5.0 includes many new features. One of the more useful additions to Excel is the ability to embed Windows tools such as scroll bars or list boxes within spreadsheets. We've taken advantage of this feature to create several *instructional templates*. Instructional templates are interactive worksheets that allow students to explore statistical concepts. In this way the student can actually work with the basic concepts of statistics rather than reading about them. Instructional templates included in this book teach the concepts of:

- Boxplots
- The Central Limit Theorem
- Confidence intervals
- Probability density functions
- Hypothesis tests
- Exponential smoothing

Chapter Summaries

Each chapter introduces problems involving real-world data as a basis of statistical exploration. Students are guided by step-by-step instructions that not only give them hands-on experience using Excel, but also feature regular signposts that help them remember where they are in the process of solving the problem. Throughout each chapter are Summary Boxes that summarize the key steps necessary to perform a statistical test using Excel. Finally, each chapter includes a set of exercises involving real data for students to practice what they learned in the chapter. The exercises give the student an opportunity to play detective in analyzing the data, to form a conclusion, and to defend the conclusion in written reports.

Chapter 1: Getting Started

Chapter 1 is intended for the new user. The student is shown how to use a mouse to navigate around Windows. Basic spreadsheet concepts are covered, such as opening and closing a workbook, entering and formatting data, using formulas and functions, and printing and saving workbooks.

Chapter 2: Entering and Manipulating Data

Chapter 2 introduces the student to different data types. The concept of data list is covered, as well as creating and using range reference names. The student will use the built-in features of Excel to sort a data list and to perform simple and complex queries. The chapter concludes with the importing of data from a text file and entering some basic formulas.

Chapter 3: Single Variable Graphs and Statistics

Chapter 3 explores basic statistics for a single variable. The mean, standard deviation, and rank statistics are introduced for the first time. The distribution of a sample is presented with frequency tables and histograms. The boxplot is introduced and explained with an instructional template.

Chapter 4: Scatterplots

Chapter 4 discusses the charting capabilities of Excel. The primary focus is on creating a scatterplot and modifying the properties of the plot. The CTI Statistical Add-Ins are introduced and used to label points on a scatterplot. The chapter concludes with a discussion of the scatterplot matrix and its application to a job discrimination data set.

Chapter 5: Probability Distributions

Chapter 5 discusses basic statistical theory, introducing the concept of samples, populations, and random variables, among other things. Several instructional templates are used in this chapter to illustrate principles of probability distributions. The second half of the chapter concentrates on the normal distribution, emphasizing the sampling distribution of the mean. The chapter concludes with a discussion of the Central Limit Theorem, explained through an instructional template.

Chapter 6: Statistical Inference

Chapter 6 continues the theoretical discussion of the previous chapter with a look at confidence intervals and hypothesis testing. Instructional templates are provided to facilitate both discussions. The one-sample and two-sample t-statistics are introduced. The chapter concludes with an analysis of paired and unpaired data using real-world examples. Wilcoxon's one-sample nonparametric test is introduced to compare with the one-sample t-test.

Chapter 7: Tables

Chapter 7 uses Excel to analyze contingency tables. Taking advantage of Excel's new Pivot Table feature, the student is shown how to create such tables from real data involving the use of computers in classrooms. Statistical tests on the tables are performed using the CTI Statistical Add-Ins. The chapter concludes by dealing with ordinal data and sparseness in tables.

Chapter 8: Correlation and Simple Regression

Chapter 8 introduces simple linear regression and correlation. Simple linear regression is performed using the Analysis ToolPak. The interpretation of regression statistics, including the coefficient of determination, R^2, is discussed. Creating correlation matrices for multiple variables is presented, along with a comparison of the Pearson correlation coefficient and its nonparametric equivalent, the Spearman rank correlation. The student then graphs these relationships with scatterplot matrices.

Chapter 9: Multiple Regression

Chapter 9 extends the discussion of simple linear regression to multiple regression involving several variables. The interpretation of the analysis of variance table is covered. Emphasis is placed upon regression diagnostics, especially normal probability plots and residual plots. The student gets a chance to see these principles put into action by analyzing sex discrimination data from a junior college.

Chapter 10: Analysis of Variance

In this chapter the student explores analysis of variance. The chapter opens by showing the student how to display the distribution of several columns of data using the Multiple Histograms command from the CTI Statistical Add-Ins. The theory of one-way analysis of variance is discussed and applied to a real-world example involving hotel prices in four U.S. cities. The presentation of one-way analysis of variance concludes with multiple comparisons using the Means Matrix command from the CTI Statistical Add-Ins. The student is introduced to the problems associated with multiple comparisons and how these can be partly resolved by the Bonferroni correction factor. The second half of the chapter looks at two-way analysis of variance for a study involving soft drinks. The student is shown how to interpret analysis of variance tables and how to check for interactions by creating an interaction plot.

Chapter 11: Time Series

Chapter 11 explores time series data by studying the Dow Jones Average in the 1980's. The concepts of lags and moving averages are introduced. The discussion of lags leads into a presentation of the autocorrelation function, which is demonstrated using the ACF Plot command. The simple one-parameter exponential smoothing model is introduced with an instructional template and is applied to the Dow Jones Average in the 1980's. The two-parameter exponential smoothing model is then explored with a discussion of the underlying theory. The second half of the chapter looks at seasonal time series data with a real-world data set of beer production in the 1980's. The two-parameter exponential smoothing model is discussed and fit to the data using an add-in command included with CTI Statistical Add-Ins.

Chapter 12: Quality Control

Chapter 12 uses Excel to analyze quality control data. The theory of quality control is followed by applying four types of quality control charts: Xbar charts, Range charts, C-charts and P-charts. The chapter concludes with the Pareto chart applied to baby powder production data.

Using Data Analysis with Excel

You will need the following hardware and software to use *Data Analysis with Microsoft Excel 5.0 for Windows*:

- An IBM or compatible microcomputer with the Intel 80386 processor or higher

- Excel 5.0, or later (to install Excel, you need a hard disk drive with 8MB of available space and 4 MB of RAM, though 8 MB is highly recommended)

- Windows version 3.1, or later

The *Data Analysis with Microsoft Excel 5.0 for Windows* instructor package includes:

- The text, which includes 12 chapters, documentation on the data sets, a reference section for Excel's statistical functions, Analysis ToolPak commands, and CTI Statistical Add-Ins commands, and a bibliography.

- Student files disk that contains data sets from nearly 50 different real-life situations, 11 instructional templates, and the CTI Statistical Add-Ins file. The instructor can install these files on a network or on stand-alone workstations to make them available to students. Chapter 1 includes instructions to students for copying the files to their own disks for use on their own computers.

- Instructor's Manual, with instructions for making the Student files accessible to your students and answers and solutions to all the exercises.

The *Data Analysis with Microsoft Excel 5.0 for Windows* student package contains only the text, but Chapter 1 includes instructions to the student for accessing the Student files.

Acknowledgments

We could not have completed this book without the support and enthusiasm of Mac Mendelsohn, Managing Editor at Course Technology. Other people at CTI who saw this book to completion include: our always-cheerful production editor, Catherine Griffin; our careful copyeditor, Joan Wilcox; proofreader, Kathy Finnegan; compositor, Kim Munsell; word processor, Cynthia Anderson; and indexer, Sherry Dietrich. Special thanks go to our reviewers who gave us valuable insights into improving the book. Any remaining mistakes or omissions are ours. The reviewers for this book are:

David Auer, Western Washington University; Sharon Hunter Donnelly, University of Tenessee at Knoxville; Richard D. Spinnetto, University of Colorado at Boulder; Wayne L. Winston, Indiana University; and Jack Yurkiewicz, Pace University.

We want to thank Bernard Gillett for expertly validating the final version of the manuscript and Robert Gillett for his careful work in testing the manuscript and working the problems. Thanks to Laura Berk, Peter Berk, Robert Beyer, David Booth, Orlyn Edge, Stephen Friedberg, Maria Gillett, Richard Goldstein, Glenn Hart, Lotus Hershberger, Les Montgomery, Joyce Nervades, Diane Warfield, and Kemp Wills for their assistance with the data sets in this book. We especially want to thank Dr. Jeff Steagall, who wrote some of the material for Chapter 12, Quality Control. If we have missed anyone, please forgive the omission.

I would like to thank my wife and sons for their support in a time-consuming effort.

Kenneth N. Berk

I want to thank my parents, who gave me a deep sense of values and personal responsibility. This book would not have become a reality without the generosity and support of Paul Reichel, Dr. Michael Kosorok, Dr. David DeMets, and Dr. George Bryan of the University of Wisconsin, who enabled me to restructure my work hours to complete this project on schedule. Thanks to John Paul, Thomas and Peter—my three sons—who kept this whole project in perspective and forced me to go outside once in a while and play. Of course, praise and thanks go to my developmental editor, Joan Carey, who in addition to being very insightful, creative, cheerful, and lovely, is a very good wife.

Patrick M. Carey

Brief Contents

Table of Contents

CHAPTER 9

MULTIPLE REGRESSION 191

CHAPTER 10

ANALYSIS OF VARIANCE 217

CHAPTER 11

TIME SERIES 244

CHAPTER 12

QUALITY CONTROL 284

EXCEL REFERENCE

GETTING STARTED

O BJECTIVES

In this chapter you will learn to:

- Navigate Windows
- Copy the Student files
- Start Excel and recognize worksheet elements
- Open an Excel workbook
- Enter data into a worksheet and apply formatting
- Scroll a worksheet and workbook
- Use formulas and functions
- Print a worksheet
- Save your work
- Exit Excel

Excel is a powerful program designed to help you organize and analyze data. Businesses worldwide use Excel to handle their computational, charting, and data management needs. One of the premier spreadsheet programs, Excel also provides extensive statistical and graphical analysis tools. The convenience of using the same package to organize data and to perform statistical analysis cannot be overrated, and you can meet many of your statistical needs with Excel's statistical capabilities. This book emphasizes those statistical capabilities, taking advantage of the almost unlimited ways you can extend them with Excel's versatile programming language and customized templates.

This book uses a hands-on approach to learning to use Excel's statistical functionality. You have the opportunity to explore data from different angles, sometimes through unexpected twists. You might find there is more going on with the data than you imagined, and some detective work will help you discover what is really happening. This book uses examples with real data to give you a feel for this discovery process.

This first chapter offers a quick overview of basic Windows techniques, and then introduces the Excel workspace.

How to Use This Book

Each chapter of this book introduces you to one or more case problems. By following the step-by-step instructions, you learn ways to analyze a data set using Excel.

Actions that you should take are listed in numbered steps and preceded by a phrase that starts "To do. . . ."

Numbered steps contain the following types of instructions:

- Keys that you press are in brackets and **boldface:** Press [**Enter**].
- Letters, numbers, and special characters that you type are in boldface: Type **Salary**.
- Items that you click with the mouse are in boldface: Click **OK**.
- Key combinations that you press simultaneously are in brackets, boldface, and joined with a plus sign (+): Press [**Alt**]+[**O**].
- Menu commands that you select are in boldface; the first word is the menu name and the second word is the name of the command on that menu: Click **File > New**. This means you click File on the main menu bar, then click New on the File menu.
- Some buttons or icons that you click appear pictorially: Click for Chart Wizard.

This book also includes summary boxes that summarize and generalize the steps you take to perform a certain procedure. These will help you after you have gone through the chapters and are ready to apply what you have learned to new data and new situations.

Using Windows

This book does not require previous Excel experience, because the steps provide complete instructions, but familiarity with basic Microsoft Windows features like dialog boxes, menus, on-line Help, and windows would make your startup time easier. This section provides a quick overview of Windows features you'll use in this book. Course Technology, Inc. publishes two excellent texts for learning Windows: *A Guide to Microsoft Windows* and *An Introduction to Microsoft Windows*, both available in current releases.

To start Microsoft Windows:

1 Turn on your computer.

2 Your computer might load Windows automatically, in which case you see the Program Manager, shown in Figure 1-1. You might also see a small icon at the bottom of the screen, as shown in Figure 1-2. If so, follow step 2 under Figure 1-2 to open the Program Manager.

If you see the DOS prompt (C:\ >), type **win** at the prompt and press [**Enter**].

The Program Manager window opens (yours might have different icons).

Program Manager

The Program Manager, shown in Figure 1-1, is the Microsoft Windows home base, from where you open program applications, manage files, work with your computer's operating system, and accomplish many other system tasks. You open Excel from Program Manager. Your Program Manager screen might look different from Figure 1-1. There might be other windows open, with fewer or more elements, depending on your computer system and how the previous user exited Windows.

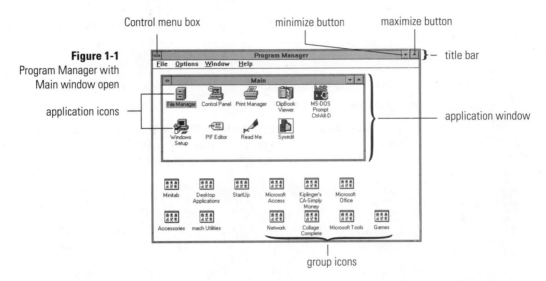

Figure 1-1
Program Manager with Main window open

The Program Manager is a **window**, a rectangular area containing an application (such as Excel) or a document (for example, a spreadsheet) in which you do your work. Table 1-1 provides a quick identification of the Windows elements shown in Figure 1-1.

Table 1-1
Windows Elements

Windows element	Icon	Purpose
title bar		identifies the window
menu bar		contains menus that list commands
Control menu box		opens a menu that lets you move, resize, and close the window
minimize button		reduces the window to an icon at the bottom of the screen
maximize button		expands the window to fill the whole screen; turns into , the restore button
icon		small graphical picture representing applications, windows, or documents
group icon		represents a group of related applications (this is the Microsoft Office group icon)
application icon		represents an application (this is the Excel application icon)
group window		contains related application icons

You work with these elements using the mouse. (You can also manipulate them using the keyboard, but this book emphasizes using the mouse.)

Mouse Techniques

There are four basic mouse techniques that you will use to communicate with the application you are using:

clicking	Move your mouse so that the tip of the pointer arrow ▷ is touching the element you want to work with. Then press and release the left mouse button.
right-clicking	Same as clicking, except you press and release the right mouse button instead of the left.
double-clicking	Point at an element as if you were going to click it, then press and release the left mouse button twice in rapid succession.
dragging	Point at an element you want to move from one place to another, then press and hold down the left mouse button while you move the mouse so that the element drags across the screen. When you have positioned the element where you want it, release the mouse button.

To practice using mouse techniques:

1   Click the **Program Manager minimize button** ▾.

The Program Manager window is reduced to a small icon at the bottom of the screen, as shown in Figure 1-2. Minimizing a window is a convenient way to get it out of the way when you are working with other applications.

Figure 1-2
Program Manager reduced
to an icon

2   Double-click the **Program Manager icon**.

The Program Manager window opens again.

3   Move the pointer arrow ▷ to the Program Manager title bar, and drag the title bar to the upper left, but not off the screen (that is, press and hold down the left mouse button while you move the mouse up and to the left).

The Program Manager moves to the upper-left corner of your screen.

4   Click the **maximize button** ▲.

Now Program Manager fills the entire screen, and the maximize button ▲ turns into a restore button ↕, as shown in Figure 1-3.

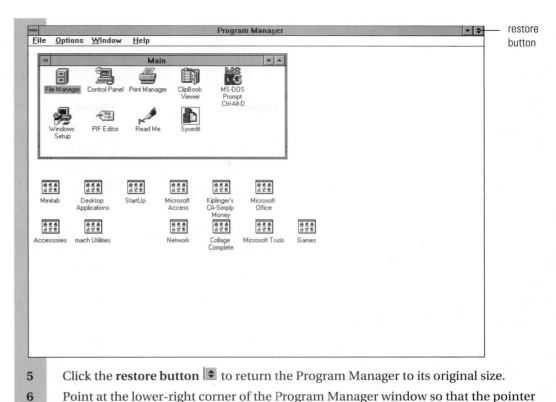

Figure 1-3
Maximized Program Manager

restore button

5 Click the **restore button** 🔽 to return the Program Manager to its original size.

6 Point at the lower-right corner of the Program Manager window so that the pointer arrow turns into a double-headed arrow 🔯, drag the window corner down and to the right, then release the mouse button.

This enlarges the Program Manager window. When you are working with more than one window at a time, you will often want to resize windows and arrange them on your screen.

Getting Help

The basic mouse techniques you just practiced are fundamental to communicating with Windows applications. If you feel you need more practice, you might want to spend some time with the Windows tutorial and with on-line Help. The Windows tutorial walks you through the basics of using Windows, whereas on-line Help gives you access to *on-line* information that appears on your screen when you request it.

To access on-line Help:

1 Click **Help** on the Program Manager menu bar to open the Help menu. See Figure 1-4.

Figure 1-4
Program Manager help menu

menu bar

Windows Tutorial menu command

Help menu

Command	Description
Contents	opens a Help window with a table of contents listing Program Manager topics you might want help with
Search for Help on	opens a Search window that displays an index of all available Help topics
How to Use Help	opens a Help window that shows you how to use the Microsoft Help window
Windows Tutorial	starts an on-line interactive tutorial that walks you through the basics of using Microsoft Windows
About Program Manager	displays information about your version of ProgramManager

The on-line Help features are extremely useful. It's well worth the time to become familiar with them. Practically every Windows package uses the same on-line Help structure, including Excel, and in Help you will find answers to many of your questions.

Accessing the Student Files

To complete the chapters and exercises in this book, you must have access to on-line Student files. Your instructor has placed the Student files that come with the instructor copy of this book on your school's computer system or made the files available to you in some other way.

This book assumes that the Student files are located on a hard drive in the directory C:\EXCEL\STUDENT. Your instructor or technical support person will provide the correct directory location for your school's computer system.

Note: If you don't understand the concepts of directory, file, and path, see your instructor or technical support person for assistance.

Using a School Computer

As you proceed through these chapters, you will want to save your work. If you are using a school computer, you will open the Student files from the directory your instructor tells you to use. However, you might need to obtain a blank, formatted, high-resolution disk (referred to as your Student Disk in this book) on which to save your work, because most schools do not allow students to save their work to school computer hard drives. The instructions later in this chapter tell you how to save your work to your Student Disk.

Using Your Own Computer

If you have your own computer and your own copy of Excel at home and you want to use the Student files on your own computer, you will need a copy of XLFILES.EXE, an executable compressed file that contains all the Student files. Your instructor will either give you a disk containing this file or will give you the location of this file on a school computer, in which case you can copy it to a disk of your own and install the Student files from that disk.

First, you will need a blank, formatted, high-density 3.5-inch or 5.25-inch disk, and you will need to find out where your instructor placed XLFILES.EXE. This book assumes XLFILES.EXE is located in the directory C:\EXCEL\STUDENT, but if your instructor gives you a different path, be sure to substitute that instead in step 6.

To copy XLFILES.EXE to your blank disk:

1   Place your blank, formatted disk in drive A or B of your school computer.

2 Start Windows if it isn't already running, and double-click the **Program Manager icon** to open Program Manager if it isn't already open.

3 Double-click the **Main group icon** in Program Manager to open the Main window if it isn't already open. See Figure 1-5.

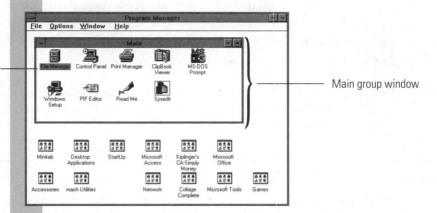

Main group window

4 Double-click the **File Manager icon** in the Main group window. File Manager opens.

5 Click **File**, then click **Copy**. The Copy dialog box opens.

6 Double-click the **From box** and type **C:\EXCEL\STUDENT\XLFILES.EXE** in the From box (substitute the path your instructor gave you if it is different).

7 Press [**Tab**] and type **A:** or **B:** in the To box.

8 Click **OK**. The copying procedure begins.

You now have your own installation disk, which you can use to install the Student files on your own computer. You must first create a directory on your hard drive that will contain the the Student files. The instructions in the *Data Analysis with Microsoft Excel 5.0 for Windows* text assume that the Student files are located in the STUDENT subdirectory of the EXCEL directory, with the path C:\EXCEL\STUDENT, but you can install them in any appropriate location.

To create the STUDENT subdirectory that will contain the Student files:

1 Open File Manager on your computer following the previous steps 1 through 4.

2 Click **File**, then click **Create Directory**.

3 Type **C:\EXCEL\STUDENT** in the Name box (or enter a different path if C:\EXCEL\STUDENT is inappropriate), then click **OK**.

This creates a STUDENT subdirectory in the EXCEL directory into which you can place the Student files. Next, you must copy XLFILES.EXE from your installation disk into the directory you just created. XLFILES.EXE contains over 50 files in a compressed format, and when you run this executable file, it "inflates" the Student files into the directory in which it is located. You will need 3.3 MB of hard disk space.

To copy XLFILES.EXE into the STUDENT subdirectory you just created and then inflate the Student files:

1 Be sure File Manager is open and the installation disk is in the appropriate disk drive.

2 Click **File** to open the File menu, then click **Copy**. The Copy dialog box opens.

3 Highlight the contents of the From box (if necessary), then type **A:** or **B:** in the From box, depending on which drive contains the installation disk.

4 Press [**Tab**], then type **C:\EXCEL\STUDENT** (or the name of the directory into which you want to place the files).

5 Click **OK**. Excel copies XLFILES.EXE into the directory you specified.

6 Click **File** to open the File menu, then click **Run**.

7 Type **C:\EXCEL\STUDENT\XLFILES.EXE C:\EXCEL\STUDENT** in the Command Line box. The first part of this command line tells File Manager to run XLFILES.EXE, and the second part tells File Manager to place the decompressed files into the STUDENT subdirectory of EXCEL.

8 Click **OK**. Wait as the EXE file inflates XLFILES.EXE into its component files.

9 Click **File**, then click **Exit** to exit File Manager.

Do not move CTI.HLP from the STUDENT subdirectory, or Excel won't be able to find the Help file. The Student files are now available in the directory you created. You can save your work to this directory or to a different one.

Launching Excel

This book assumes that you have Excel loaded on the computer you are using. If you are using a school computer on a network, you might need to make some adjustments when you start Excel or work with files. You will need Excel version 5.0 for Windows in order to follow the steps in this book.

Excel is a computerized spreadsheet program. A **spreadsheet** is a business tool that helps you organize and evaluate data. Before the era of computers, a spreadsheet was simply a piece of paper with a grid of rows and columns to facilitate entering and displaying information, as shown in Figure 1-6.

Figure 1-6
Sales spreadsheet

you added these numbers ———

to get this number ———

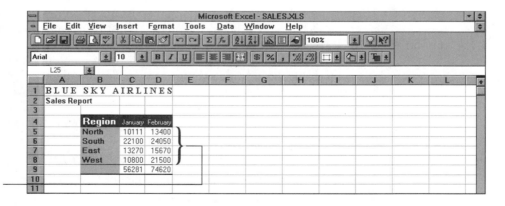

Computer spreadsheet programs use the old hand-drawn spreadsheets as their visual model, as you can see from the Excel Worksheet window, shown in Figure 1-7.

Figure 1-7
Sales spreadsheet in Excel

Excel adds these numbers
for you

Computerized spreadsheets, however, are highly sophisticated; they allow you to perform complex operations on the data contained in the rows and columns. This chapter shows you the basics of getting around in Excel, but the focus will soon shift to statistics. To learn more about Excel's spreadsheet capabilities, consult the Excel manual.

To start Excel:

1   Launch Windows so that Program Manager is open and all group windows are minimized.

2   Double-click the **Microsoft Office icon** or the **Excel group icon**. See Figure 1-8.

Figure 1-8
Program Manager

group windows are
minimized to icons

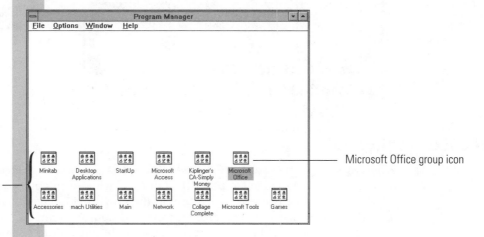

Microsoft Office group icon

Note: If you can't find the Microsoft Office icon or the Excel group icon, your computer's copy of Excel might be located in a different group window. Consult your instructor or technical support person for assistance.

The Microsoft Office or Excel group window opens. See Figure 1-9. Microsoft offers Excel alone or, more commonly in today's market, as part of a suite of applications packaged together as Microsoft Office, shown in Figure 1-9.

Figure 1-9
Microsoft Office group
window

Microsoft Excel application
icon

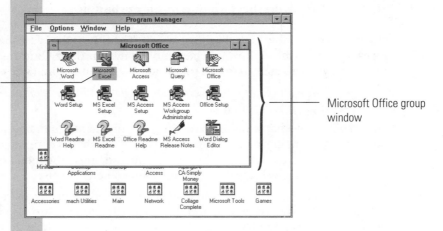

Microsoft Office group
window

3   Double-click the **Microsoft Excel application icon**.

Excel starts up and displays the opening screen. See Figure 1-10.

Excel window Control menu box

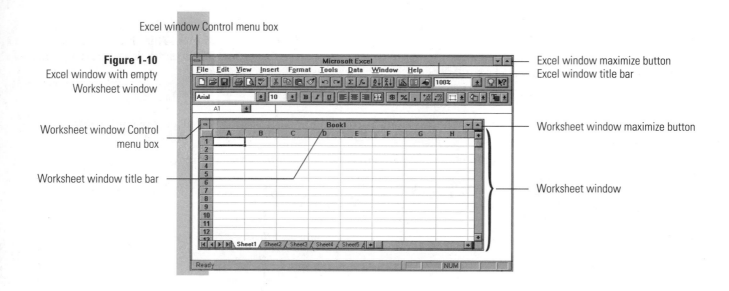

Figure 1-10
Excel window with empty
Worksheet window

Excel window maximize button
Excel window title bar

Worksheet window Control
menu box

Worksheet window maximize button

Worksheet window title bar

Worksheet window

Using the Excel Window

When you first start Excel, the Excel window opens and, within it, an empty Worksheet window appears, as shown in Figure 1-10. Your Excel window might look different. Each of these windows shares the same features as the Program Manager window: the title bar, menu bar, minimize and maximize buttons, and Control menu box all function the same way (see the Program Manager section earlier in this chapter to review these features).

Note: If your window doesn't look like the one shown in Figure 1-10, your version of Excel might have started with both windows already maximized or sized differently.

To maximize the Excel window and the Worksheet window if they aren't already maximized:

1   Click the **Excel window maximize button**.

2   Click the **Worksheet window maximize button**.

Now the Excel window looks like Figure 1-11. The title of the Worksheet window, Book1, moves into the Excel window title bar, and the two maximize buttons turn into restore buttons,which you can click to return the windows to their original sizes.

Note: Depending on your monitor type, you might be able to see more or less of the window shown here and elsewhere in this book.

You can tell Excel what you want to do by using the menus or the buttons on the toolbars. The menu bar contains names of menus, each of which lists commands that accomplish certain tasks. To open a menu and choose a command, move the mouse pointer on the menu name, click the mouse button to open the menu, and click the command you want to use.

Excel offers 13 different toolbars that let you work more quickly and easily. When you first start Excel, the Standard toolbar and the Formatting toolbar are displayed directly below the menu bar, as shown in Figure 1-11. Each toolbar contains rows of buttons that are shortcuts for selecting menu commands. For example, if you wanted to open a file, you could do so in one of two ways: using the File menu Open command or using the Open button on the Standard toolbar.

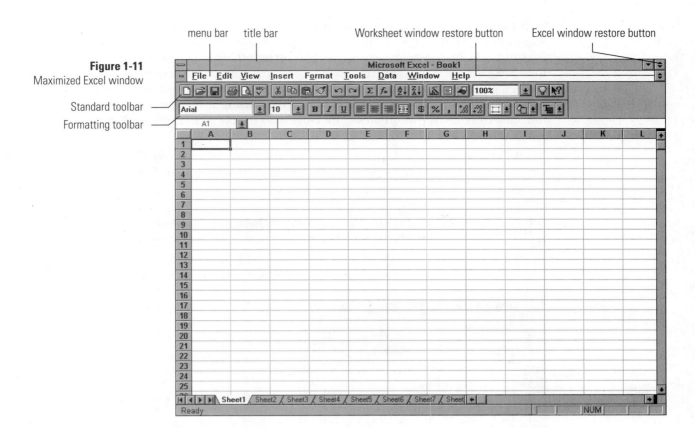

menu bar   title bar   Worksheet window restore button   Excel window restore button

Figure 1-11
Maximized Excel window

Standard toolbar
Formatting toolbar

To practice both methods of opening a file:

1   Click **File** on the menu bar to open the File menu.

2   Click **Open** (the second command).

A dialog box appears, letting you specify the file you want to open. For now, close this dialog box without opening a file. Now use the toolbar to display this dialog box.

3   Click **Cancel** in the Open dialog box.

4   Move the pointer over the second button on the Standard toolbar 📂 and let it rest there for a second.

The word "Open" appears, telling you the name of the button.

5   Click the **Open button** 📂.

The Open dialog box appears.

6   Click **Cancel** in the Open dialog box.

The toolbar buttons are quick substitutes for the menu commands or for other common options in the dialog boxes. Excel displays different toolbars depending on the task you are performing. In addition, you can customize a toolbar by selecting which buttons you want to display on the toolbar; for that reason, your toolbars might look different from the ones shown in this book.

Opening an Excel Workbook

Figure 1-11 shows "Book1" in the title bar. Excel documents are called workbooks and are made up of individual sheets. The worksheet is one type of sheet, but Excel has several sheet types, including chart sheets, which you'll learn about later.

To introduce you to the basics of opening and working with workbooks, take a look at an Excel workbook that might be used by the chief of interpretation at Kenai Fjords National Park in Alaska. Park employees keep records on monthly public-use data (the number of visitors in a given period). They can use these data, for example, to help them decide when to hire temporary rangers during the busy periods. The data are stored in an Excel workbook, PARK.XLS. PARK is the name of the file, and XLS is the file extension (a three-letter suffix to a filename with file type information) that identifies the file as an Excel workbook. Let's take a look at this workbook.

Note: The next set of instructions assumes that the Student files are on drive C in the STUDENT subdirectory of Excel. Substitute the correct path information if appropriate.

To open the PARK.XLS workbook:

1   Click the **Open button** 📂 or click **File > Open** (this is a shortcut way of saying "click Open from the File menu," as explained in the previous section).

The Open dialog box appears. See Figure 1-12. Your dialog box might display different directories and might not have a Network button.

Figure 1-12
Open dialog box

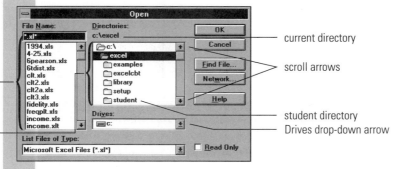

2   Click the **Drives drop-down arrow** to open the Drives list.

If your current directory already reads C:\EXCEL, you can skip steps 3-5 and proceed to step 6.

3   Click **C:** in the Drives list (or substitute the drive containing the Student files).

4   Double-click the **C:** folder in the directory tree.

5   Double-click the **EXCEL** folder in the directory tree (you might have to scroll to see it by clicking the up or down scroll arrows in the directory tree, shown in Figure 1-12, to move through the available directories; you'll learn more about scrolling later in this chapter).

6   Double-click the **STUDENT** folder in the directory tree (you might have to click the down scroll arrow to see it).

7   Click the **down scroll arrow** in the File Name list box until you see PARK.XLS, then double-click **PARK.XLS** to open the PARK workbook. See Figure 1-13.

Figure 1-13
PARK.XLS workbook

cell reference

active cell

row heading

sheet tabs

sheet tab for active worksheet

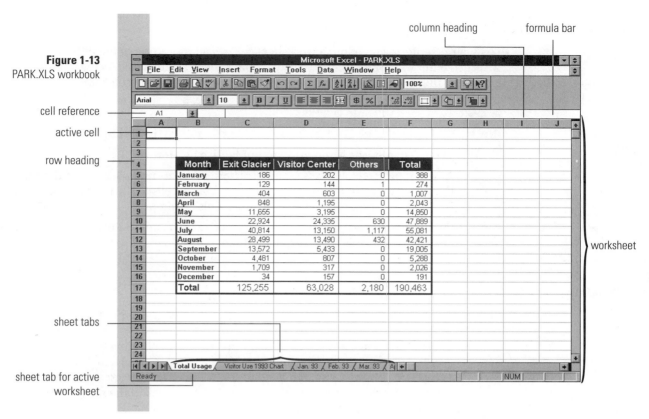

column heading

formula bar

worksheet

A single workbook can have as many as 255 sheets. The sheet names appear on tabs at the bottom of the workbook window. In the PARK.XLS workbook, the first sheet name is "Total Usage." It is in boldface because it is the active sheet. The second tab is "Visitor Use 1993 Chart," and the rest of the tabs are "Jan. 93," "Feb. 93," "Mar. 93," and so on. You'll see later how to move to different sheets.

Note: Your screen might not show as many sheet tabs as are shown in Figure 1-13 because you might have a different monitor type. You'll learn later how to scroll through the workbook to view all the sheet tabs.

The first sheet in the workbook contains a table that shows visitor counts by month. Park staff tallied the number of visitors at the Exit Glacier site (an information booth at the foot of an immense glacier) and at the Visitor Center in the nearby town of Seward. Visitors attending other sites are included in the Others column. A glance over the data shows that the peak visitor months are May through September.

Using the Worksheet Window

You will work most often with the worksheet, the grid of rows and columns. The rows are identified by their **row headings** (1, 2, 3, and so on), and the columns by their **column headings** (A, B, C, and so on). The intersection of a row and a column is called a **cell**. The **active cell** accepts any entry you type: it is identified by the double border. Each cell has its own **reference**, which is its address. The address is the letter of its column followed by the number of its row; the cell reference is displayed in the **cell reference box**. In Figure 1-13, the active cell is A1 (the intersection of column A and row 1), and that reference is displayed in the cell reference box on the left side of the formula bar. The formula bar displays cell entries.

If you move the mouse pointer around the screen, you'll notice that it changes from an arrow ↖ to a cross ✛ when you move it onto the worksheet. The mouse pointer changes shape depending on the task you are trying to accomplish. You used ↖ earlier in the "Mouse Techniques" section of this chapter to select and move items. In the Worksheet window, you will use ✛ to select cells and ranges. You will use other mouse pointer shapes throughout this book.

Entering Data into the Active Cell

To illustrate entering data into a worksheet cell, try adding a title to the public-use worksheet. To enter information into a cell, you make the cell active by clicking it; then type your entry, and then press [Enter] to accept the entry and move to the cell below.

To enter a title into cell B2:

1   Click cell **B2**. Now B2 is the active cell, as identified by the double border.

2   Type **Kenai Fjords National Park Public Use by Month, 1993**.

As you type, Excel displays your entry both in the formula bar and in the cell. Notice that the display of the text spills over into empty cells to the right of cell B2, but cell B2 actually contains the entry; the other cells are empty.

3   Press [**Enter**] to accept your entry. The active cell is now B3 (note that the cell reference in the formula bar changed to B3).

Formatting Cell Entries

This book is not intended to teach comprehensive spreadsheet formatting, but when you present data or statistical analysis to others, formatting can help your worksheet look more professional. For instance, in the public-use worksheet, try centering the title you just entered over the five columns of the table, and give it a larger, bold font. To center the title over a space of five columns, you first need to select the **range**, the group of the cells that you want to center over. To highlight a range, click the first cell in the range, drag the mouse over the extent of the range, and then release the mouse button.

To format the title:

1   Click cell **B2** and hold down the mouse button, and while the button is pressed, drag to cell F2.

This selects, or highlights, the range B2 to F2 by enclosing the entire range in a double border and reversing the colors (the first cell in a selected range is not highlighted), as shown in Figure 1-14. A range is usually identified by the first and last cell, separated by a colon (:). The range you selected is B2:F2.

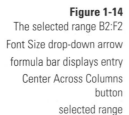

Figure 1-14
The selected range B2:F2
Font Size drop-down arrow
formula bar displays entry
Center Across Columns
button
selected range

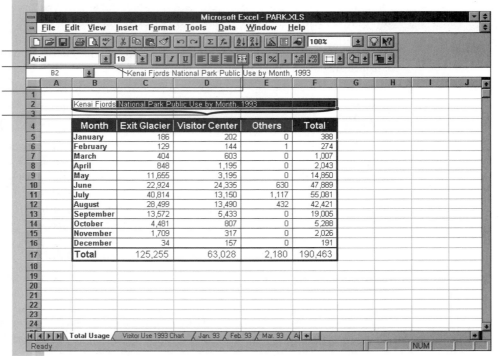

2 Click the **Center Across Columns button** 🔳.

Excel centers the title across the range B2:F2. Now change the font size and make the title bold:

3 Click the **Bold button** **B**.

4 Click the **Font Size drop-down arrow** to open the Font Size list box.

5 Click **12** to increase the point size to 12 points (a **point** is a size measurement of a typeface: 1/72").

6 Click cell **A1** to deselect the range and see the results of your formatting. See Figure 1-15.

Figure 1-15
Formatted title

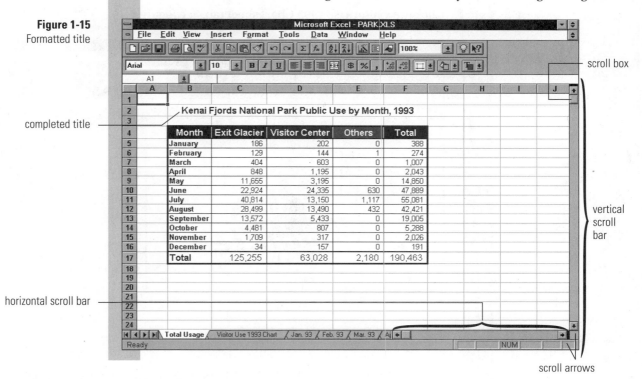

completed title

scroll box

vertical scroll bar

horizontal scroll bar

scroll arrows

Scrolling a Worksheet

When a window contains more information than can fit within it, the window scrolls; that is, earlier information rolls off the top of the screen to make room for more.

To scroll a window one line or column at a time, click the arrows at the extreme ends of the worksheet scroll bars, shown in Figure 1-15, located along the bottom and right edges of the window. To scroll one page (window) at a time, click the scroll bar anywhere between the scroll box (the box inside the scroll bar) or a scroll arrow. The window scrolls in the direction of the scroll arrow. To make larger leaps, drag the scroll box to a new position.

To scroll down the worksheet:

1 Click the **down scroll arrow** repeatedly and notice how the screen scrolls one row at a time.

2 Drag the scroll box about halfway down the vertical scroll bar. Excel shows additional empty columns and rows.

3 Press and hold down the [**Ctrl**] key, and then press [**Home**] to return to cell A1.

This is an example of a key combination, using two keys in combination as a short-cut, to quickly move around the Worksheet window. In this book, key combinations are indicated by the key names joined by a plus sign (+), such as [Ctrl]+[Home].

Scrolling a Workbook

The PARK.XLS workbook also contains a chart of visitor count data on the Visitor Use 1993 Chart worksheet. This chart is on the second worksheet in the workbook.

To look at the Visitor Use 1993 chart:

1   Click the **Visitor Use 1993 Chart sheet tab.** See Figure 1-16.

Figure 1-16
Total Usage chart

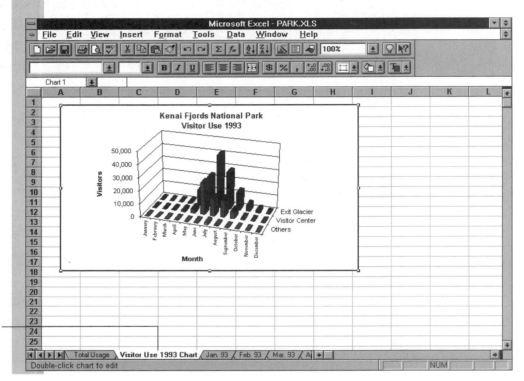

sheet tab for Visitor Use
1993 Chart

Note: If a small collection of buttons with the word "Chart" in its title bar appears, click its Control menu box to close it. This is another Excel toolbar, one you don't need right now.

The vertical bars in this chart (a histogram, which you'll learn about in Chapter 3) show the number of visitors for each month. The number of visitors appears on the vertical axis whereas the month appears on the horizontal axis. You can easily see from the chart that visitor usage is heaviest from May through September, because the bars are highest for these months. You'll learn how to create charts like this in the next chapter.

Now take a look at the next sheet in the workbook:

1   Click the **Jan. 93** sheet tab.

The form that appears resembles the form that the Kenai Fjords staff uses to record public use: it contains park information, number of visits, visitor hours, and other information. See Figure 1-17.

Figure 1-17
Third worksheet in the
PARK.XLS workbook

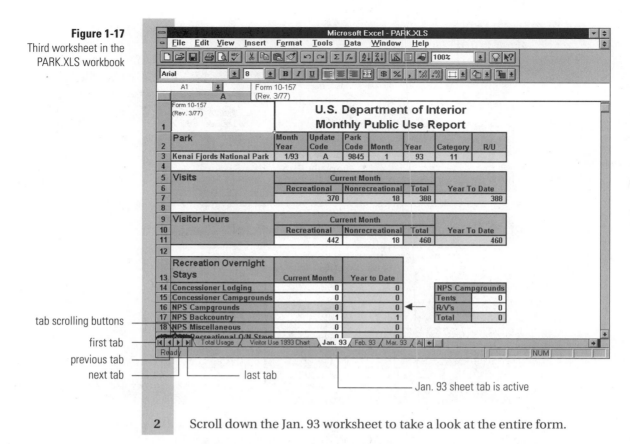

tab scrolling buttons

first tab

previous tab

next tab

last tab

Jan. 93 sheet tab is active

2   Scroll down the Jan. 93 worksheet to take a look at the entire form.

Clearly the PARK.XLS workbook is complex: its pages contain many cells with a variety of information that is interrelated. This book will not cover all the spreadsheet techniques used to create a workbook like this one, but you should be aware of some of the formatting and computational possibilities that exist.

To reach other pages of the workbook, you can use the tab scrolling buttons, shown in Figure 1-17. Clicking these buttons moves you to either the first page, the previous page, the next page, or the last page in the workbook.

Formulas and Functions

Formulas and functions are crucial Excel tools that you enter into cells to perform calculations. A formula in Excel always begins with an equals sign (=) and can contain numbers or cell references. Most formulas contain mathematical operators such as +, -, *, /, or ^, for addition, subtraction, multiplication, division, and exponentiation, respectively. For example, the formula =20*2 tells Excel to multiply 20 by 2, and returns a value of 40 (called the result of the formula). The formula =C5+C6 tells Excel to add the contents of cells C5 and C6. If you want to view a formula in a cell, click the cell to make it active. The formula bar shows the formula and the cell shows the result.

To get a feel for formulas, take a look at how this spreadsheet handled the computations involved in tracking public use:

1   Click **Edit > Go To**.

2   Type **F7** in the Reference box.

3   Press [**Enter**].

The Go To command moves you to the cell you specify, in this case cell F7, shown as the active cell in Figure 1-18.

Figure 1-18
Contents of cell F7

formula bar displays formula

cell F7 displays results

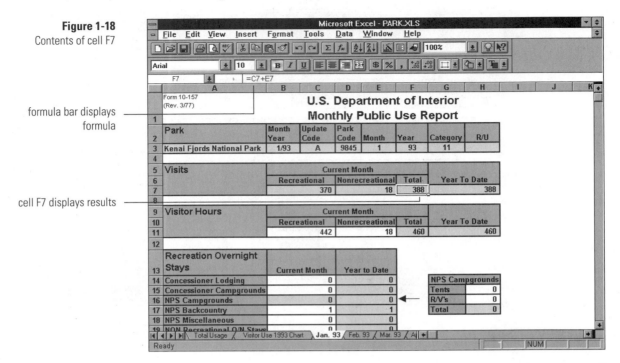

The formula bar contains the entry "=C7+E7." The workbook uses a formula in cell F7 that adds the contents of C7 (370) and E7 (18). The equals sign (=) indicates that cell F7 contains a formula that automatically updates if you change the contents of the cells making up the formula. The cell itself displays the result of the formula, whereas the formula bar shows the formula. The formula in cell F7 includes only the addition operator (+), but formulas can use many different operators.

Excel formulas can perform computations across cells in a single worksheet, across sheet pages in a workbook, or even across workbooks. The Total Usage table, for example, adds all the monthly totals.

To look at the total usage formula:

1   Click the **Total Usage sheet tab**.

2   Click cell **C5**.

Notice that the formula bar shows a formula that adds the contents of cells B23, G23, and B26 in the Jan. 93 worksheet.

Many formulas use **functions**, that is, calculation tools that perform operations automatically. Excel provides a wide variety of functions that perform many different types of calculations. For example, you can use the SUM function to create the formula =SUM(C5:C11) instead of having to type the longer formula =C5+C6+C7+C8+C9+C10+C11. Excel offers a myriad of functions (such as square roots or averages) in a number of categories (statistical, financial, mathematical, and so on).

To view a function:

1   Click cell **C17**.

Now the formula bar shows a formula that uses a function: =SUM(C5:C16). The function SUM adds the contents of all the cells indicated within the parentheses, in this case, the range C5 to C16, the monthly totals for the Exit Glacier site. When you enter a function, you type the function name and, in parentheses, the arguments that the function operates on. **Arguments** can be cell references, numbers, or other functions. For example, if you are taking the absolute value (represented by the function ABS) of the average of the values of the cells in the range C1:C15, you would enter =ABS(AVERAGE (C1:C15)).

Every function has its own syntax, which tells you how to enter the function and its arguments. For example, the AVERAGE function syntax is AVERAGE(range). More complicated functions have more complicated syntaxes.

The summary box that appears below, "Entering Formulas," lists the general steps you take when you want to enter a formula. Do not perform these steps now. Use this box as a reference. Summary boxes appear throughout this book.

··

Entering Formulas

- Click the cell that will display the formula values.

- Type = to begin the formula.

- Type the operation you want to perform (such as C6+C10 or 100∗F2). If the operation includes a function, type the function name in either uppercase or lowercase, and then type the cells or numbers on which you are operating using the correct syntax. You can also use the Function Wizard to enter functions, as described in the next section.

- Press [Enter].

··

Function Wizard

To save you the effort of looking up function names and their arguments, Excel provides a Function Wizard. The **Function Wizard** is a series of dialog boxes from which you can choose a function. Try using the Function Wizard to calculate the average Exit Glacier site usage in 1993.

To calculate the average and place it in cell C18:

1   Click cell **B18**, type **Average**, and press **[Tab]**; the active cell is now C18.

Note: If pressing [Tab] moves you an entire screen to the right rather than just to the next cell, you need to change an Excel setting. Choose Tools > Options, and then click the Transition tab. In the Settings area, there is a check box labeled Transition Navigation Keys. If it is selected, click it to deselect it, then click OK. [Tab] should work properly now.

2   Click the **Function Wizard button** [fx] (or click **Insert > Function**).

The Function Wizard - Step 1 of 2 dialog box appears. See Figure 1-19.

Figure 1-19
Function Wizard - Step 1 of 2 dialog box

Function Category list box ———

description of selected function

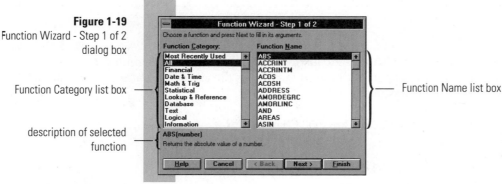

——— Function Name list box

Excel divides functions into categories such as Financial, Date & Time, Math & Trig, Statistical, and so on. This list appears in the Function Category list box. The Function Name list box contains the function names that fall under each particular category. Below the list boxes, Excel displays the highlighted function's name, the arguments you enter to use the function, and a description of the function itself.

3   Click **Statistical** in the Function Category box.

Excel lists function names alphabetically in the Function Name list box. The first statistical function is AVEDEV, or the average of the absolute deviations of data points from their mean, as the description tells you.

4   Click **AVERAGE** in the Function Name list box.

Now the description tells you that this function returns the average of its arguments.

5   Click **Next >**.

The next dialog box displays all the options for the AVERAGE function. Note from the description that the AVERAGE function can have up to 30 arguments. In this case you need only one—the range C5:C16, the Exit Glacier usage per month in 1993.

6   Type **C5:C16** in the number1 box, as shown in Figure 1-20.

Figure 1-20
Function Wizard - Step 2 of 2
dialog box

number1 box

Value box

Note that the dialog box shows the value of the average in the Value box in the upper-right corner of the dialog box.

7   Click **Finish**.

The value 10,438 now appears in cell C18. This value is the average monthly attendance at the Exit Glacier site in 1993. In this book, when you are instructed to enter a function, you can use the Function Wizard or type the function name directly, whichever you prefer. Using the Function Wizard, Excel inserts the function name for you and prompts you to enter the argument values, whereas when you simply type the function, you must remember the function name and the appropriate format for the arguments.

Printing Worksheets

In the Kenai Fjords example, the chief of interpretation could use this workbook to display the months of heaviest visitor use to justify the hiring of a seasonal employee. It is useful to print a hard copy (on paper) of the chart that shows the visitor use patterns. First take a look at a preview of the worksheet.

To preview the Visitor Use 1993 Chart:

1   Click the **Visitor Use 1993 Chart sheet tab**.

2   Click the **Print Preview button** 🔍.

The Print Preview window opens with a new set of buttons on top and the worksheet as it will look when printed on paper. See Figure 1-21.

Figure 1-21
Print Preview window

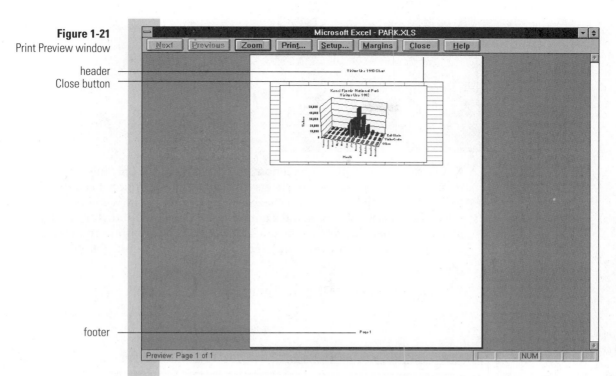

header

Close button

footer

3   Move the pointer onto the page. It turns into a magnifying glass ⊕ , which indicates that if you click the mouse button, you will zoom in on the page section where you clicked.

4   Click the chart area of the page. The chart is enlarged so that you can read it.

5   Click the page again. The page zooms back out.

Notice that Excel places a header (line at the top) and footer (line at the bottom) on the page with useful identification information. The header contains the name of the sheet, and the footer contains a centered page number.

Note: If a user has customized your version of Excel to include a different header or footer, your Print Preview window might look different.

6   Zoom in on the footer to see the information it contains.

7   Click the **Close button** to close the Print Preview window.

You can change or remove the header or footer (as well as other aspects of the printed page) using the Setup dialog box (available by clicking the Setup button).

The quickest way to print a worksheet is to use the Print button 🖨 on the toolbar. However, using the Print button, you won't have access to the Print dialog box. Although some toolbar buttons function the same way as menu commands (Open, for example, produces the same dialog box as the File > Open command), many buttons bypass the dialog boxes or bring up quicker versions with fewer options. The Print button bypasses the Print dialog box altogether. This should be fine once you know what printer you are using and are comfortable with the default printer settings, but for now, take a quick look at the Print dialog box.

To print the worksheet:

1   Click **File > Print**. The Print dialog box appears. See Figure 1-22.

Figure 1-22
Print dialog box

current printer

Notice that you can print a selection (you could highlight only the range you wanted to print), the selected sheet or sheets (this is the default option, because you have the Visitor Use 1993 Chart worksheet selected), or the entire workbook. You can also select the number of copies and a page range. The other option buttons let you make more specific printer selections. The dialog box also shows you the current printer. If you need to select a different printer, use the Printer Setup button.

2   Click **OK**.

Your printout should resemble Figure 1-23.

Figure 1-23
Hard copy of Visitor Use
1993 Chart worksheet

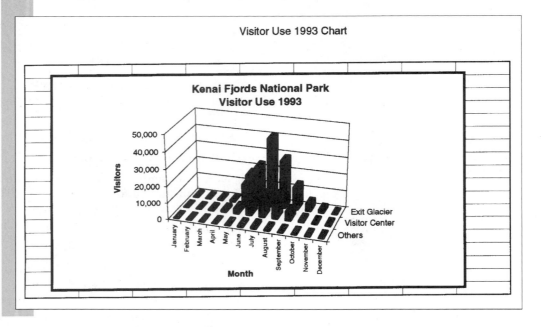

This hard copy effectively shows why the chief of interpretation opts to hire the seasonal employee for the months of May, June, July, August, and September.

Note: Whether or not your hard copy shows gridlines (the lines that appear behind the chart) depends on whether you have the Gridlines check box in the Page Setup dialog box selected. Click File > Page Setup; then click the Sheet tab in the Page Setup dialog box to view this setting.

Saving Your Work

You should periodically save your work when you make changes to a workbook or are entering a lot of data, so that you won't lose much work if there is a power failure. Excel offers two options on the File menu for saving your workbook changes: the Save command, which saves the file under the same name, and the Save As command, which you use to save a file under a new name or to save a file for the first time.

So that you do not change the original files you are working with (and can go through the chapters again with unchanged files, if necessary), this book instructs you to save your work under a new file name, made up of the original file name with a chapter-number prefix. To save the changes you made to the PARK.XLS workbook, save the file as 1PARK.XLS. If you are using your own computer, you can save the workbook to your hard drive. If you are using a school computer, you should obtain a blank, formatted, high-density 3.5-inch disk on which you will save all your work. This book refers to this disk as your Student Disk, and the saving instructions assume you are saving your work to this Student Disk.

To save the 1PARK.XLS workbook on your Student Disk:

1 Click **File > Save As** to open the Save As dialog box.

2 Insert your Student Disk into drive A (or the drive you use on your computer).

3 Click the **Drives drop-down arrow**.

4 Click **a:** in the list of drives that appears.

Note: If your Student Disk is in a different drive, click that drive name instead. If you are saving to a hard drive, select the correct directory in the directory tree by double-clicking the appropriate yellow folders. The Save As dialog box displays the current directory. If you don't understand the directory structure, consult your Windows documentation.

5 Double-click the **File Name box**.

6 Type **1PARK** in the File Name box. The Save As dialog box should look like Figure 1-24.

Figure 1-24
Save As dialog box

File Name box

current directory

Drives drop-down arrow

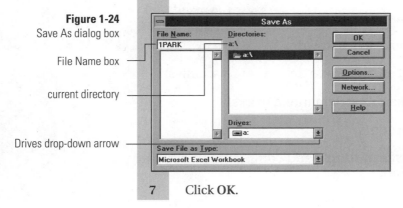

7 Click **OK**.

Excel automatically adds the XLS file extension. The Excel title bar changes to reflect the new filename.

Exiting Excel

When you are finished with an Excel session you should exit the program so that all the program-related files are properly closed.

To exit Excel:

1 Click **File > Exit**.

If you have unsaved work, Excel asks if you want to save it before exiting. If you click No, Excel closes and you lose your work. If you click Yes, Excel opens the Save As dialog box so that you can save your work, and then it closes. You can also click Cancel, which returns you to the Excel window.

You are back in Program Manager.

CHAPTER 2

ENTERING AND MANIPULATING DATA

OBJECTIVES

In this chapter you will learn to:

- Recognize data types and enter data

- Apply styles to data

- Create names to define your variables

- Insert new data into a worksheet

- Sort data

- Perform simple and complex queries using filters

- Import data from a text file

- Create calculated variables

- Format your spreadsheet using freeze panes

Variable Types

Statisticians study **data**, which are collections of measurements for objects. Quantities that change and can be measured and studied in statistics are called **variables**. For example, the time spent by several individuals reading the first chapter in this book is a variable. **Quantitative variables** measure quantities, like age or weight. **Qualitative** (or **categorical**) variables indicate categories or classes. If you can order a qualitative variable in some way it is called an **ordinal** variable. Examples of ordinal variables include preferences (greatly disliked, mildly disliked, liked, greatly liked) or education (grade school, high school, college, graduate school). If the qualitative variable does not have an inherent order, such as a car brand or color, then it is known as a **nominal** variable. In performing statistical analyses, it is important to understand what type of variables you are working with so that you can apply the appropriate statistical test.

Entering Data

Table 2-1 contains average daily gasoline sales and other (non-gasoline) sales for each of 10 service station/convenience franchises in a store chain in a western city. There are three variables in this table: Station, Gas, and Other. The Gas and Other variables are quantitative variables because they measure quantities (sales in dollars). The Station variable is a categorical variable, but it is not clear by simply observing the data whether Station is ordinal or not. The service station number could indicate the order in which each service station was built if they were constructed as part of a franchise plan. In that case, if you were interested in how the newer service stations compare with the older ones, it would be important to consider Station as an ordinal variable. However, in this case the numbers are simply labels and do not imply any specific order.

Table 2-1
Service station data

Station	Gas	Other
1	$3,415	$2,211
2	$3,499	$2,500
3	$3,831	$2,899
4	$3,587	$2,488
5	$3,719	$2,111
6	$3,001	$1,281
7	$4,567	$8,712
8	$4,218	$7,056
9	$3,215	$2,508

To practice entering data, in the next section you'll type these values into the rows and columns of the Worksheet window.

Using Lists

A rectangular block of data in table format, like the service station data, with a row of labels across the top (in this case, Station, Gas, and Other) is known in Excel as a list. A **list** is a labeled series of rows that contain similar data. In the service station data, each row contains a single record: all the information for a single service station, including its number, its gas sales, and its sales for other merchandise.

To create the first column of the service station list:

1   Launch Excel as described in Chapter 1 (remember to maximize the Excel window and the Worksheet window).

Excel shows an empty workbook with Microsoft Excel—Book1 in the title bar.

Note: If you do not see an empty workbook, click File > New or click the New Workbook button 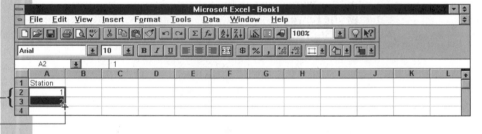 *to create a new workbook.*

2　Click cell **A1** to make it the active cell if it isn't already, type **Station**, and press [**Enter**].

Cell A1 accepts your entry and moves the active cell down one row to cell A2.

3　Type **1** into cell A2 and press [**Enter**].

4　Type **2** into cell A3 and press [**Enter**].

At this point you could continue to type the service station numbers through number nine. However, Excel provides a quick way of filling in the values from a series (in this case, the first nine integers) called **AutoFill**. To fill in the rest of the service station numbers using AutoFill:

5　Select the range **A2:A3** by clicking A2, holding down the mouse button, and dragging the mouse down to highlight A3. When both A2 and A3 are highlighted, release the mouse button.

Notice the small black box at the lower-right corner of the double-border around the selected range. This is called a **fill handle**; you can select and move it around to work with the selected range.

6　Move the mouse pointer over the selected range's fill handle. You know that the pointer is on the fill handle when the pointer changes from a cross ✛ to a crosshair +. See Figure 2-1.

Figure 2-1
Crosshair on the fill handle
of the selected range

selected range

crosshair on fill handle

7　Click the **fill handle** and drag it down so that the range **A2:A10** is selected.

8　Release the mouse button and watch as Excel finishes entering the nine integers in cells A2:A10 for you. See Figure 2-2.

Figure 2-2
Results of using AutoFill

selected range is now
A2:A10

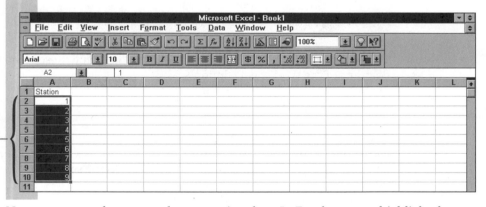

Now you are ready to enter the gas station data. In Excel, you can highlight the range of cells you want to use. When you press [Tab] to accept an entry and move to the next cell, you move to the next cell within the selected range rather than to a cell outside the range.

To enter the gasoline and other sales data in the cell range B1:C10:

1　Select the range **B1:C10** so that your Worksheet window looks like Figure 2-3 (use the dragging technique described in Chapter 1).

Figure 2-3
Selected range for
data entry

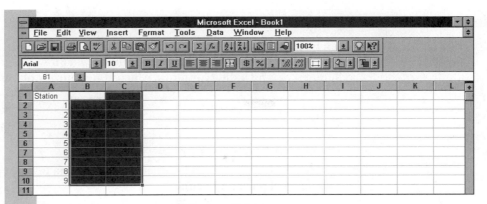

2　　Type **Gas** in cell B1 and press [**Tab**].

3　　Type **Other** in cell C1 and press [**Tab**].

　　Cell B2 should be highlighted. Pressing [Tab] at the end of a selected row takes you to the beginning of the next row. This is the advantage of highlighting B1:C10 before entering the data.

4　　Type **3415** in cell B2 (you do not need to type the $ yet) and press [**Enter**].

　　Cell B3 is now the active cell. Pressing [Enter] moves you down one row. If you make a mistake in entering data, you can return to the previous row by pressing [Shift]+[Enter] or to the previous column by pressing [Shift]+[Tab].

5　　Type **3499** in cell B3 and press [**Enter**].

6　　Continue typing values from Table 2-1 into Sheet1. When you get to cell B10 and are ready to enter the data for the Other column, press [**Enter**] twice to reach C2, and then start entering the Other data.

7　　Click any cell to remove the highlighting and view the list, which looks like Figure 2-4.

Figure 2-4
Completed service station
list

	Station	Gas	Other
1			
2	1	3415	2211
3	2	3499	2500
4	3	3831	2899
5	4	3587	2488
6	5	3719	2111
7	6	3001	1281
8	7	4567	8712
9	8	4218	7056
10	9	3215	2508
11			

You have now completed entering your first Excel list.

Data Entry Tips

- Select the cell or cell range into which you want to enter data, type the data, then press [Enter]. Press [Enter] to move through a range from top to bottom, [Shift]+[Enter] to move from bottom to top, [Tab] to move from left to right, and [Shift]+[Tab] to move from right to left.

- To enter the same data into several cells at once, select the cells into which you want to enter data, type the data, then press [Ctrl]+[Enter].

- Use AutoFill to enter sequences. Type the first few entries, then highlight those entries. Drag the crosshair pointer to the bottom of the range. When you release the mouse button, Excel fills in the sequence.

Applying Styles to Data

You did not type any dollar signs ($) when entering the sales information. You can do that now by specifying a display style for the data. Styles do not alter the values or variable type. Excel offers different styles, including Currency, which automatically places a $ before a value.

To format the sales data:

1   Select **B2:C10**.

2   Click the **Currency Style button** ⊞ .

Excel adds a $ before each number, adds a comma to separate thousands, and adds two decimal places for cents.

Note: Depending on your monitor type, Excel might not have room to display the new entries, so it displays ##### instead. The next step should take care of this problem.

3   Click the **Decrease Decimal button** ⊞ twice to remove the cents information.

4   Click cell **A1** to deselect the range and get a better look at the list. See Figure 2-5.

Figure 2-5
Service station list with dollar values in Currency style

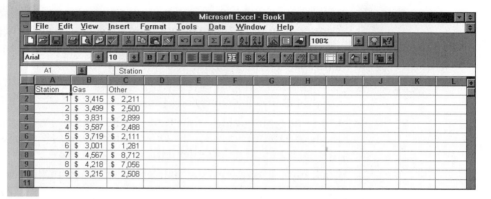

Defining Range Names

When using Excel for statistical analysis, it is often easier to refer to variable values by a range name (Gas) rather than by the range of cells containing the data (B2:B10). Currently, each column of variables has a label in row 1. However, although the Gas label in B1 tells you what cells B2:B10 contain, the label doesn't tell Excel that it applies only to that range. To do that, you must assign the labels as range names for the variables they identify; only then can Excel use the range name rather than the range of cells in its computations.

Using a range name can be an advantage; let's say you wanted to find the average value of gas sales in your nine franchises using the AVERAGE function. You could enter AVERAGE(B2:B10) or AVERAGE(GAS). It's easier to identify the data by a variable range name than by a range of cell references.

To create range names that represent ranges of cells in your workbook:

1   Select the range **A1:C10**.

2   Click **Insert > Name > Create** to open the Create Names dialog box. See Figure 2-6.

Figure 2-6
Create Names dialog box

Excel will create range names for the variables based on where you have entered the data labels.

3 Verify that the **Top Row check box** is selected. Leave the other check boxes unselected.

4 Click **OK**.

Although Sheet1 (the current worksheet, as indicated by the active sheet tab) doesn't look any different, Excel has created three range names: Service, Gas, and Other. These range names refer to the cell ranges A2:A10, B2:B10, and C2:C10.

To see a list of defined range names and the cell ranges they represent:

1 Click **Insert > Name > Define**.

Figure 2-7 shows the Define Name dialog box. The list box lists all names and objects defined in your workbook (an **object** is any element in Excel that you can manipulate: a chart, a cell, a range of cells, a label, a formula, and so on). You've defined only the three range names so far.

The text box at the bottom of the dialog box displays the reference of the currently selected block. The block A1:C10 on Sheet1 is still selected, so it is the reference that appears.

Figure 2-7
Define Name dialog box

Name drop-down arrow —

list box —

reference of selected block —

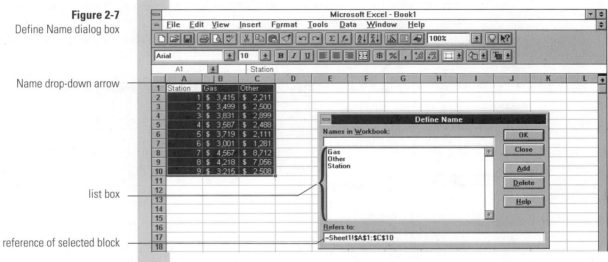

2 Click **Gas** and watch how the reference changes to display the Gas variable data.

You can add or delete range names from this dialog box.

3 Click **Close** to close the dialog box.

4 Click cell **A1** to deselect the list.

You can use the Name drop-down list box to quickly select a variable.

5 Click the **Name drop-down arrow** to the right of the cell reference box.

6 Click **Gas**.

Excel highlights the Gas data.

Cell References

A dollar sign ($) appears as part of the cell reference in the dialog box shown in Figure 2-7. When a cell reference includes a $, the cell reference is absolute. There are three kinds of cell references: relative, absolute, and mixed. A **relative reference**, written without a $, identifies a cell location based on its position relative to other cells (B1). An **absolute reference**, written with a $ before the column and row references, always points to the same cell ($B$2). A **mixed reference** combines the two. It can have an absolute column reference, written with the $, but a relative row reference, written without the $ ($B2), or vice versa (B$2).

Cell references are an important concept. For now, simply notice that the $ indicates a cell reference type. Later you'll have an opportunity to work with cell references more closely.

Inserting New Data

In the gas station example, let's say the company recently purchased a tenth service station whose data need to be entered into the sales workbook. You could type the new data into cells A11:C11, as indicated in the previous section, and then redefine the Station, Gas, and Other range names to correspond to cell range A2:C11 of Sheet1 in the workbook. A second method is to insert the new row within the cell range A2:C10 and take advantage of Excel's ability to automatically redefine the range corresponding to your variable range names.

To add a new station to the list of service stations:

1   Select the cell range **A10:C10**.

Excel offers quick access to common formatting and data manipulation tasks with a shortcut menu that opens when you click an element in the worksheet, such as a range, cell, or chart, with the right mouse button.

2   Point at the cell range and click the right mouse button (this is called right-clicking) to open the shortcut menu. See Figure 2-8.

Figure 2-8
Shortcut menu

Shortcut menu

Insert command

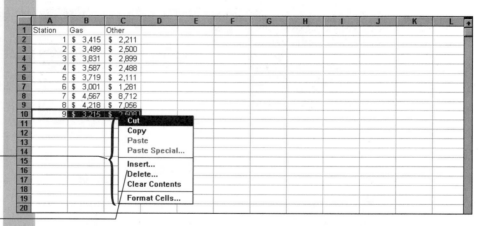

3   Click **Insert**.

4   Click the **Shift Cells Down option button**.

5   Click **OK**.

Excel shifts the values in cells A10:C10 to A11:C11 and inserts a new blank row in cells A10:C10.

6   Type **10** in cell A10 and press [**Tab**]. (This row is now out of order, but you'll sort the list in a moment.)

7   Type **3995** in cell B10 and press [**Tab**].

8   Type **1938** in cell C10 and press [**Enter**].

Notice that Excel automatically formats the new values for the gas and other sales data as currency. This is because you entered the data *within* the formatted range. Now check to see that Excel also redefined the range names you created to include the new entry.

9   Click the **Name drop-down arrow**, then click **Gas**.

Excel highlights the range B2:B11. It is important to remember that if you had entered the data at the bottom of the list, Excel would not have updated the range names and applied the styles as it did. When you want a new entry to be part of an existing variable, be sure you insert it within the range defined for that variable and not outside it.

To save your service station workbook with the name 2GAS:

1   Click **File > Save As**.

2   Insert your Student Disk into drive A (if that is the drive you plan to use).

3   Click the **Drives drop-down arrow**.

4   Click **a:** in the list of drives that appears for the A drive.

5   Double-click the **File Name text box**.

6   Type **2GAS** in the File Name text box.

7   Click **OK**.

Note: Because this is the first time you are saving this workbook, Excel might prompt you to enter summary information about the file. You can fill in the Summary Info dialog box if you want, indicating comments about the file. Otherwise, click OK.

Sorting Data

You might have noticed that the data are no longer in order. You can quickly reorder the data by either selecting the list you want to sort or by clicking a cell within the list and letting Excel sort it. Excel detects and excludes column labels from the data you are sorting.

To sort the data by service station number:

1   Click any cell within the range **A1:C11**.

2   Click **Data > Sort**. Notice that Excel selects the entire range A2:C11, which is your data range.

3   Click the **Sort By drop-down arrow**, click **Station**, then verify that the **Ascending option button** is selected. Also verify that the **Header Row option button** is selected. See Figure 2-9.

Figure 2-9
Sort dialog box

Station is selected

Header Row is selected

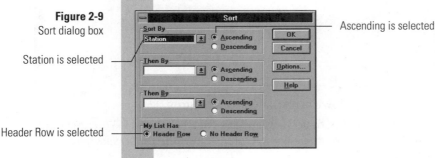

Ascending is selected

4   Click **OK**.

Excel sorts the data by service station number. Before sorting, you used the Save As command to save your file with a new name. Now, update the 2GAS.XLS file using the Save command, which saves your changes to the file you designated when you used Save As.

5   Click the Save button ■ to save your file as **2GAS.XLS** (the drive containing your Student Disk should already be selected).

Querying Data

Statisticians are often interested in subsets of data rather than in the complete list. For instance, a manufacturing company trying to analyze quality-control data from three work shifts might be interested in looking only at the night shift. An employment firm interested in salary data might want to consider just the salary subset of those making between $25,000 and $35,000. These subsets of a data list fulfill certain criteria. Excel can select only the records that fit your criteria by using a **data filter**. Another term for filtering (often used in database applications) is **querying**, because you are asking the database to discover which records fulfill your criteria. There are two types of criteria you can use with Excel:

- **Comparison Criteria**, which compare variables to specified values or constants (for example, which service stations have gas sales of over $3,500)

- **Calculated Criteria**, which compare variables to calculated values, which are themselves usually calculated from other variables (for example, which service stations have gas sales that exceed the average for all franchises)

Excel provides two ways of filtering data. The first method is the **AutoFilter**, which is primarily used for simple queries with comparison criteria. For more complicated queries and those involving calculated criteria, Excel provides the **Advanced Filter**.

Querying Data Using AutoFilter

Let's say the service station company plans a massive advertising campaign to boost sales for the service stations that are reporting sales of $3,500 or less. You can construct a simple query using comparison criteria to have Excel display only service stations with Gas ≤ $3500.

To query the service station list:

1 Make sure one of the cells in the range **A1:C11** is still selected.

2 Click **Data > Filter > AutoFilter**. Notice that Excel adds drop-down arrow buttons to row 1, which contains the column labels. See Figure 2-10.

Figure 2-10
List with AutoFilter enabled

filter drop-down arrows

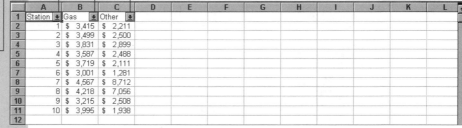

3 Click the **Gas drop-down arrow**, then click **Custom**.

The Custom AutoFilter dialog box appears. See Figure 2-11. You are going to enter a comparison statement that says, in effect, "Show me the rows where Gas ≤ $3500."

Figure 2-11
Custom AutoFilter dialog box

variable you are filtering

comparison operator drop-down arrow

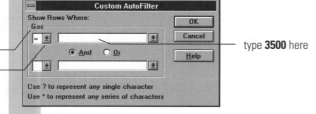

type **3500** here

4 Click the **comparison operator drop-down arrow**, then click **<=**.

5 Press [**Tab**], then type **3500** in the text box to the immediate right.

6 Click **OK**.

Excel modifies the list of service stations to show stations 1, 2, 6, and 9 in rows 2, 3, 7, and 10. See Figure 2-12.

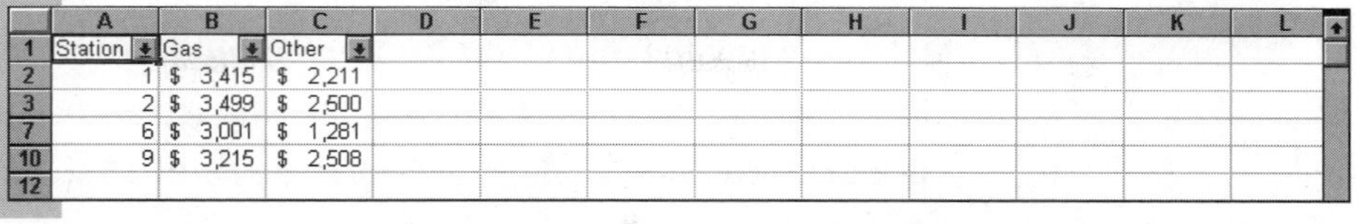

Figure 2-12
Filtered service station list

	A	B	C	D	E	F	G	H	I	J	K	L	
1	Station	Gas	Other										
2	1	$ 3,415	$ 2,211										
3	2	$ 3,499	$ 2,500										
7	6	$ 3,001	$ 1,281										
10	9	$ 3,215	$ 2,508										
12													

The service station data for the other stations have not been lost, but merely hidden. You can retrieve the data by choosing the All option from the Gas drop-down list.

Adding a Second Filter

Let's say you needed to add a second filter that filters out service stations selling less than $2,500 worth of other products. This filter does not negate the AutoFilter you just inserted; instead, it adds to the filter criteria.

To add a second filter:

1 Click the **Other drop-down arrow**, then click **Custom**.

2 Enter **<= 2500** in the Custom AutoFilter dialog box, as you did for the first filter.

3 Click **OK**.

This kind of filter is known as an **and filter**: the list of service stations now includes only stations 1, 2, and 6, the service stations with gasoline sales less than or equal to $3,500 *and* other sales less than or equal to $2,500. You can see that you should concentrate your advertising campaign on these three service stations.

When a list displays only filtered data, Excel places a checkmark next to the AutoFilter command. You can remove filtering from the list by clicking the command to remove the checkmark.

To remove filtering from the data list:

1 Select **Data > Filter > AutoFilter** (notice the checkmark; if you open the Filter submenu again, the checkmark will not appear next to the command).

The drop-down lists disappear and Excel displays all the data again.

Querying Data Using Advanced Filter

There might be situations where you want to use more complicated criteria to select data. Such situations include criteria:

- that require several *and/or* clauses
- that involve the calculations of variables

In such cases you can query the data on your list using the Advanced Filter. To use the Advanced Filter, enter the selection criteria into cells on the worksheet and then refer to those cells in the Advanced Filter dialog box.

To recreate the query you just performed using the Advanced Filter:

1 Click cell **B13**, type **Advanced Filter Criteria**, then press [**Enter**].

2 Type **Gas** in cell B14, then press [**Enter**].

3 Type **<=3500** in B15, click cell **C14**, type **Other** in C14, then press [**Enter**].

4 Type **<=2500** in C15, then press [**Enter**].

Now, apply these criteria to the service station data. Tell Excel the range containing the list you are filtering and the range containing the filtering criteria (B14:C15).

5   Click one of the cells in the range **A1:C11**.

6   Click **Data > Filter > Advanced Filter**.

7   Make sure that the **Filter the List, in-place option button** is selected.

8   Make sure that the range **$A$1:$C$11** is entered as the List Range.

9   Click the **Criteria Range text box** and type **B14:C15** so that the dialog box looks like Figure 2-13.

Figure 2-13
Completed Advanced Filter
dialog box

list range A1:C11

criteria range B14:C15

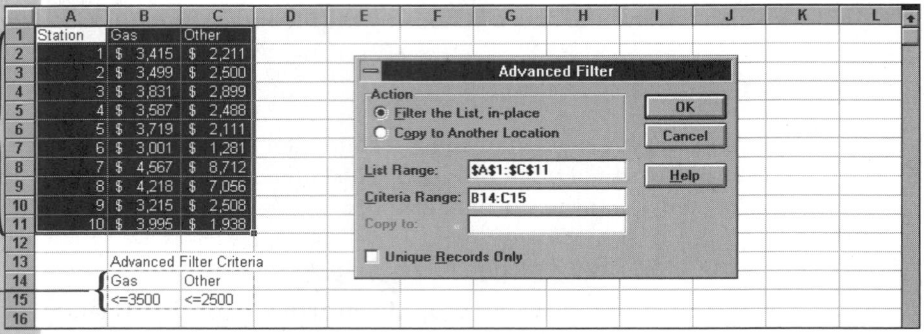

10   Click **OK**.

Excel shows service stations 1, 2, and 6, as before.

What if you want to look at only those service stations with *either* gasoline sales less than or equal to $3,500 *or* other sales less than or equal to $2,500? Entering *or* criteria between columns is not possible with AutoFilter. However, using the Advanced Filter, you can link criteria together by an *or* clause (Gas ≤ $3500 *or* Other ≤ $2500) by placing the criteria in different columns and in different rows. The filter you just performed used criteria from different columns but in the same row (B15 and C15), which is an example of an *and* filter. Now try putting these criteria in different rows to produce an *or* filter.

To perform an *or* filter:

1   Click **C15**, which contains the <=2500 criterion.

2   Move the mouse pointer to one of the borders of C15 so that it turns from a cross ✛ to an arrow ⬉ and the tip of the arrow is on the border.

3   Press and hold down the mouse button and drag the value to C16. The criteria are now in different rows. See Figure 2-14.

Figure 2-14
Criteria in different rows

different rows

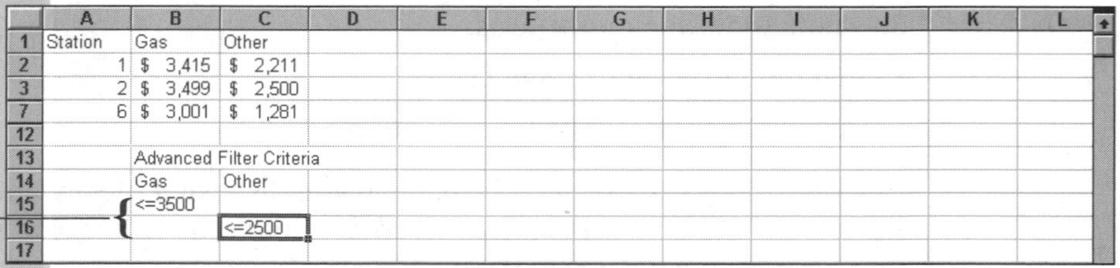

4   Click any cell in the range **A1:C11** (it does not matter if some of the rows are hidden).

5   Click **Data > Filter > Advanced Filter** from the menu.

6   Make sure the range **$A$1:$C$11** is entered in the List Range.

7   Enter the range **B14:C16** as the Criteria Range (if the dialog box still contains the previous entry, you should only have to delete the final 5 and change it to a 6; it doesn't matter in this example if the references are absolute or not).

8   Click **OK**.

Service stations 1, 2, 4, 5, 6, 9, and 10 are now shown. These are the service stations with either gasoline sales ≤ $3,500 *or* other sales ≤ $2,500.

Querying Data Using Calculated Values (Optional)

The Advanced Filter also lets you specify criteria based on calculated values. Let's suppose you wanted to query the data to see only those service stations whose gasoline sales were greater than the average gasoline sales of all the service stations, so that you could award a bonus to those franchise managers? You could calculate the average gasoline sales and enter this number explicitly into the criterion, but it's easier to include the calculation in the criterion itself if you have a good understanding of formulas and cell references (which is why this section is optional).

To select those service stations with gasoline sales higher than or equal to the average gasoline sales of all the service stations:

1   Click cell **E13**, type **Highgas**, and press [**Enter**].

2   Type **=B2>= AVERAGE($B$2:$B$11)** in cell E14 and press [**Enter**] so that FALSE appears in the cell (you'll see why in a moment).

3   Select a cell in the range **A1:C11**.

4   Click **Data > Filter > Advanced Filter**.

5   Make sure the range **$A$1:$C$11** is entered in the List Range.

6   Select the contents of the Criteria Range box, type **E13:E14** as the Criteria Range, then click **OK**.

Service stations 3, 5, 7, 8, and 10 are shown as having gasoline sales higher than average. They've earned the bonus.

7   View all the service station data again by clicking **Data > Filter > Show All**.

A few comments about the formula entered into cell E14 might help you understand Excel's requirements for the filtering formula. Recall that you typed a single cell in the formula to begin with (=B2). Then you typed the operator (>=), then the function (AVERAGE), and finally, the range the function is operating on. Excel evaluates the first cell (B2) to determine whether the calculated criterion is TRUE, and then evaluates the rest (B3:B11), one at a time (Excel evaluates all rows in the list).

Other requirements for the filtering formula are:

• A calculated criteria formula must refer to at least one cell in the list. Use a relative reference to tell Excel to adjust the formula for each cell in the column it is evaluating.

• Your criteria formula must produce a logical value of TRUE or FALSE (Excel displays only those rows that produce a value of TRUE). In this example, the formula produced the value FALSE, because the value of B2 is less than the average gasoline sales.

• Do not use the name of the column label (in this example, Gas) in the calculated criterion, because it will produce erroneous results when Excel looks for a single value. This example used the criterion label of Highgas for just that reason. For the same reason, the formula used an absolute reference ($B$2:$B$11) rather than the defined range name (Gas). You could have defined a new range name such as Gasdata to refer to cells B2 through B11, which you could then use in the formula. As long as the range name doesn't conflict with the label, the formula will work.

Note: You don't have to enter criteria on the same worksheet as the data list. You could put all the criteria of interest to you on a separate sheet and then refer to them as you please.

	To save your work:
1	Click the **Save button** 🖫 to save your file as **2GAS.XLS** (the drive containing your Student Disk should already be selected).
2	Select **File > Close**.

Querying Data

- Click any cell in the list you want to query.

To use AutoFilter, which queries one column at a time:

- Click Data > Filter > AutoFilter.
- Click the filter drop-down arrow to open a list of possible criteria.
- Either select one of the listed criteria, in which case Excel performs the query immediately, or click Custom to open the Custom AutoFilter dialog box.
- Enter the criteria expression that determines which data are displayed in the Custom AutoFilter dialog box; then click OK.

To use Advanced Filter, which lets you specify complex criteria:

- Enter the criteria into a range of cells in the workbook. If your criteria are in the same row, Excel displays only the data that meet *all* the criteria. If they are in different rows, Excel displays the data meeting *any* of the criteria.
- Click Data > Filter > Advanced Filter.
- Enter the list range and the criteria range in the Advanced Filter dialog box; then click OK.

Importing Data from a Text File

When you receive data from other sources, it often will be in formats other than the XLS files that are designed to be used by Excel. Excel can read data from a variety of formats including database files, other spreadsheet files, and text files. Text files usually have the file extension TXT or DAT; they provide one of the most common ways of sharing data because almost all applications can read them.

For example, a young family-owned bagel shop might gather data on wheat products on the market that people eat as snacks or for breakfast for comparison with their own products. The data are contained in a text file, WHEAT.TXT, shown in Table 2-2. These data were obtained from the nutritional information on the packages of competing wheat products.

Table 2-2
Wheat data

Brand	Food	Price	Package_oz	Serving _oz	Calories	Protein	Carbo	Fat
Anderson	Pretzel	$1.55	14	1	110	3	23	1
Uncle B	Bagel	$0.99	15	1.5	120	5	25	0.5
Bays	Eng Muffin	$1.09	12	2	140	5	25	2
Thomas	Eng Muffin	$1.69	12	2	130	4	25	1
Quaker	OHs Cereal	$2.49	12	1	120	1	24	2
Nabisco	Grah Cracker	$2.65	16	0.5	60	1	11	1
Wheaties	Cereal	$3.19	18	1	100	3	23	1
Wonder	Bread	$1.31	16	1	70	3	14	1
Brownberry	Bread	$1.29	16	0.84	60	2	11	1
Pepperidge	Bread	$1.59	16	0.89	80	2	14	2

Importing a text file into Excel is a little different from opening an Excel file because a text file doesn't contain cells. It contains just characters and spaces. Columns are determined by the number of spaces or by a special character such as a [Tab] or comma. During the importing process, Excel tries to determine where one column ends and the next begins so that it can place the data in the correct cells.

To import WHEAT.TXT:

1   Click the **Open button** 📖.

2   Click the **List File of Type drop-down list arrow** and select **Text Files** (you might have to scroll to see it). Be sure you are in the directory containing the Student files (as described in Chapter 1). WHEAT.TXT should now appear in the File Name list box.

3   Double-click **WHEAT.TXT**.

Excel displays the Text Import Wizard to help you import the text properly. See Figure 2-15. Excel contains a number of wizards that walk you through the steps of accomplishing a task; the Text Import Wizard is one of them.

Figure 2-15
Text Import Wizard – Step 1 of 3 dialog box

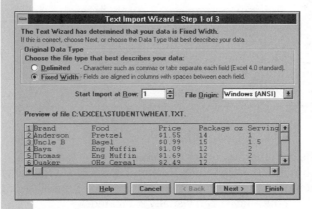

Excel automatically determines that the data are aligned in columns with spaces delimiting each field. By moving the horizontal and vertical scroll boxes, you can see the whole data set. The data should look like Table 2-2.

4   Click the **Next button**.

The second step of the Text Import Wizard allows you to define borders between data columns. See Figure 2-16. Note that the right border for **Package oz** cuts the column label in half.

Figure 2-16
Text Import Wizard – Step 2 of 3

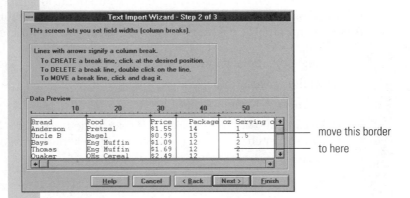

move this border
to here

5   Click and drag the right border for **Package oz** to the right so that it aligns with the left edge of the Serving oz column.

Don't worry that the left border isn't exactly aligned with the left edge of the Package oz column. As long as the border doesn't cut through any data or labels, the importing process will be successful.

6   Click **Next**.

The third step of the Text Import Wizard allows you to define column formats and to exclude specific columns from importing. General, the default format for each column, properly formats the data except in the most special cases.

7   Click **Finish**.

Excel imports the wheat data and places them into a new workbook. See Figure 2-17.

Figure 2-17
WHEAT.TXT workbook

Range Select All button ——

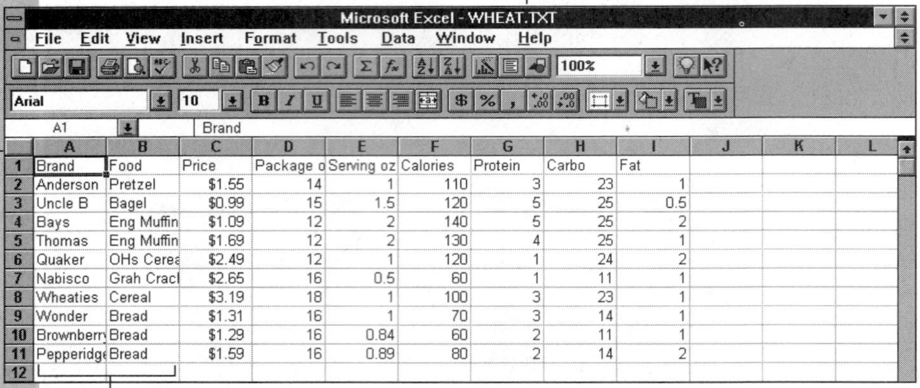

first two columns are truncated

Notice that the data for the first two columns don't look anything like the preview from the Text Wizard; but don't worry. When Excel imports a file, it formats the new workbook with the standard column width of about nine characters, regardless of column content. The data are still there, but they have been visually truncated.

To format the column widths to show all the data:

1   Click the **Range Select All button** in the upper-left corner of the worksheet. See Figure 2-17.

2   Double-click the right border of one of the column headers in the selection; the pointer changes from ⇩ to ✛ when you point at the border.

Excel changes the column widths to match the width of the longest cell in each column. Because you'll be applying statistical analysis techniques to the columns in this workbook, create range names for the columns in this data list as you did for the service station data.

To create the range names:

1   Select the range **A1:I11**.

2   Select **Insert > Name > Create**.

3   Make sure the **Top Row** check box is selected in the Create Names dialog box. Be sure the other check boxes are deselected (click them if necessary to deselect them).

4   Click **OK**.

5   Click **A1** to deselect the range and view the worksheet.

Creating New Variables

Now that you have imported the data successfully, you can examine these data to see how bagel products compare with other wheat products. How do the different products compare in terms of food value per dollar? For example, how do you think the breakfast cereals compare with the breads? Because the packages come in various sizes, it is hard to make cost comparisons, but you can make the comparison more equivalent by computing a new variable.

To find the ounces per dollar spent on each product, divide the package weight by the price:

1   Type the label **Oz per Dol** in cell J1 and press [**Enter**].

2   Highlight the range **J2:J11**.

3   With the range still highlighted, type =**Package oz/Price** in cell J2 and press [**Enter**].

Excel computes the division for you (indicated by the /). Notice that you had to type an underscore (_) in place of the space in the label name contained in D1. When Excel creates range names from labels, it converts all spaces to underscores.

4   Click **Edit > Fill > Down** (or press [Ctrl]+[D]).

Excel enters the ounces per dollar amount into each cell in the column. See Figure 2-18.

Figure 2-18
Results of ounces per dollar computation

Excel applies formula to each cell in selection

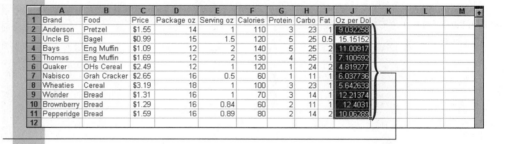

	A	B	C	D	E	F	G	H	I	J	K	L	M
1	Brand	Food	Price	Package oz	Serving oz	Calories	Protein	Carbo	Fat	Oz per Dol			
2	Anderson	Pretzel	$1.55	14	1	110	3	23	1	9.032258			
3	Uncle B	Bagel	$0.99	15	1.5	120	5	25	0.5	15.15152			
4	Bays	Eng Muffin	$1.09	12	2	140	5	25	2	11.00917			
5	Thomas	Eng Muffin	$1.69	12	2	130	4	25	1	7.100592			
6	Quaker	OHs Cereal	$2.49	12	1	120	1	24	2	4.819277			
7	Nabisco	Grah Cracker	$2.65	16	0.5	60	1	11	1	6.037736			
8	Wheaties	Cereal	$3.19	18	1	100	3	23	1	5.642633			
9	Wonder	Bread	$1.31	16	1	70	3	14	1	12.21374			
10	Brownberry	Bread	$1.29	16	0.84	60	2	11	1	12.4031			
11	Pepperidge	Bread	$1.59	16	0.89	80	2	14	2	10.06289			
12													

5   Click some of the cells in **column J** and notice that the formula in each cell of column J is the same.

If you've used Excel before, you might be used to seeing cell references such as =D2/C2 in the second row, =D3 / C3 in the third row, and so on. As long as you have given the data a range name and the data are part of a list, you can use range names in your formulas.

To name the column you just created:

1   Select **J1:J11**.

2   Click **Insert > Name > Create**.

3   Verify that **Top Row** is selected.

4   Click **OK**.

The results in your new variable show you how much you get per dollar spent, but the results are misleading because some foods have fewer nutrients per ounce. A half-ounce serving of graham crackers, for example, has a nutritional content similar to a 0.84-ounce serving of Brownberry bread.

To make the comparison more equivalent, compute the calories per dollar. Create a new variable in column K called Cal per Dollar in the range K2:K11 with the formula in each cell =Oz_per_Dol*Calories/Serving_oz. You can save some typing by having Excel paste the variable range name into the formula.

To create the calories per dollar variable:

1   Type the label **Cal per Dol** in cell K1 and press [**Enter**].

2   Highlight the range **K2:K11**.

3   Type = in cell K2 to begin the formula.

4   Click the **Names drop-down arrow** and click **Oz_per_Dol** from the drop-down list. The range name **Oz_per_Dol** appears in cell K2.

5   Type * (for multiplication).

6   Click the **Names drop-down arrow** and click **Calories** from the drop-down list. Excel appends the range name **Calories** to the formula in cell K2.

7   Type / (for division).

8   Click the **Names drop-down arrow** and click **Serving oz** in the drop-down list; then press [**Enter**] to accept the formula.

9   Press [**Ctrl**]+[**D**] to fill down the rest of the cells.

The calories per dollar values are now entered in column K. See Figure 2-19.

Figure 2-19
Worksheet with the calories per dollar variable

	A	B	C	D	E	F	G	H	I	J	K	L	M
1	Brand	Food	Price	Package oz	Serving oz	Calories	Protein	Carbo	Fat	Oz per Dol	Cal per Dol		
2	Anderson	Pretzel	$1.55	14	1	110	3	23	1	9.032258	993.5484		
3	Uncle B	Bagel	$0.99	15	1.5	120	5	25	0.5	15.15152	1212.121		
4	Bays	Eng Muffin	$1.09	12	2	140	5	25	2	11.00917	770.6422		
5	Thomas	Eng Muffin	$1.69	12	2	130	4	25	1	7.100592	461.5385		
6	Quaker	OHs Cereal	$2.49	12	1	120	1	24	2	4.819277	578.3133		
7	Nabisco	Grah Cracker	$2.65	16	0.5	60	1	11	1	6.037736	724.5283		
8	Wheaties	Cereal	$3.19	18	1	100	3	23	1	5.642633	564.2633		
9	Wonder	Bread	$1.31	16	1	70	3	14	1	12.21374	854.9618		
10	Brownberry	Bread	$1.29	16	0.84	60	2	11	1	12.4031	885.9358		
11	Pepperidge	Bread	$1.59	16	0.89	80	2	14	2	10.06289	904.5297		
12													

10   Highlight the range **K1:K11**, click **Insert > Name > Create**, be sure only **Top Row** is selected, and click **OK** to create a range name for your new variable.

The preceding computation adjusts the calories per dollar by including the number of calories per ounce. Of course, this comparison is not perfect either, because calories are not necessarily the best measure of food value. Instead of measuring the energy value in the form of calories, it might be more useful to measure protein.

To measure protein:

1   Type **Protein per Dol** in cell L1 and press [**Enter**].

2   Enter the formula =**Oz_per_Dol*Protein/Serving_oz** in the L column, using the techniques you just learned. Type the variable range names in the cell or use the Names drop-down list, as you prefer.

Freeze Panes

The completed worksheet now contains a number of columns that the bagel company can use to compare its competitors' products with its own products. You might have to increase the widths of some of the columns in order to see each entire column label. In addition, if you find nine decimal places in the calculated variables too much, you can format the appropriate columns to display fewer decimal places.

To display only two decimal places:

1 Highlight the range **J2:L11**.

2 Click the **Decrease Decimal button** 🔢 on the formatting toolbar until the calculated variables are shown to two decimal places. See Figure 2-20.

Figure 2-20
Worksheet with truncated decimals

	C	D	E	F	G	H	I	J	K	L	M	N	O
1	Price	Package oz	Serving oz	Calories	Protein	Carbo	Fat	Oz per Dol	Cal per Do	Protein per Dol			
2	$1.55	14	1	110	3	23	1	9.03	993.55	27.10			
3	$0.99	15	1.5	120	5	25	0.5	15.15	1212.12	50.51			
4	$1.09	12	2	140	5	25	2	11.01	770.64	27.52			
5	$1.69	12	2	130	4	25	1	7.10	461.54	14.20			
6	$2.49	12	1	120	1	24	2	4.82	578.31	4.82			
7	$2.65	16	0.5	60	1	11	1	6.04	724.53	12.08			
8	$3.19	18	1	100	3	23	1	5.64	564.26	16.93			
9	$1.31	16	1	70	3	14	1	12.21	854.96	36.64			
10	$1.29	16	0.84	60	2	11	1	12.40	885.94	29.53			
11	$1.59	16	0.89	80	2	14	2	10.06	904.53	22.61			
12													

Although Figure 2-20 shows all 12 variables, some computer monitors might display fewer columns. When you want to view two parts of a large worksheet simultaneously, you can create **freeze panes**, which allow you to scroll one part of the worksheet while keeping another part always in view. For example, you might want to view the brand and food variables and the calculated variables as a group.

To create freeze panes that let you see the brand and food variables and the calculated variables at the same time:

1 Scroll left, then click cell **C2**.

2 Click **Window > Freeze Panes**.

A dividing line appears between the first and second row and the second and third column, indicating that the first two columns and the first row are frozen in place on the screen.

3 Click the **right scroll arrow** on the horizontal scroll bar until **column J** is aligned with the vertical freeze pane. See Figure 2-21.

Figure 2-21
Freeze panes

horizontal freeze pane border

vertical freeze pane border

	A	B	J	K	L	M	N	O	P	Q	R
1	Brand	Food	Oz per Dol	Cal per Do	Protein per Dol						
2	Anderson	Pretzel	9.03	993.55	12.03						
3	Uncle B	Bagel	15.15	1212.12	18.48						
4	Bays	Eng Muffin	11.01	770.64	13.51						
5	Thomas	Eng Muffin	7.10	461.54	9.10						
6	Quaker	OHs Cereal	4.82	578.31	5.82						
7	Nabisco	Grah Cracker	6.04	724.53	8.04						
8	Wheaties	Cereal	5.64	564.26	8.64						
9	Wonder	Bread	12.21	854.96	15.21						
10	Brownberry	Bread	12.40	885.94	14.78						
11	Pepperidge	Bread	10.06	904.53	12.31						
12											

It is now easier to study the variables you've created. The bagels are first in weight per dollar (cell J3), calories per dollar (cell K3), and protein per dollar (cell L3). Perhaps the bagel company could advertise their bagels as one of the quickest sources of energy and protein for the money when compared with other wheat products.

The products in rows 6, 7, and 8 are all expensive relative to what they provide. Row 6 contains the OHs breakfast cereal, which is the most expensive relative to its weight (that is, it has the lowest Oz per Dol figure, as shown in cell J6), and its calories tend to be nutritionally empty. It has very little protein, which is not surprising considering that, according to the information on the package, its second and third ingredients are sugar and brown sugar, respectively.

The last three products in the list are breads, and they provide more protein per dollar than the cereals. Wonder Bread is highest in protein per dollar among the three breads, which might surprise those who think of it as an "air bread."

The pretzels (Row 2) and the Bays English Muffins (Row 4) have patterns similar to the breads, while the Thomas English Muffins (Row 5) are similar to the breakfast cereals.

Statistics should always be considered in context. Who eats bread, bagels, or breakfast cereals with nothing on them? Based on these data, you could advocate toast over breakfast cereal, but the story could be different when butter, cream cheese, peanut butter, and milk are factored in! The pretzels look good from this point of view, because they can be eaten plain.

To remove the freeze panes, save your data, and exit Excel:

1   Click **Window > Unfreeze Panes**.

2   Click **File > Save As**.

3   Click the **Save File as Type drop-down arrow**, then click **Microsoft Excel Workbook**.

4   Select the appropriate drive and directory containing your Student Disk.

5   Double-click the **File Name text box**, type **2WHEAT**, then click **OK**.

6   Fill out the Summary Info dialog box if you want, then click **OK**. Notice that the title bar changes from WHEAT.TXT to 2WHEAT.XLS: the file is now saved in an Excel format.

7   Click **File > Exit**.

E X E R C I S E S

In the Exercises, you will be instructed to save your work with an "E" prefix (for Exercises) followed by the chapter number and descriptive name.

1. The file POLU.XLS contains air-quality data collected by the Environmental Protection Agency. The data show the number of unhealthy days (heavy levels of pollution) per year for 14 major US cities in the years 1980 and 1985-1989.

    a. Define range names for each of the variables and then investigate the change in the number of unhealthy days from 1985 to 1989 by creating a new column of the difference between 1989 unhealthy days and 1985 unhealthy days. Name the new column Diff.

    b. Sort the data in ascending order, based on the change in unhealthy days between 1985 and 1989.

    c. Use AutoFilter to view only those cities that showed an increase in the number of unhealthy days between 1985 and 1989.

    d. Repeat steps a through c using a new column of the ratio of the 1989 count of unhealthy days to the 1985 count.

    e. (Optional) Using the Advanced Filter, show only those cities that had an increase in the number of unhealthy days from 1985 to 1989 *without* using the newly created columns of differences and ratios between the two years (Hint: You'll have to use a calculated criterion).

    f. Write a paragraph summarizing the change from 1985 to 1989. Discuss the difference and also the ratio for the 14 cities.

    g. Save the file as E2POLU.XLS.

2. Import the ASCII text file BREWER.TXT into Excel (the file is shown in the following table) using the Text Import Wizard. Columns are delimited by tabs. Save the file as E2BREWER.XLS.

Table 2-3
Brewer Data

Company	Yr 1981	Yr 1985	Yr 1989
Anhser-Busch	54.5	68.0	80.7
Miller	40.3	37.1	41.9
Coors	13.3	14.7	17.7
Stroh	23.4	23.4	18.4
Heileman	14.0	16.2	13.0
Pabst	19.2	9.1	6.6
Genesee	3.6	3.0	2.4
Others	10.9	7.0	2.1
Imports	5.2	7.9	8.7

This file has four variables and nine cases. The first variable is the name of the company. The other three variables are their sales in millions of barrels for the years 1981, 1985, and 1989, as given in the *Beverage Industry Annual Manual 91/92* (page 37). Define range names for the four columns.

3. Explore the change from 1981 to 1989 in the E2BREWER.XLS data you saved in Exercise 2. Both the actual changes and the relative changes (ratio of sales) are interesting. Sort the data by the 1981 sales, and list the data, including the difference and ratio, between 1981 and 1989.

    a. In the listing, is there a relationship apparent between the size of a company and its changes?

    b. Could you say that the big are getting bigger?

    c. Write a paragraph discussing your conclusions.

4. The following table shows data for the Big Ten universities. It includes the freshman class size in 1984, percentage of students graduating by 1989, male athletes recruited in 1984, percentage of male athletes graduating by 1989, female athletes recruited in 1984, percentage of female athletes graduating by 1989, average freshman score on the ACT test, and percentage of freshmen who finished among the top 10% in high school. Penn State declined to respond to the survey.

Table 2-4
Big Ten Data

College	Freshmen 1984	Percent by 1989	Males 1984	Percent by 1989	Females 1984	Percent by 1989	ACT	Percent Top Ten
Illinois	5881	76.2	71	62.0	28	71.4	27	58
Indiana	5602	57.6	79	54.4	25	52	24	34
Iowa	3594	55.4	61	59.0	34	64.7	24	25
Michigan	4627	76.5	82	54.9	27	77.8		69
Mich State	6440	59.7	54	55.6	20	75	23	26
Minnesota	4685	27.0	77	24.7	37	59.5		27
N. Western	1833	86.0	55	83.6	17	88.2	28	82
Ohio State	7709	46.2	95	47.4	60	61.7	22	23
Purdue	4435	66.7	68	58.8	26	73.1	23	35
Wisconsin	5608	59.8	86	54.7	29	62.1		33

Chronicle of Higher Education, March 27, 1991, pages A39-A44 and *U.S. News and World Report*, "America's Best Colleges 1992."

Type the data into Excel and define names for the columns. Save the data as E2BIGTEN.XLS.

5. For the data in Exercise 4, find the difference and ratio between male and female athlete graduation rates. Make a general statement comparing the male and female graduation rates.

6. For the data in Exercise 4, examine the difference and ratio between the female graduation rate and the overall university graduation rate.

   a. How does the female athlete percentage compare with the overall rate?

   b. Does one university stand out? Use AutoFilter to display only the university with the highest female to overall graduation rate ratio.

7. For the POLU.XLS data:

   a. Calculate the average number of unhealthy days for each city from 1985 through 1989.

   b. Calculate the ratio of this average to the number of unhealthy days in 1980.

   c. Sort the list by this ratio. How is New York ranked? Summarize your conclusions.

8. Open the WHEAT.XLS file (this file contains the same data you imported in this chapter and saved as 2WHEAT.XLS):

   a. Compute a new variable, Cal Oz, the ratio of calories per unit weight. Do this by dividing the calories by the serving size in ounces.

   b. Use AutoFilter to discover which foods have 100 calories or more per ounce. What is it about these foods that causes them to have more calories? (Hint: These foods have only a small amount of a particular ingredient that has weight but no calories.)

   c. Save the file as E2WHEAT.XLS.

9. The ALUM.XLS file contains measurements on eight aluminum chunks from a high school chemistry lab. Both the mass in grams and volume in cubic centimeters were measured for each chunk. In the worksheet, compute the density (the ratio of mass to volume).

   a. Is there an outlier (an observation that stands out as being different from the others)? Ignoring the outlier, the remaining values of density should be all the same constant, except for observation errors (the value given in chemistry books is 2.699).

   b. Save the file as E2ALUM.XLS.

10. The ECON.XLS file has seven variables related to the US economy from 1947 to 1962. The Deflator variable is a measure of the inflation of the dollar, and it is arbitrarily set to 100 for 1954. The last variable, Total, contains total employment.

    a. Notice that the Population variable shows increases in population each year. Use the Sort command to find out for which other variables this is true by sorting a variable and then checking to see if the years are still in ascending order.

    b. There is an upward trend to GNP, although it does not increase every year. What about GNP per person? Calculate the ratio of GNP to the Population. Call the new variable "GNPPOP." Sort the data set to find out if it shows an increasing trend. Print the data set and summarize your results.

    c. Save the file as E2ECON.XLS.

CHAPTER 3

SINGLE VARIABLE GRAPHS AND STATISTICS

O BJECTIVES

In this chapter you will learn to:

- Create frequency tables and histograms

- Calculate basic summary statistics

- Create and interpret boxplots

- Use add-ins, templates, and macros to expand Excel's capabilities

- Interpret and transform distribution shapes

- Calculate and interpret descriptive statistics

Statistics and Distributions

In Chapter 2, you used Excel's filter and sorting capabilities to gather information about a data list. In this chapter you will learn about numbers called **statistics** that describe data. For example, the **average**, or **mean**, value is a commonly used statistic that describes the typical value in a data set. Another important statistic is the **standard deviation**, which indicates how widely the data are dispersed about the mean.

Before applying these and other statistics to data, it is important to first look at the **distribution** of the data, or how the data are spread over a range of values. In some distributions the data tend to cluster around certain values, whereas for other distributions the data are spread evenly through a range of values. Although you might be interested in the values of certain statistics, you might find after looking at the distribution of the data that those statistics are no longer pertinent.

For example, you might be interested in the average salary at a business. However, computing the mean salary is not meaningful if the boss makes a lot of money and everyone else is near the minimum wage. The mean is somewhere between the boss's salary and what others make, so it might look impressively high when, in reality, it is high only because the distribution of data from which it is calculated includes one very high salary. Nobody would actually have a salary near the average. In this distribution, the mean is nearly useless for describing a typical wage.

Frequency Tables and Histograms

Two of the main tools a statistician uses to explore the distribution of data are frequency tables and histograms. **Frequency tables** are tables that divide data into ranges and show the count, or frequency, of all values in that range. A **histogram** is a bar chart of these frequencies in which the length of the bar for each range is proportional to the number of values within that range. Let's use these two tools to investigate the distribution of data from a data list involving personal computers.

In 1993, a consumer information service studied and tested 191 prominent 486 PCs and published the information in a popular computing journal (*PC Magazine*, Vol. 12, #21, pp.148-149). The workbook PCINFO.XLS contains the price, performance, make, and model of these PCs.

For the rest of this book, whenever you open a workbook, you will be instructed to save it immediately to your Student Disk, adding the prefix of the chapter number. This prevents you from saving your work over the original file. If you want to go back through the chapter using the original files, they are still available to you.

To open PCINFO.XLS, be sure that Excel is started and the windows are maximized, and then:

1 Open **PCINFO.XLS** (be sure you select the drive and directory containing your Student files).

2 Click **File > Save As**, click the drive containing your Student Disk, then save your workbook on your Student Disk as **3PCINFO.XLS**.

The workbook opens to the first sheet, PC Data (as you can see from the sheet tab), which contains a list of 192 rows, or records, and six columns, or fields. See Figure 3-1.

Figure 3-1
PCINFO.XLS worksheet

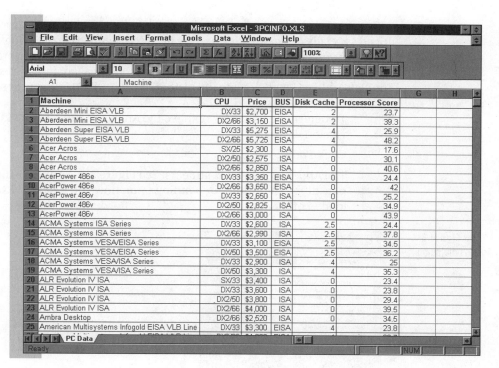

Range names for the individual columns of data have already been created for you, and a freeze pane has been added using the procedures described in Chapter 2. The variables in the list include those described in Table 3-1:

Table 3-1
PC Data Variables

Range Name	Range	Description
Machine	A1:A192	The brand and model of the PC.
CPU	B1:B192	The central processing unit, or chip, that handles the computing; note that there may be several models of a single machine, each outfitted with a different CPU.
Price	C1:C192	The cost in dollars of the PC.
BUS	D1:D192	The connection, or *electrical highway*, through which information within the computer is passed. There are two BUS types in this list: ISA (Industry Standard Architecture) and EISA (Extended Industry Standard Architecture).
Disk Cache	E1:E192	The size (in megabytes) of a memory chip designed to speed the reading and writing of data from the computer disk.
Processor Score	F1:F192	Test results of the speed of the PC's processor. Higher scores indicate better CPU performance.

If you were thinking of purchasing one of these PCs, your first question might be: "What is the typical price for a 486 machine?" Another question close on the heels of the first could be: "If I plan to spend between X and Y dollars for a 486, how many of these machines will be in that price range?" And if you're still asking questions, a third might be: "What is an unreasonable price to pay for a 486?" In other words, which prices should you regard as unusual when compared with others? You want to get a feeling not just for the average price, but also for the distribution of the prices. This is where a frequency table and histogram become useful.

The Frequency Table and Histogram Add-Ins

Excel does not have the ability to create frequency tables and histograms in its default configuration (that is, as it is originally installed), but you can add this capability through the use of add-ins. As the name suggests, **add-ins** are extra features added to Excel. Excel comes with several add-ins that you can choose to include. Third-party developers have also produced specially designed add-ins, one of which you'll use in this book. The features added to Excel give you, among other things, the ability to create special reports, design slide shows, and, most importantly for this book, perform statistical analyses. Specifically, the Analysis ToolPak Add-Ins let you create the histograms described in the next section.

To determine whether you have the Analysis ToolPak Add-Ins installed with your version of Excel:

1   Click **Tools > Add Ins**.

The Add-Ins dialog box appears. See Figure 3-2. Your Add-Ins dialog box might display different options, depending upon how you installed Excel and whether you've loaded other add-ins.

Figure 3-2
Add-Ins dialog box

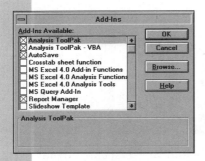

2   Verify that the Analysis ToolPak Add-Ins check box is selected; click it if it is not.

If you are not able to click the Analysis ToolPak check box, this set of add-ins might not have been included with the original installation. If you are using your own computer, use the Excel installation disks to install the add-ins. If you are running Excel from a school computer or network server, ask your instructor or technical support person to install these add-ins for you.

3   Click **OK** to close the Add-Ins dialog box.

The Analysis ToolPak Add-Ins add the Data Analysis command to the Tools menu. See Figure 3-3.

Figure 3-3
Tools menu

Data Analysis command

Creating a Frequency Table and Histogram

You are ready to analyze the distribution of the price variable using Excel's histogram add-in.

To create a frequency table and histogram of price:

1   Click **Tools > Data Analysis**.

2   Click **Histogram** in the Analysis Tools list box. See Figure 3-4.

Figure 3-4

Data Analysis dialog box

Analysis Tools list box

Histogram

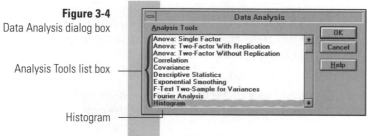

3   Click **OK**.

4   Type **Price** in the Input Range box (the name has already been defined for you to represent the range C1:C192 on the PC Data sheet).

5   Click the **Labels check box** because the range name Price includes the column label.

6   Click the **New Worksheet Ply box** and type **Price Histogram** so that Excel will create a new worksheet named Price Histogram and place the histogram there.

7   Click the **Cumulative Percentage** check box to calculate cumulative percentages (you'll see why these are important in a moment).

8   Click the **Chart Output check box** to create the histogram chart. The Histogram dialog box should look like Figure 3-5.

Figure 3-5

Histogram dialog box

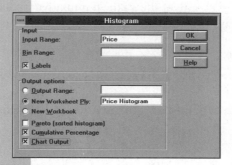

9   Click **OK**.

Excel creates a new worksheet in your workbook called Price Histogram. See Figure 3-6.

Figure 3-6

Price Histogram worksheet

frequency table

histogram chart

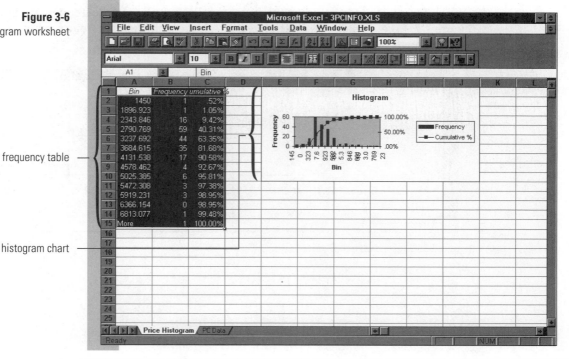

Creating a Frequency Table and Histogram

- Click Tools > Data Analysis (if this command does not appear, you must install the Analysis ToolPak Add-Ins).

- Click Histogram in the Analysis Tools list box.

- Enter the range containing the data for the frequency table and histogram. If you are using customized bin values, enter the range for the bin values.

- Select the Cumulative Percentage check box to produce cumulative percent values. Select the Chart check box to create the histogram.

- You can send the frequency table and histogram output to a cell, new worksheet, or new workbook.

The frequency table appears in the range A1:C15. Column A displays the **bin values**, or boundaries for each range of values (the next section explains how Excel calculates these boundaries). Column B shows the frequency of values within these bins (the bin values given in column A are the upper boundaries for each bin). Column C indicates the **cumulative percentage**: the percentage of values within or below the bin value. The histogram chart in the range E1:J10 shows these values graphically. The bars display the frequency values, and the curve (also known as an **ogive**) shows the cumulative frequency. Before continuing, format the frequency table output so that you can interpret it more easily.

To format the table:

1. Widen column **C** (as described in Chapter 2) so that the label **Cumulative %** is not truncated.

2. Select the range **A2:A14** and click the **Currency Style button** 🖻.

3. Click the **Decrease Decimal button** 🖾 twice to format the price values in whole dollar amounts.

4. Decrease the width of column **D** to show more of the histogram chart in the active window (if your monitor requires this).

5. Click cell **A1** to remove the highlighting and view the list and the chart.

Excel reformats the bin values in the histogram list as well as in the frequency table. See Figure 3-7. Excel's charts are *linked* to values on the worksheet; therefore, when you change data or data formats, the charts change as well.

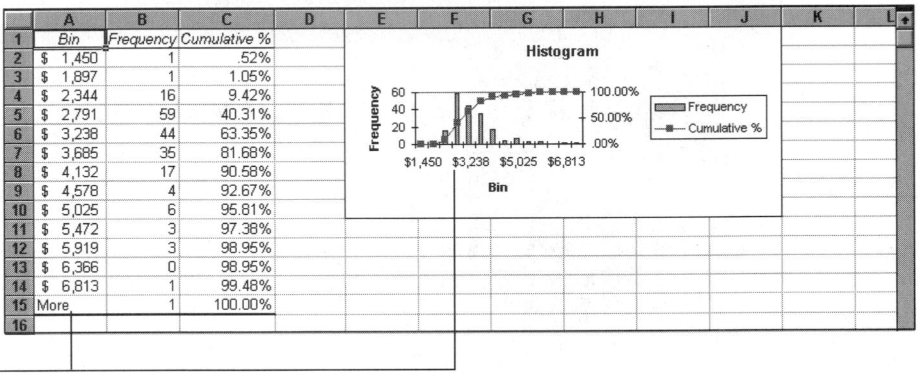

Figure 3-7
Formatted Price Histogram worksheet

bin values

50   Frequency Tables and Histograms

Interpreting the Frequency Table and Histogram

Excel creates a set of evenly distributed bins between the data's minimum and maximum values (the number of bins depends on the number of data points). The frequency of each bin is the number of observations less than or equal to the bin value and greater than the previous bin value. For example, in Figure 3-7 the information in cell B6 tells you that there are 44 prices less than or equal to $3,238 (from cell A6) and greater than $2,791 (from cell A5). The final row in the bin column, row 15, contains the value "More," with a frequency of 1, which indicates that there is one price greater than $6,813.

From the frequency table you can see at a glance that roughly 90% of the prices are less than or equal to $4,132 (from cell C8). If, for example, you were planning to spend from $1,900 to $2,800 on a 486 computer, about 75 (16 + 59 from cells B4 and B5) models would be within that price range.

Changing the Bin Intervals

The default bin ranges Excel chose resulted in bin boundaries that are not even numbers. If you want your bin boundaries to fall on even values, such as in intervals of $500 from $1,000 to $1,500, you can recreate the frequency table with your own bin values.

To create a histogram with the new bin ranges, first create a column that contains the bin ranges you want; then create the histogram as before:

1   Click the **PC Data sheet tab**.

2   Click cell **H1**, type **New Bin**, and press **[Enter]**.

3   Type **1000** in cell H2, press **[Enter]** and type **1500** in cell H3.

4   Select cells **H2** and **H3**, click the fill handle and drag it down to cell **H15**, then release the mouse button.

Excel creates a column containing the bin ranges you want in your frequency table and histogram chart.

5   Click **Tools > Data Analysis**, then double-click **Histogram**.

6   Type **Price** in the Input Range box and press **[Tab]**.

7   Select **H2:H15** on the PC Data sheet (move the dialog box out of the way to see this column, if necessary). This enters the selected range automatically into the Bin Range box. (You can also simply type the range.)

8   Click the **Labels check box** to select it, if necessary.

9   Change the name in the **New Worksheet Ply box** to Price Histogram 2, then click **OK**.

The histogram appears on a new page, Price Histogram 2.

10 Format the histogram list as you did for the first histogram so that the Price Histogram 2 worksheet looks like Figure 3-8.

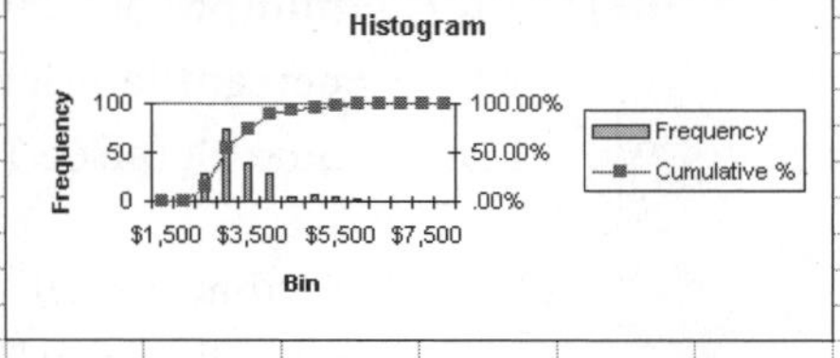

Figure 3-8
Histogram with customized bin ranges

	A	B	C	D	E	F	G	H	I	J	K
1	Bin	Frequency	Cumulative %								
2	$ 1,500	1	.52%								
3	$ 2,000	1	1.05%								
4	$ 2,500	29	16.23%								
5	$ 3,000	73	54.45%								
6	$ 3,500	39	74.87%								
7	$ 4,000	29	90.05%								
8	$ 4,500	5	92.67%								
9	$ 5,000	6	95.81%								
10	$ 5,500	4	97.91%								
11	$ 6,000	2	98.95%								
12	$ 6,500	1	99.48%								
13	$ 7,000	0	99.48%								
14	$ 7,500	1	100.00%								
15	More	0	100.00%								
16											

These new bin values are easier to interpret. In Figure 3-8, cell C5 shows that over half (54.45%) of the PCs in the list cost $3,000 or less. The frequency of prices from $2,500 to $3,000 (73, in cell B5) is by far the largest of the 15 bins in this table. You might conclude that a typical price for a 486 computer ranges from $2,500 to $3,000. But what is the price that is in the center of the distribution so that half the prices are lower and half the prices are higher than it? This statistic is called the **median**. From the frequency table you might guess that the median is close to $3,000, but you can't be sure of this because you don't know how the prices within each bin are distributed. With Excel you can calculate the median value as well as other statistics describing the distribution.

Statistics Describing a Distribution

Excel offers more than 70 statistical functions that aid in data analysis. Functions often used to describe the distribution of data include:

Table 3-2
Excel statistical functions

median	the middle value when the data are lined up in order; if there is an even number of observations, then the median is the average of the middle two.
percentile	the pth percentile has p% of the data below it; the median is the 50th percentile.
first quartile	the 25th percentile, which has 25% of the data below it
second quartile	the 50th percentile, which is the same as the median
third quartile	the 75th percentile, which has 75% of the data below it
minimum	the lowest value of the data list
maximum	the highest value of the data list

Let's use these functions on the current data set to create a table of statistics describing the distribution of the price data.

To calculate the descriptive statistics:

1 Select the range **A19:A25** on the Price Histogram 2 worksheet, then enter the following row labels into this range, pressing [**Enter**] after each one:

A19	**Minimum**
A20	**5th Percentile**
A21	**25th Percentile**
A22	**Median**
A23	**75th Percentile**
A24	**95th Percentile**
A25	**Maximum**

2 Select the range **B18:B25**, type **Price** in cell B18, and press [**Enter**].

3 Enter the following formulas into the selected range, pressing [**Enter**] after each one:

B19	**=min(price)**	calculates the minimum price
B20	**=percentile(price,0.05)**	calculates the price at the 5th percentile
B21	**=quartile(price,1)**	calculates the price at the first quartile (the 25th percentile)
B22	**=median(price)**	calculates the median price
B23	**=quartile(price,3)**	calculates the price at the 3rd quartile (the 75th percentile)
B24	**=percentile(price,0.95)**	calculates the price at the 95th percentile
B25	**=max(price)**	calculates the maximum price

Your values should look like those shown in Figure 3-9 (resize column A to see the complete row labels).

Figure 3-9
Cells A18:B25

18		Price							
19	Minimum	1450							
20	5th Percentile	2250							
21	25th Percentile	2600							
22	Median	2950							
23	75th Percentile	3510							
24	95th Percentile	4875							
25	Maximum	7260							

Creating this table gives you values for specific percentiles in the distribution of the price data. In Figure 3-9, the median price is $2,950, verifying your hunch that the typical price for a 486 PC is around $3,000. The prices range from a minimum of $1,450 to a maximum of $7,260, but 90% of prices fall in the range $2,250 to $4,875, a much smaller interval. Finally, the central 50% of the data ranges from $2,600 (the 25th percentile) to $3,510 (the 75th percentile). The width of this range, $910, is known as the **interquartile range**.

You now have a good idea of how the PC data are distributed, and you know what the typical price of a 486 computer is, as well as the price you can expect for the cheapest and most expensive brands. In addition to price, there are other factors, such as performance, that you want to take into consideration when comparing PCs. You'll explore this factor in future chapters.

For now, save and close the computer workbook. You'll use the Save command rather than the Save As command. Recall from Chapter 2 that Save As lets you select a new location and filename for your file. You've already renamed this file 3PCINFO and saved it to your Student Disk, so you can use the Save command to save this version with the same name to the same location.

To save and close the 3PCINFO workbook:

1 Click the **Save button** 🖫 to save your changes to the 3PCINFO.XLS file (the drive containing your Student Disk should already by selected).

2 Click **File > Close**.

Boxplots

An important tool for viewing a distribution is a boxplot, described in John Tukey's influential 1977 book, *Exploratory Data Analysis*. The **boxplot** is a useful way of displaying the quartiles and extreme values (largest and smallest) of the distribution. Excel does not produce boxplots nor does it include boxplots with its add-ins, but the Student files that come with this book include both a boxplot template that introduces you to boxplot feature, and a boxplot menu command that lets you create boxplots.

The Boxplot Template

Several instructional templates are included with your Student files to allow you to interactively explore basic statistical concepts. One of these templates shows you how to create boxplots. A **template** is a special workbook you can use as a pattern to create other workbooks of the same type. Templates can contain text, graphics, formatting styles, page layout instructions, formulas, and macros. When you open a template, Excel creates a copy of the template for you to work with. This leaves the original template document intact. No matter what changes you make to the copy, you can always get the original version of the template back by reopening the document.

To open the boxplot template:

1 Click the **Open button** 🖼️.

2 Select the drive and directory containing the Student files.

3 Double-click **BOXPLOT.XLT** (templates have the file extension XLT).

The empty boxplot template opens in your Worksheet window. See Figure 3-10.

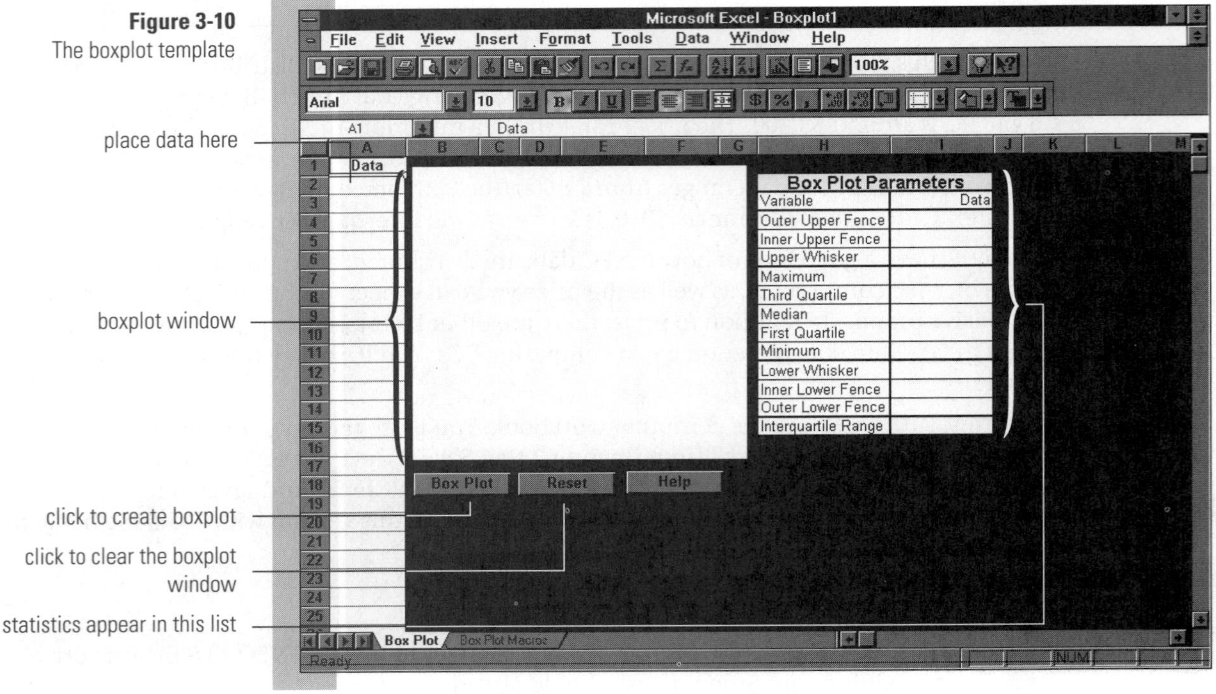

Figure 3-10
The boxplot template

place data here →

boxplot window →

click to create boxplot →

click to clear the boxplot
window →

statistics appear in this list →

Note that the name on the title bar is not BOXPLOT.XLT but Boxplot1. This is a copy of the original template file BOXPLOT.XLT. Any changes you make in this document will not affect that source document.

The boxplot template creates a boxplot from data entered into column A, which appears in the boxplot window (now empty). The boxplot shows the distribution of the data; and statistics describing the distribution, such as the median, first and third quartiles, and the interquartile range, appear in the Box Plot Parameters table. The three buttons beneath the boxplot window have the following functions: Box Plot creates the boxplot using the data in column A; Reset clears the boxplot window to its startup position, removing any data in the process; and Help opens on-line Help.

To enter a set of numbers to see how the boxplot displays a distribution of data:

1 Highlight the range **A2:A16**.

2 Type the numbers **0, 1, 2, 2, 3, 3, 3, 3, 4, 4, 4, 5, 5, 6, 7** into cells A2:A16 (one integer for each row), pressing **[Enter]** after each entry.

3 Click the **Box Plot button**.

The macro included in this template runs through a predefined series of steps to create the boxplot, which appears as shown in Figure 3-11.

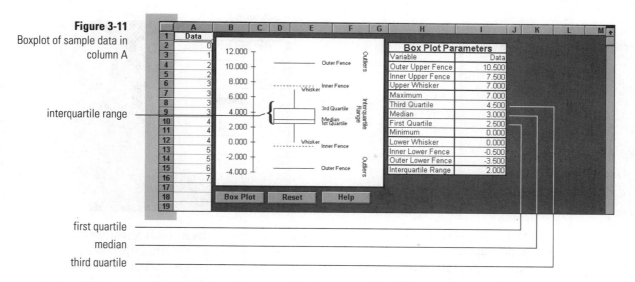

Figure 3-11
Boxplot of sample data in column A

interquartile range

first quartile

median

third quartile

The box in the boxplot indicates the center of the distribution. The bottom of the box is drawn at the first quartile, and the top at the third quartile. For these data, the box shows that the central 50% of the data covers the range 2.5 to 4.5 (check the Box Plot Parameters statistics to verify these numbers). The height of the box is the interquartile range, in this example, 2.0. The median value, 3, appears as a solid line across the box (note that the median is not necessarily halfway between the first and third quartiles).

The lines extending from the box are known as **whiskers**; they indicate the upper and lower range of the data that lie outside the box. The length of a whisker is limited; it can be no more than 50% longer than the interquartile range. The locations at these maximum distances are called **inner fences**. Therefore, the inner fence is a distance of 1.5 times the interquartile range from the first and third quartiles, that is:

> inner fences = first quartile – 1.5(interquartile range) and third quartile + 1.5(interquartile range)

Beyond the inner fence is the outer fence. The outer fence is a distance of 3.0 times the interquartile range from the first and third quartiles, that is:

> outer fences = first quartile – 3.0(interquartile range) and third quartile + 3.0(interquartile range)

Almost all the values in the distribution fall between the two inner fences (boxplots usually do not show the fences; they are included here for instruction). Data values that fall beyond the inner fences are known as **outliers** (extreme values). Outliers can affect the validity of some of the statistics you use to describe your data—particularly the mean. The boss's salary discussed in the chapter introduction is an example of an outlier. The boxplot is an excellent way to discover the presence of outliers.

To see how the boxplot displays outliers, enter some extreme values in column A:

1   Select the range **A15:A16**.

2   Type **9** in A15 and press [**Enter**].

3   Type **12** in A16 and press [**Enter**].

Figure 3-12
Boxplot with mild and
severe outliers

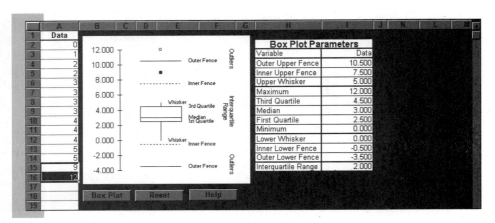

Note from Figure 3-12 that Excel automatically updates the boxplot as you change the source data. Observations that lie beyond the inner fence but within the outer fence are called **moderate outliers**, and appear in the boxplot with the • symbol. Values that lie beyond the outer fence are called **severe outliers**, and appear with the o symbol. One of the more important uses of the boxplot is to highlight extreme observations. Usually you should look carefully at an outlier to see why it differs so much from the other data points.

On your own, try changing a few more values and see how the boxplot changes shape. If you want to start with a new set of data, click the Reset button. Because the boxplot is a simple plot that displays the values of several important statistics, it has widespread use. In the next section you'll use an actual menu command, not a template, to create a boxplot using real data.

To close the boxplot template file when you are finished experimenting:

1   Click **File > Close**.

2   Click **No**, because you don't need to save changes you made to the template copy.

The CTI Statistical Add-Ins

Let's apply the boxplot to some real data. Neither Excel nor the Analysis ToolPak provides a menu command to create boxplots, but you can create them using the CTI Statistical Add-Ins provided with your Student files.

To open the CTI Statistical Add-Ins:

1   Click the **New Workbook button** 🗋 to open a new workbook.

2   Click **Tools > Add Ins**.

3   If the **CTI Statistical Add-Ins check box** appears in the list box as in Figure 3-13, click it to select it, if necessary (you might have to scroll to see it), then click **OK** (you can then skip steps 4 through 8). If the check box does not appear, complete steps 4 through 8.

Figure 3-13
Add-Ins dialog box

CTI Statistical Add-Ins
check box

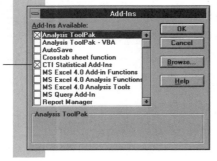

If you can't find the CTI Statistical Add-Ins check box, these add-ins might not have been installed with the version of Excel you are using. Depending upon your institution, you might have to install the add-ins yourself, or your instructor or technical support person might do it for your class. You can install the add-ins in any directory on the hard disk, but once they are installed, do not move them to another directory or else workbook files you've created using the add-ins functions will not work, and will have to be relinked with the add-ins in their new location. Once the add-ins have been installed, you do not need to reinstall them. Further information on the CTI Statistical Add-Ins is included in the appendix at the end of this book.

Continue with steps 4 through 8 to install the CTI Statistical Add-Ins from your Student files if they are not already installed. These instructions assume that your CTI.XLA file is located in the C:\EXCEL\STUDENT directory, as specified in Chapter 1.

4   Click **Browse** in the Add-Ins dialog box.

5   Select the drive and directory containing your Student files and the add-ins file (although you can run the add-ins off a disk, performance is better if the Student files and the add-ins are on the hard drive).

6   Double-click **CTI.XLA** (the XLA extension stands for Excel Add-Ins).

7   The CTI Statistical Add-Ins should now appear in the Add-Ins list box, so click the **CTI Statistical Add-Ins check box** to select it if it is not already selected. If a message appears asking if you want to store a copy of the add-ins in the LIBRARY directory of Excel, click **No** so that the CTI.XLA file remains in your Student dictory. If you need further assistance, ask your instructor or technical support person about the appropriate location for the add-ins on your hard drive.

8   Click **OK**.

A new menu called CTI appears on the menu bar (if it wasn't there already).

Applying the Boxplot to Real Data

Let's apply the boxplot menu command to data on the top 50 women-owned businesses from the Wisconsin State Department of Development. The data are entered into an Excel workbook named WBUS.XLS.

To open WBUS.XLS and then save it as 3WBUS.XLS:

1   Open **WBUS.XLS** (be sure you select the drive and directory containing your Student files).

2   Click **File > Save As**, select the drive containing your Student Disk, then save your workbook on your Student Disk as **3WBUS.XLS**.

The workbook opens, displaying one worksheet, Business Data, with a data list of fifty rows and six columns ordered by sales. See Figure 3-14.

Figure 3-14
WBUS.XLS worksheet

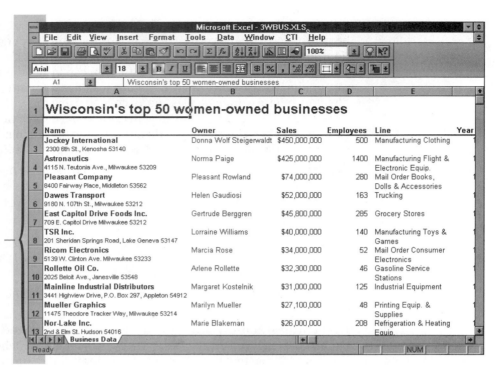

rows are ordered by sales,
from highest to lowest

Range names and the freeze pane have already been created for you. The variables in the data list are in Table 3-3:

Table 3-3
Business Data variables

Range Name	Range	Description
Name	A2:A52	The name and address of the business
Owner	B2:B52	The owner or owners of the business
Sales	C2:C52	The total sales in 1993 of the business
Employees	D2:D52	The number of persons employed by the business
Line	E2:E52	The focus of the business
Year	F2:F52	The year the business was started

Let's first examine the distribution of the number of employees in these 50 businesses.

To create a boxplot of the employee data showing the number of persons employed:

1 Click **CTI > Boxplots** to open the Create Box Plots dialog box.

2 Type **Employees** (the range name for the employee data) in the Input Columns box.

3 Verify that the **Selection Includes Header Row check box** is selected, because the range name includes a column label.

4 Click the **New Sheet option button**.

5 Type **Emp Boxplot** in the New Sheet box so that the Create Box Plots dialog box looks like Figure 3-15.

Figure 3-15
Completed Create Box Plots
dialog box

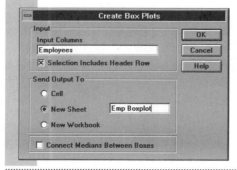

6   Click **OK**.

Note: The CTI Boxplots command requires that the data be organized into columns.

Creating a Boxplot

- Click CTI > Boxplots (if this command does not appear, you must install the CTI Statistical Add-Ins).

- Enter the range containing the data for the boxplot (indicate whether or not the range contains a header row).

- Enter the output destination: cell, new sheet, or new workbook.

A boxplot showing the distribution of the number of employees appears on the Emp Boxplot worksheet. The first thing you notice is that the distribution has four outliers, three of which are severe outliers. See Figure 3-16. The interquartile range as shown by the dimensions of the box covers a range from near zero to around 150 employees, but the range of all the data covers from close to zero to around 1,400 employees. The median appears to be around 50 employees.

To calculate the exact values of these statistics:

1   Select the range **A18:B23**.

2   Enter the following row labels into A18:A23:

    A18   **Minimum**

    A19   **25th Q**

    A20   **Average**

    A21   **Median**

    A22   **75th Q**

    A23   **Maximum**

3   Enter the following formulas into B18:B23:

    B18   **=min(employees)**

    B19   **=quartile(employees,1)**

    B20   **=average(employees)**

    B21   **=median(employees)**

    B22   **=quartile(employees,3)**

    B23   **=max(employees)**

Your boxplot and values should look like Figure 3-16.

Figure 3-16
Boxplot and table of descriptive statistics

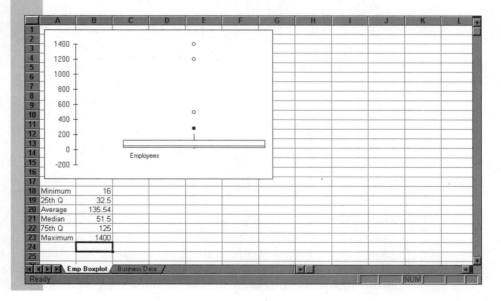

There is a great discrepancy between the median number of employees (51.5 in cell B21) and the mean number of employees (135.54 in cell B20) because of the outliers. In fact, the effect of the business with the large number of employees is so severe that the mean value is greater than the 75th percentile. This is an example of positively skewed data.

Distribution Shapes

Histograms and boxplots can give you a lot of information on the shape of the distribution, which plays an important role in determining the appropriate statistical analysis to perform. If data are distributed symmetrically with most of the values near the center, then the mean is near the center of the distribution and is useful in describing the distribution. But as you've seen, if the data are **skewed**, as they are in the employee example for the women-owned businesses, the mean might not be very useful and could, in fact, be misleading.

Figure 3-17 shows three possible distribution shapes of sample data in histogram and box-plot form. As a general rule for determining skewness:

Mean > median accompanies positive skewness
Mean = median accompanies symmetry
Mean < median accompanies negative skewness

Figure 3-17
Histograms and boxplots showing distribution shapes

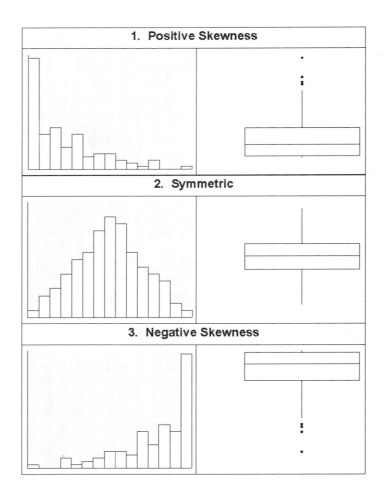

If you plan further statistical analysis on data distributions showing negative skewness (long tails stretching down on a boxplot) or positive skewness (long tails stretching up on a box-plot), you should be concerned about the skewness. Sometimes you can transform the shape of the distribution when this is the case to create a more symmetric distribution. Data involving counts, such as the number of employees, are often skewed in the positive direction because, by definition, the counts cannot be negative and low values often cluster near zero. Transforming the variable might bring it into balance. Specifically, a logarithm or square-root transformation might shorten the long tail, making the distribution more symmetric.

Consider using a logarithmic transformation for the employee data. Because these data extend through several orders of magnitude, it is natural to classify a business size by powers of 10 (that is, think of groups of companies with 10-100 employees, 100-1,000 employees, 1,000-10,000 employees, and so on). You are more likely to group a pair of companies with employee sizes of 5,000 and 5,500 together rather than two companies with 50 and 500 employees, even though the arithmetic difference is less for the second pair. Analyzing the logarithm of employee size is a way of mathematically expressing how you naturally compare company sizes.

To transform the employee data:

1 Click the **Business Data sheet tab** to return to the worksheet containing the data.

2 Click cell **H2**, type **Log(Employees)**, then press [**Enter**].

3 Select the range **H3:H52** (notice how the freeze panes are in effect so that you can still see the column labels).

4 Type **=log(employees)** in cell H3, then press [**Enter**] (you can paste the name from the Name drop-down list if you want to save typing).

5 Press [**Ctrl**]+[**D**] to apply the formula to the entire range, then click any cell to deselect the range.

The base-10 logarithms of the employee counts are now entered into cells H3:H52, as shown in Figure 3-18.

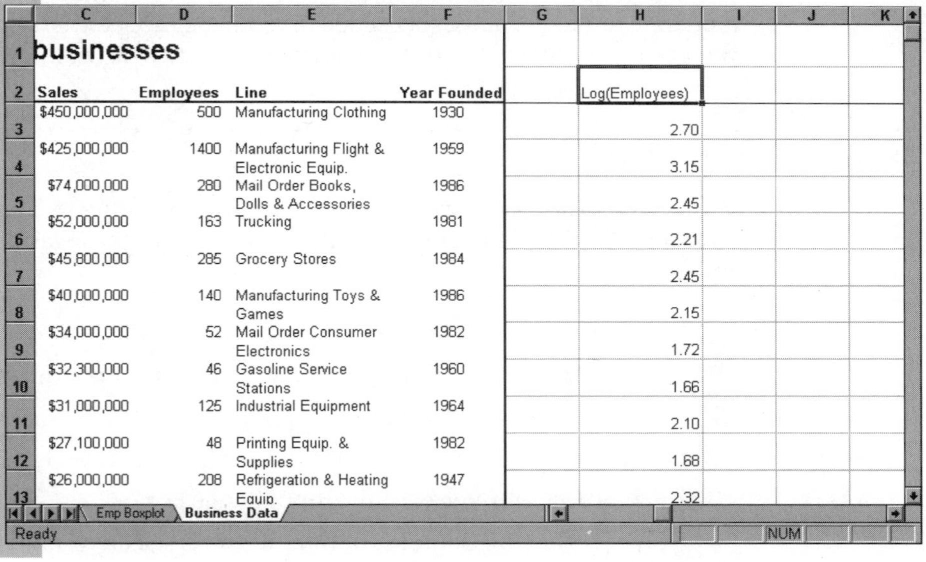

	Sales	Employees	Line	Year Founded	Log(Employees)
1	businesses				
3	$450,000,000	500	Manufacturing Clothing	1930	2.70
4	$425,000,000	1400	Manufacturing Flight & Electronic Equip.	1959	3.15
5	$74,000,000	280	Mail Order Books, Dolls & Accessories	1986	2.45
6	$52,000,000	163	Trucking	1981	2.21
7	$45,800,000	285	Grocery Stores	1984	2.45
8	$40,000,000	140	Manufacturing Toys & Games	1986	2.15
9	$34,000,000	52	Mail Order Consumer Electronics	1982	1.72
10	$32,300,000	46	Gasoline Service Stations	1960	1.66
11	$31,000,000	125	Industrial Equipment	1964	2.10
12	$27,100,000	48	Printing Equip. & Supplies	1982	1.68
13	$26,000,000	208	Refrigeration & Heating Equip.	1947	2.32

Figure 3-18
Column H, showing transformed values

Now use a trick to create a range name for the transformed variable you just created. You can use the cell reference box to specify the range name rather than using the Insert menu as you did in Chapter 2.

To create a range name for the transformed data and column label:

1 Select **H2:H52**.

2 Click the **cell reference box** (currently showing the cell reference H2).

3 Type **log_employees**, then press [**Enter**]. Excel creates a new range name for the range H2:H52.

You are now ready to check the transformed data by viewing their distribution in a boxplot.

To create the boxplot of the transformed data:

1   Click **CTI > Boxplots**.

2   Type **log_employees** in the Input Columns box.

3   Verify that the **Selection Includes Header Row check box** is selected (because you've included the column label in the range).

4   Click the **New Sheet option button** to select it.

5   Type **Log Emp Boxplot** in the New Sheet box, then click **OK**.

The boxplot appears as shown in Figure 3-19.

Figure 3-19
Boxplot of logarithmic
transformation of number of
employees

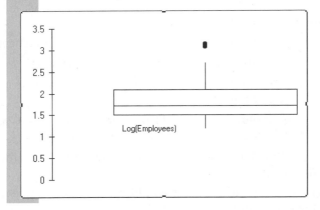

The logarithmic transformation eliminated the severe outliers, leaving you with only two moderate outliers. You've succeeded in reducing the skewness of the data. As a final step, you can compare the descriptive statistics of the two distributions.

Other Descriptive Statistics

The Analysis ToolPak provided by Excel includes a feature to create a table of basic statistics. You'll use this feature to create a table of statistics for the logarithm of the number of employees.

To create a table of descriptive statistics:

1   Click the **Business Data sheet tab** to return to the worksheet containing the data.

2   Click **Tools > Data Analysis**.

3   Click **Descriptive Statistics** in the Analysis Tools list box, then click **OK**.

4   Type **log_employees** in the Input Range box.

5   Verify that the **Columns option button** is selected and that the **Labels in First Row check box** is selected.

6   Click the **Summary statistics check box** to select it.

7   Click **New Worksheet Ply**, then click the corresponding box and type **Log Emp Stats**. The Descriptive Statistics dialog box should look like Figure 3-20.

Figure 3-20

Descriptive Statistics dialog
box

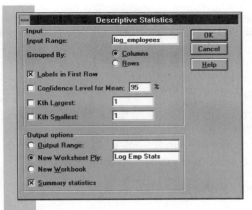

8   Click **OK**.

Excel creates a new worksheet titled Log Emp Stats and displays the summary statistics.

9   Format column **A** to fit the longest entry.

10   Click cell **A1** to remove the highlighting and view the list. The Log Emp Stats worksheet should look like Figure 3-21.

Figure 3-21

LogEmps Stats worksheet

	A	B	C	D	E	F	G	H	I	J
1	*Log(Employees)*									
2										
3	Mean	1.839479								
4	Standard Error	0.061317								
5	Median	1.711787								
6	Mode	2.146128								
7	Standard Deviation	0.433577								
8	Sample Variance	0.187989								
9	Kurtosis	1.313708								
10	Skewness	1.105934								
11	Range	1.942008								
12	Minimum	1.20412								
13	Maximum	3.146128								
14	Sum	91.97393								
15	Count	50								
16	Confidence Level(95.000%)	0.120179								
17										

Creating a Table of Descriptive Statistics

- Click Tools > Data Analysis (if this command does not appear, you must install the Analysis ToolPak Add-Ins).

- Click Descriptive Statistics in the Analysis Tools list box.

- Enter the range containing the data for the descriptive statistics table and indicate whether the variables are arranged in rows or columns. Also indicate whether the first row (if the data are arranged in columns) or first column (if the data are arranged in rows) contains the variable label.

- Select the Summary Statistics check box to produce a table of descriptive statistics.

- You can send the frequency table and histogram output to a cell, new worksheet, or new workbook.

Measures of Central Tendency: Mean, Median, and Mode

The Descriptive Statistics command returns a table of basic statistics that includes three measures of **central tendency** (statistics that describe the most "typical" value of a data list): the mean, median, and mode. See Figure 3-21. The mean and median have already been discussed in this chapter; you calculated them using formulas containing functions. Note that in Figure 3-21, in contrast to the raw employee counts, the mean (1.839 in cell B3) and the median (1.712 in cell B5) of the transformed data are close together, indicating that you have achieved some degree of symmetry in the distribution. If you want to express the mean and median in terms of the number of employees, you would raise the transformed values to the power of ten: Median: $10^{1.1712} = 51.498$; Mean: $10^{1.8395} = 69.100$.

Note that the median of the transformed data raised to the 10th power is the same as the median of the untransformed data. This is to be expected, because taking the logarithm does not change the order of the values. Transforming the mean back to the original scale is equivalent to another measure of central tendency called the **geometric mean**, usually expressed as the nth root of the product of the data values, or:

$$\text{Geometric mean} = \sqrt[n]{(X_1)\,(X_2)\,(X_3)...(X_n)}$$

where $X_1, X_2, X_3, \ldots X_n$ are the values of the data list. In this case the geometric mean = $(500*1400*280*\ldots 25)^{\frac{1}{50}}$. The geometric mean is often used instead of the arithmetic mean when the data are in the form of proportions (values ranging from 0 to 1).

Another summary measure of values in the data list is the **mode**, which is the most frequently occurring value in a distribution. In Figure 3-21, the mode is 2.146, or in terms of the number of employees, $10^{2.146} = 140$. The most common number of employees in the 50 women-owned businesses is 140.

Measures of Shape: Skewness and Kurtosis

The skewness, as you have seen, is a measure of the lack of symmetry of the distribution. Recall that a positive skewness indicates that the tail extends up on a boxplot, whereas a negative skewness indicates a long tail extending down; a perfectly symmetric distribution has a skewness value of 0. In Figure 3-21, the skewness value for the transformed data is 1.106 (cell B10), confirming what you saw in the boxplot—a relatively symmetric distribution when you compare it to the value for the untransformed data, as you'll see in the exercises.

Kurtosis (cell B9) indicates the heaviness of the tails of the data distribution. Distributions with much of the data located in the tails (heavy tails) have a high kurtosis value. In Figure 3-21, the kurtosis value (1.314) indicates that this data list does not have long tails.

By transforming the employee data, you've taken your first step in preparing this data list for such statistical analyses. Many statistical analyses that you will perform later in this book expect the data to be at least symmetric and do not perform well with heavily skewed data. In the exercises at the end of this chapter, there are comparisons of the skewness and kurtosis statistics for the untransformed employee counts.

Measures of Variability: Range, Standard Deviation, and Variance

Variability is the extent to which data values differ from each other; in other words, how widely the data are spread out. A simple measure of this is the **range** (maximum value – minimum value). The range is easy to calculate but is also heavily influenced by outlying values and is, therefore, considered a crude measure of variability. More commonly used is the standard deviation.

The **standard deviation** summarizes how far the data values are from the average, or mean, value. The difference between a data value and the mean is called the **deviation**. For the log(employees) values the mean is 1.8395 and the deviations are {2.699 – 1.8395 = 0.86; 3.146 – 1.8395 = 1.31; ... 1.398 – 1.8395 = –0.44}. The sum of the deviations is the sum of the differences from the mean and is always equal to zero, so the average deviation is also zero and does not give you any useful information regarding the variability of the data. Statisticians commonly use the standard deviation to summarize the values of the deviations where:

$$\text{Standard deviation} = \sqrt{\frac{(d_1^2 + d_2^2 + ... + d_n^2)}{n-1}}$$

and d_1, d_2, . . . d_n are the deviations. For the log(employees) data, the value of the standard deviation is:

$$\text{Standard deviation} = \sqrt{\frac{(0.86^2 + 1.31^2 + \ldots + (-0.44)^2)}{49}} = 0.4336$$

You can interpret the standard deviation as the "typical" deviation of a data value from the mean. In Figure 3-21, the typical, or standard, deviation is 0.4336 units (cell B7) from the mean value, 1.8395. Like the mean, the standard deviation is susceptible to the influence of outliers and is very large if severe outliers are present.

The **variance** is the square of the standard deviation. For the log(employees) data, the variance is 0.188 (cell B8).

To save the 3WBUS.XLS workbook and exit Excel:

1. Click the **Save button** to save your changes to the 3WBUS.XLS file (the drive containing your Student Disk should already be selected).

2. Click **File > Exit**.

E X E R C I S E S

1. A data distribution has a median value of 22, a first quartile of 20, and a third quartile of 30. Six values lie outside the interquartile range with values of 17, 18, 18, 40, 50, and 70.

   a. Draw the boxplot for this distribution.

   b. Is the skewness positive, negative, or zero?

2. Open the file 3WBUS.XLS, which you saved at the end of Chapter 3.

   a. Create a table of descriptive statistics using the Analysis ToolPak Add-Ins for the untransformed employee data.

   b. What are the skewness and kurtosis values and how do they compare with the statistics for the logarithm of the number of employees? What does this tell you about the relative shapes of the two distributions?

   c. The **coefficient of variation** is a measure of the standard deviation adjusted for the size of the mean value, specifically: coefficient of variation = (standard deviation) ÷ mean.

   You can think of the coefficient of variation as a measure of the variability as a fraction of the average. Using information you obtained from the descriptive statistics tables, compute the coefficient of variation for the employee and log(employee) data. What does the coefficient of variation tell you about the relative variability of the two data sets?

   d. The coefficient of variation statistic is useful only for data sets with values always greater than zero. What would happen to the coefficient of variation if some of the data values were negative and resulted in the data set having a mean value of zero?

   e. Save your file as E3WBUS.XLS.

3. Create a new variable for the E3WBUS.XLS workbook equal to the total sales per $1,000, divided by the number of employees (sales ÷ 1,000 ÷ employees). Give the new column of data the range name sales_per_emp.

   a. Create a boxplot for the total sales of each business.

   b. Create a boxplot for the total sales per $1000 per number of employees.

   c. What has adjusting the sales data for the number of employees done for the shape of the distribution? Explain why the shape changes.

   d. Save your changes to the E3WBUS.XLS file, then close the file.

4. The BASE.XLS data have salary values for 263 major-league baseball players at the start of the 1988 season.

   a. Obtain a histogram and boxplot for the salaries of the players.

   b. Calculate the 10th and 90th percentiles for the salaries.

   c. Using the value for the 90th percentile, create an AutoFilter of the data list and determine the players that were in the upper 10% of the salary range.

   d. Save your file as E3BASE.XLS.

5. Repeat Exercise 4 using the logarithm of salary.

   a. Is its distribution more symmetric, and does it have less skewness?

   b. Is the mean closer to the median?

   c. Save your changes to the E3BASE.XLS file, then close the file.

6. In Chapter 2 you looked at the Diff variable in the POLU.XLS data set and saved the result in the file E2POLU.XLS. Diff was the difference between the 1985 to 1989 average pollution and the 1980 pollution.

   a. Obtain distribution graphs (histogram and boxplot) and summary statistics for Diff. Is the mean useful to summarize the differences? What about the median?

   b. Is there a problem in considering both the difference for Boston, which started out with low pollution, and the difference for New York, which started out with high pollution? Are the 14 cities too diverse to summarize with one number?

7. Repeat Exercise 6 using instead the ratio of the 1985 to 1989 average pollution to the 1980 pollution.

   a. Is New York less of an outlier here?

   b. Is the distribution skewed?

   c. Does the median give a better summary of the data than does the mean? Does it make more sense to deal with all 14 cities in terms of the ratio?

   d. Could you say that there were roughly half as many polluted days per year during the period 1985 to 1989 as there were during 1980?

   e. Save your file as E3POLU.XLS.

8. Repeat Exercise 6 using the logarithm of the ratio.

   a. Is it more symmetric? Is the skewness lower?

   b. Is the median closer to the mean? It often makes sense to look at ratios and logarithms when you are looking for a percentage (fractional) change, as opposed to the simple difference. This is especially true when there is a wide initial range.

9. You can use the CTI Statistical Add-Ins to create boxplots of multiple variables on one plot. Try this with the E3POLU.XLS data.

   a. Select the range B2:G16 and create a boxplot using the CTI Statistical Add-Ins Boxplots menu command (make sure you select the Selection Includes Header Row check box). Select the Connect Medians Between Boxes check box.

   b. What additional insight does this plot give you?

   c. What information is lost with this plot as compared to your previous efforts?

SCATTERPLOTS

O BJECTIVES

In this chapter you will learn to:

- Plot one variable against another using a scatterplot

- Edit a chart to change its scale

- Label data points

- Plot several variables on the y-axis

- Identify points on a chart with a grouping variable

- Create, interpret, and edit a scatterplot matrix

In Chapter 3 you looked at some basic statistics and charts in order to display the distribution of one variable. More often you'll be interested in how variables relate to each other. The best place to start assessing this relationship is with charts. In this chapter, you'll use Excel to create charts that show several data series plotted against each other.

Creating Scatterplots with the Chart Wizard

To show the basics of plotting, let's begin with a simple **scatterplot**, a chart that illustrates the relationship between two variables.

The Excel file BIGTEN.XLS contains information on students who enrolled in and graduated from Big Ten universities. As Table 4-1 shows, the columns include:

Table 4-1 Big Ten information	Range Name	Range	Description
	University	A1:A11	The name of the college
	Top Ten Percent	B1:B11	The percentage of freshmen who were in the top 10% of their high school graduating class
	ACT	C1:C11	The average ACT score for freshmen enrolling in 1984
	Freshman	D1:D11	The number of freshmen enrolled in 1984
	Grad Percent	E1:E11	The percentage of those freshmen graduating by 1989
	Male Recruit	F1:F11	The number of male athletes recruited in 1984
	Male Percent	G1:G11	The percentage of male athletes who graduated by 1989
	Female Recruit	H1:H11	The number of female athletes recruited in 1984
	Female Percent	I1:I11	The percentage of female athletes who graduated by 1989
	Computer.	J1:J11	The ratio of the number of students per computer.

To open the BIGTEN.XLS workbook, be sure that Excel is started and the windows are maximized, and then:

1   Open **BIGTEN.XLS** (be sure you select the drive and directory containing your Student files).

2   Click **File > Save As**, select the drive containing your Student Disk; then save your workbook on your Student Disk as **4BIGTEN.XLS**.

The workbook opens to the first sheet, Big Ten, which contains a list of 11 rows, or records, and 10 columns, or fields. See Figure 4-1. Range names based upon the column labels of each column have already been created for you. Freeze panes have also been created so that you can view the row and column labels as you move around the worksheet.

As you look over the data list, notice that the graduation percentage for Ohio State (cell E9) ranks ninth in the Big Ten and it is one of only two schools in which less than 50% of the 1984 freshman class graduated by 1989. You might wonder what makes this school so different. Perhaps the composition of the 1984 freshman class was different from the other universities in the Big Ten?

Let's start investigating the Big Ten data by charting the relationship between the percentage of freshmen who graduated in the top 10% of their high school graduating class (Top Ten Percent) and the percentage of freshmen who graduated within five years of having started college (Grad Percent).

Figure 4-1
4BIGTEN.XLS workbook

Name drop-down arrow ———

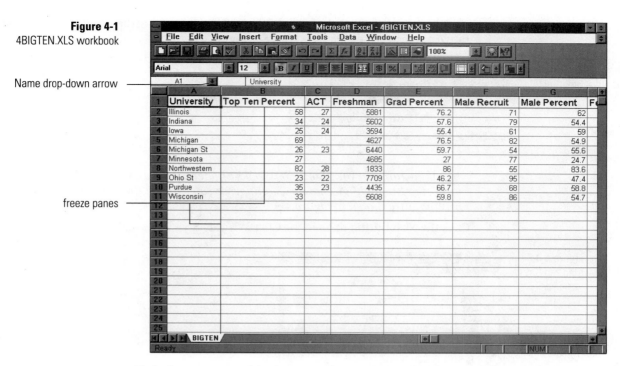

freeze panes ———

To help you create this chart, Excel provides a tutor called the Chart Wizard that walks you through a set of five steps. To use the Chart Wizard to plot Top Ten Percent against Grad Percent, first highlight the data you want to chart, then start the Chart Wizard and follow its instructions.

To create a scatterplot showing the relationship between Top Ten Percent and Grad Percent:

1 Click the **Name drop-down arrow** (shown in Figure 4-1), click **Top_Ten_Percent**, press and hold down the [Ctrl] key, then click the **Name drop-down arrow** again and click **Grad_Percent**. (You've learned several ways to highlight columns; this method uses the range names and might be the quickest method.)

The Worksheet window looks like Figure 4-2.

Figure 4-2
Highlighted columns for creating a scatterplot

	A	B	C	D	E	F	G
1	University	Top Ten Percent	ACT	Freshman	Grad Percent	Male Recruit	Male Percent
2	Illinois	58	27	5881	76.2	71	62
3	Indiana	34	24	5602	57.6	79	54.4
4	Iowa	25	24	3594	55.4	61	59
5	Michigan	69		4627	76.5	82	54.9
6	Michigan St	26	23	6440	59.7	54	55.6
7	Minnesota	27		4685	27	77	24.7
8	Northwestern	82	28	1833	86	55	83.6
9	Ohio St	23	22	7709	46.2	95	47.4
10	Purdue	35	23	4435	66.7	68	58.8
11	Wisconsin	33		5608	59.8	86	54.7
12							

2 Click the **ChartWizard button** on the Standard toolbar.

The pointer arrow turns into a crosshair + which you use to designate the area on the worksheet where you want to place the chart.

3 Drag the crosshair to select a rectangle from cell **B13** (be sure you start to the right of the freeze pane) to cell **E24** so that the chart will appear in the range B13:E24.

The ChartWizard - Step 1 of 5 dialog box appears, showing the two columns whose range names you selected to chart. See Figure 4-3.

Figure 4-3
ChartWizard — Step 1 of 5
dialog box

range addresses

Range box

Note: If the range in your Range box looks different, you can either click Cancel and repeat step 1 to highlight the correct range, or you can edit the Range box until it contains the entry shown in Figure 4-3.

4   Click the **Next button** to open the ChartWizard - Step 2 of 5 dialog box. See Figure 4-4.

Figure 4-4
ChartWizard — Step 2 of 5
dialog box

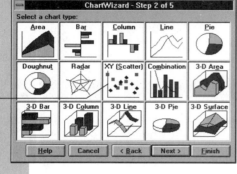

XY (Scatter)

5   Double-click **XY (Scatter)** to select a scatterplot as your chart type and to move to the next step.

6   Click **1** in the ChartWizard - Step 3 of 5 dialog box to show data values with no connecting lines or gridlines, then click **Next**.

Excel opens the ChartWizard - Step 4 of 5 dialog box, which includes a preview of what your chart will look like with the options you've selected.

7   Make sure that the **Columns option button** is selected, then click the **Use First Column(s) up spin arrow** so the Use First Column(s) For X Data spin box contains **1** (a spin box accepts numbers that you select by clicking the up or down spin arrow to scroll through a series of numbers). This tells Excel to use the first highlighted column, B1:B11, as the data for the x-axis of the scatterplot.

8   Make sure that the Use First Row(s) for Legend Text spin box contains **1** (this tells Excel to consider the first row of the selected columns as your column labels).

The ChartWizard - Step 4 of 5 dialog box should look like Figure 4-5.

Figure 4-5
ChartWizard — Step 4 of 5
dialog box

spin boxes

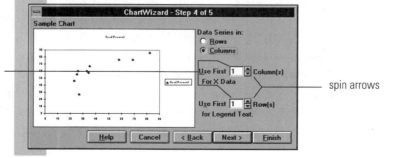

spin arrows

9   Click **Next** to open the ChartWizard - Step 5 of 5 dialog box, and click the **No option button** under Add a Legend? Because there is only one y-variable, no legend is necessary.

10   Click the **Chart Title box**, type **Big Ten Graduation Percentage**, press [**Tab**], type **Top Ten Percent** in the Category (X) box, press [**Tab**], then type **Grad Percent** in the Value (Y) box to add titles to your chart.

The ChartWizard - Step 5 of 5 dialog box should look like Figure 4-6.

Figure 4-6
ChartWizard — Step 5 of 5
dialog box

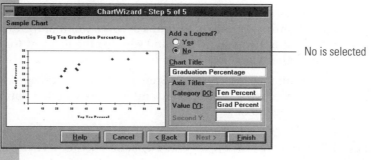

11   Click **Finish**.

The scatterplot appears in the worksheet range you selected, B13:E24, as shown in Figure 4-7 (you might have to scroll to see it). A small Chart toolbar might appear with buttons that you can use to create and format charts. The Chart toolbar shown in Figure 4-7 appears only when the chart is selected, and even then it depends on how your version of Excel is configured.

Figure 4-7
Worksheet window with
chart in B13:E24

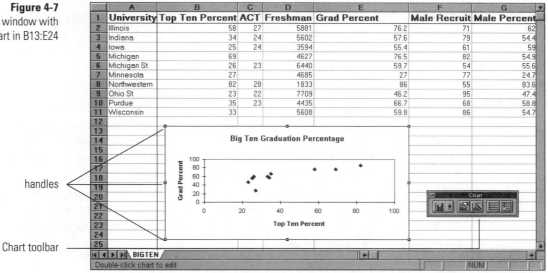

	A	B	C	D	E	F	G
1	University	Top Ten Percent	ACT	Freshman	Grad Percent	Male Recruit	Male Percent
2	Illinois	58	27	5881	76.2	71	62
3	Indiana	34	24	5602	57.6	79	54.4
4	Iowa	25	24	3594	55.4	61	59
5	Michigan	69		4627	76.5	82	54.9
6	Michigan St	26	23	6440	59.7	54	55.6
7	Minnesota	27		4685	27	77	24.7
8	Northwestern	82	28	1833	86	55	83.6
9	Ohio St	23	22	7709	46.2	95	47.4
10	Purdue	35	23	4435	66.7	68	58.8
11	Wisconsin	33		5608	59.8	86	54.7

handles

Chart toolbar

The chart shows the relationship between Grad Percent and Top Ten Percent. As the percentage of incoming freshmen who graduated in the top ten percent of their high school class increases, the percentage of freshmen who graduated from college within five years generally increases.

Creating a Scatterplot

- Select the range of data you want to plot.

- Click the ChartWizard button ▣ on the Standard toolbar and drag the cross hair over the area in which you want to place the chart.

- Select the scatterplot chart type in the ChartWizard - Step 2 of 5 dialog box.

- Select the scatterplot format in the ChartWizard - Step 3 of 5 dialog box.

- In the ChartWizard - Step 4 of 5 dialog box, indicate whether the data are organized by columns or rows, whether the first column (or row) is used for the x-axis, and whether the first row (or column) is used for the legend text.

- In the ChartWizard - Step 5 of 5 dialog box, enter the chart, and the x-axis and y-axis titles, and indicate whether you want to show the legend box.

Note: When you are working on your own with the ChartWizard to create scatterplots, you might try to chart two noncontiguous columns. In this case, the Chart Wizard assumes that both variables are y–variables and charts them against the row number. To tell Excel to treat the first column as the x–variable, be sure that you set the Use First Column(s) For X Data spin box to 1 in the ChartWizard - Step 4 of 5 dialog box.

Editing a Chart

The chart you've just created is an **embedded chart**, which is a chart that Excel places on the worksheet. The charts you created in Chapter 3 were also embedded in the worksheet. You can place a chart on a different kind of sheet, called a **chart sheet**, that is specially designed for displaying charts. Later in this chapter you'll see how to create a chart sheet. For now, take a look at the scatterplot that the Chart Wizard created for you.

Resizing a Chart

Notice that the chart you just created is automatically selected, as you can tell by the eight handles (the small squares on the chart's border, as shown in Figure 4-7). The Chart Wizard automatically selects the chart after creating it; but if it is not selected, no handles appear, and you can select it by clicking anywhere within its border.

You use the handles to resize the chart. As you move the pointer arrow over the handles, you'll see the pointer change to a double-headed arrow of various orientations, ↕, ↔, ↗, or ↘, indicating that you can resize the chart in the direction indicated by the pointer. Try making your chart a little bigger.

To enlarge the chart:

1. Move the cross pointer ✛ over the lower-middle handle; the pointer changes to a vertical double-headed arrow pointer ↕.

2. Drag the lower-middle handle to cell **E29**.

   The chart now covers the range B13:E29. Save your work.

3. Click the **Save button** ▣ to save your file as **4BIGTEN.XLS** (the drive containing your Student Disk should already be selected).

Activating a Chart

If you want to modify the chart, you must first activate it. Activating a chart gives you access to commands that modify different chart elements such as the title, axis, or plot symbol.

To activate your chart:

1 Double-click the **chart** anywhere within its borders.

The chart border changes from a thin line to a thick, colored or gray border, indicating that the chart is active. See Figure 4-8.

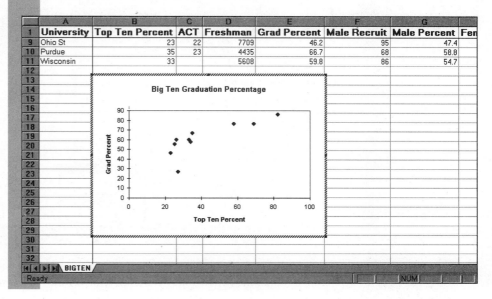

Note: If your chart appears in its own window and does not look like Figure 4-8, don't worry. The chart is too big for your monitor but you can still use it. You can also click the Zoom Control arrow and select a reduced zoom factor.

When you activate a chart, new commands appropriate to charts become available on the menus. Selecting Insert from the menu bar, for example, opens a menu containing commands for inserting objects such as titles, data labels, and legends. See Figure 4-9. If a cell were active, however, this menu would let you insert cells, rows, columns, and so on.

Figure 4-9
Insert menu for worksheet with active chart object

Changing a Chart's Scale

One of the first things you might have noticed about the chart is that the default scale creates a blank space in the lower-left corner of the data point area. Even though the graduation percentage and the percentage of freshmen from the top ten percent of their high school class are both above 20%, Excel created the chart with the scales for the two axes starting at 0%. There are situations where you want the axes formatted to show a complete range of possible values and other times where you want to concentrate only on observed values. In this case, let's rescale the axes to display only the observed values.

To change the scale for the horizontal (x) axis:

1   Right-click the **horizontal axis** so that the name **Axis 2** appears in the formula bar chart reference and the shortcut menu appears. See Figure 4-10.

Figure 4-10
Chart with horizontal axis selected and shortcut menu

Axis 2, horizontal axis, is selected

border indicates chart is active

horizontal axis

shortcut menu

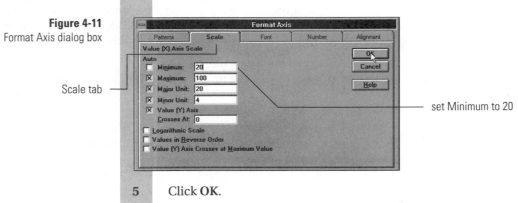

Note: When the active object in the worksheet is a cell, the formula bar displays a cell reference. When the active object is a chart, the formula bar displays a chart reference.

2   Click **Format Axis** on the shortcut menu.

3   Click the **Scale tab** in the Format Axis dialog box to display scale options.

4   Double-click the **Minimum box**, type **20** to set the Value (X) Axis scale minimum to 20 (Figure 4-11).

Figure 4-11
Format Axis dialog box

Scale tab

set Minimum to 20

5   Click **OK**.

The scale for the horizontal axis now ranges from 20 to 90 percent (Excel automatically reduced the maximum to 90 when you selected the minimum of 20).

To modify the scale for the vertical (y) axis:

1   Double-click the **vertical axis**.

2   Click the **Scale tab** in the Format Axis dialog box if it's not already in the forefront.

3   Double-click the **Minimum box**, type **20** to set the Value (Y) Axis scale minimum to 20, then click **OK**.

Note: You right-clicked the x-axis to modify its scale, then you double-clicked the y-axis to modify its scale. Either method gives you access to the Format Axis dialog box.

Your chart now looks like Figure 4-12.

Figure 4-12
Chart with modified x-axis
and y-axis scales

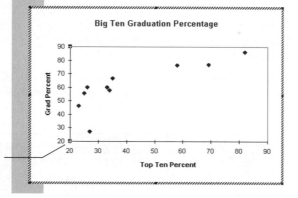

new scale minimum

···

Modifying the Scale of a Scatterplot Axis

• Double-click the chart to activate it if it is not already active.

• Double-click the axis to open the Format Axis dialog box.

• Click the Scale tab and enter the new range for the axis in the minimum and maximum scale boxes, then click OK.

···

You can also modify other elements of the chart, such as the font or size of the title or axis label text, the plot symbols, or the background color, by right-clicking the object and choosing the appropriate shortcut menu command or by double-clicking the object. Now, let's interpret the scatterplot.

···

Working with the Data Series

The points on the scatterplot are known as the **data series**. In this chart there is only one data series—the series of values that describes the relationship between Grad Percent and Top Ten Percent. Each of its points comes from the values of the Grad Percent and Top Ten Percent columns. For example, you could scroll up the worksheet to see that the point on the upper right is Northwestern, where Grad Percent = 86 and Top Ten Percent = 82 (that is, 82% of the students in the 1984 freshman class at Northwestern graduated in the top 10% of their high school class, and 86% graduated from college in five years or less). Excel plots the point at 82 horizontally and 86 vertically. (The y–variable is plotted vertically and the x–variable is plotted horizontally.) Ohio State, with a Top Ten Percent of 23 and a Grad Percent of 46.2, appears roughly in line with other Big Ten schools, even though it is at the low end of the scale.

However, there is a point at the lower left that departs from the trend. It is so far out that there could conceivably be an error here. Which university is it? You could continue to compare the worksheet values with the chart to find out, although you certainly wouldn't want to keep scrolling up and down to identify data points, especially for larger data sets. You can label each data point with its corresponding university name to avoid having to scroll between the chart and the data list.

Labeling a Single Data Point

To label the lowest data point with its university name, select just that data point and use the Insert Data Label command (available through the menu bar or the shortcut menu).

To label the lowest data point with its university name:

1 Make sure the chart is active (indicated by the thick, colored border or its own window).

2 Click the lowest data point. Excel highlights all the data points in the series, and the name **S1** appears in the formula bar chart reference if you have selected the plot symbols correctly.

3 Press and hold down the [**Ctrl**] key, and while it is pressed, click the lowest data point to select it, as shown in Figure 4-13. The pointer arrow turns into a four-pointed arrow ✛.

Figure 4-13
Chart with lowest data point selected

formula bar chart reference is S1P6

pointer selecting lowest data point

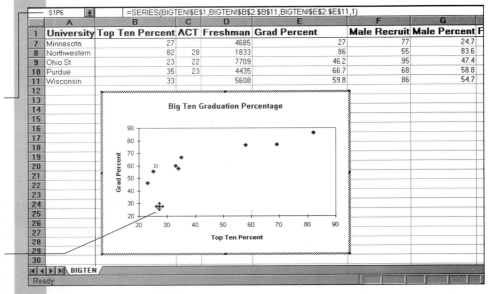

Note: Be sure S1P6 (data series 1, point 6) appears in the formula bar chart reference. If it doesn't, you might have mistakenly selected the entire data series. Repeat steps 2 and 3, being careful to click only the lowest data point with the [Ctrl] key pressed.

4 Right-click the **lowest data point**, which you just selected to open the shortcut menu.

5 Click **Insert Data Labels** on the shortcut menu to open the Data Labels dialog box. See Figure 4-14.

Figure 4-14
Data Labels dialog box

Excel provides you with three options to attach a label to this point:

Data Labels dialog box option	Result
None	clears any data labels previously entered for this point
Show Value	labels the point with the corresponding y-value (in this case, the graduation percentage)
Show Label	labels the point with the corresponding x-value (in this case, the top ten percentage)

6   Click **Show Value**, then click **OK**.

Excel attaches the y-axis value (27, the percentage of students graduating within five years) to the point. To replace this value with the name of the university, you can use a formula that tells Excel where to find the university name for that data point. Note from the chart reference that this is the sixth point (P6) in the data series. Recall that the university name is in column A. Because the first row of column A contains the data labels, the sixth point is contained in the seventh row of column A, so you will enter A7 as your worksheet reference (that is, the cell whose text you want Excel to link to this data point).

To enter the formula that specifies the university name location:

1   Verify that a selection box surrounds the text (27) you want to modify.

2   Type **='BIGTEN'!A7** (be sure to start with an equals sign and include the single quotes around the worksheet name as well as the exclamation point that separates the sheet name and the cell reference). What you type appears in the formula bar.

3   Press [**Enter**].

The selection box now contains the university name Minnesota, indicating that the point on the chart corresponds to the graduation and top ten percentages from the University of Minnesota. See Figure 4-15.

Figure 4-15
Selection box showing
university name

chart reference —

formula —

selection box —

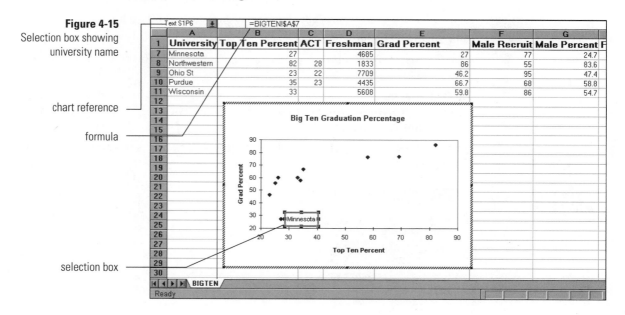

Labeling Multiple Data Points

In a large data set it would be tedious to use this procedure to label every data point individually. Because the labels are already stored in a column, it would be better to enter all the data labels by simply directing Excel to that column. Excel does not provide this capability directly, but your Student files include a macro that does this for you. The macro comes with the CTI Statistical Add-Ins which you used in Chapter 3.

To open the CTI Statistical Add-Ins:

1   Click **any cell** outside the chart to select the worksheet. Now look at your menu bar. If you see a menu named CTI, the CTI Statistical Add-Ins have already been installed on your computer and you can skip steps 2 and 3.

2   Click **Tools > Add Ins**.

3   If the **CTI Statistical Add-Ins check box** appears in the list box, as shown in Figure 4-16, click it (you might have to scroll to see it), then click **OK**. If the check box doesn't appear, refer to Chapter 3 for instructions on accessing the CTI Statistical Add-Ins and adding them to the list of add-ins.

Figure 4-16
Add-Ins dialog box

CTI Statistical Add-Ins is
selected

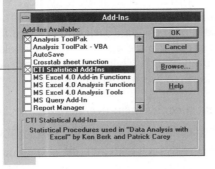

The CTI menu name should now appear on the menu bar.

To attach the appropriate names to each data point:

1   Double-click the **scatterplot** to activate the chart.

2   Select the **data series** by clicking one of the data points. The name **S1** appears in the formula bar chart reference, indicating that you have selected the entire data series.

3   Right-click the **data series**. The data series shortcut menu opens.

4   Click **Insert ID Labels**.

5   With the Insert ID Labels dialog box active, click outside the chart to activate the worksheet.

6   Select the range **A1:A11** on the worksheet. You may need to click the sheet to de-select the chart, then scroll up to select the range. The range **BIGTEN!$A$1:$A$11** appears in the Input box.

7   Verify that the **Selection Includes Header Row check box** is selected and that the dialog box looks like Figure 4-17.

Figure 4-17
Completed Insert ID Labels
dialog box

8   Click **OK**.

Watch while Excel moves from data point to data point, labeling each point with its corresponding university name. When all the labels are in place, click any cell outside the chart twice to deactivate and then deselect the chart, which now looks like Figure 4-18.

Figure 4-18
Chart of Big Ten data with
university name labels

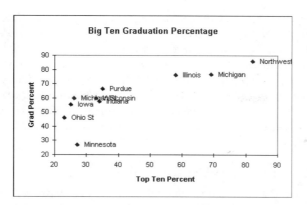

When you label every data point, there is often a problem with crowding. Points that are close together tend to have their labels overlap, as you can see here with Wisconsin and Michigan State. This is not necessarily bad if you are interested mainly in the outliers. Minnesota certainly stands out at the bottom of the chart, perhaps enough to make you want to write to the University of Minnesota to verify the Grad Percent value. However, if you were to check other sources for information on graduation percentage rates, you would find values that are close to this one, so it is very likely correct.

> *Labeling Points on a Scatterplot with Values from Another Variable*
>
> - Make sure that the CTI Statistical Add-Ins are installed.
> - Double-click the chart to activate it if it is not already active.
> - Right-click the chart series that you want to add the labels to, and select Insert ID Labels from the shortcut menu.
> - With the Insert ID Labels dialog box still active, click outside the chart to activate the worksheet and drag the mouse over the range of data containing the ID labels; then click OK.

You might still be wondering why the Grad Percent for the University of Minnesota is so much lower than the rates for the other universities. Perhaps it is because Minneapolis–St. Paul is the largest city among Big Ten towns, and students might have more distractions there. Columbus is the next largest city, and Ohio State is next to last in graduation rate, which seems to verify this hypothesis. On the other hand, Northwestern is in Evanston, right next door to Chicago, the biggest Midwestern city, so you might expect it to have a low graduation rate, too. However, Northwestern might be different because it is a private school and has an elite student body. These facts could account for it having the highest graduation rate of the ten universities. Knowing something about its students, you might have predicted that it would stand out at the top of the plot.

Plotting Several Y's

It is possible to plot more than one variable on the y-axis at the same time. As an illustration, let's plot Grad Percent, Female Percent, and Male Percent simultaneously on the y-axis and Top Ten Percent on the x-axis, with different symbols representing the different series. This allows comparison of the overall university graduation rates with the rates for female and male athletes. Are graduation rates different for the two sexes?

You placed the previous chart right on the BIGTEN worksheet. Excel also provides chart sheets, sheets in the workbook devoted exclusively to charts. When you create a chart sheet, Excel inserts it into the current workbook to the left of the worksheet it is based on. Use the same Chart Wizard options that you used before, but this time place your scatterplot on a chart sheet instead of a worksheet.

Note: Excel uses the left-most column from a contiguous range as the independent, or x–variable in its Chart Wizard. In this case, because Top Ten Percent is left-most, Excel automatically charts it on the x-axis. When your independent variable is to the right of your dependent variables, you need to move or copy the column so that it is the left-most of the variables you are charting. Select the data range containing the independent variable, click Edit > Copy, click the column label of the column that is currently the left-most of the dependent variables, click Insert > Copied Cells, click Shift Cells Right, then click OK. Excel copies the independent variable to the left of the dependent variables, and you can now use the Chart Wizard successfully. You'll try this in the exercises.

To plot the three variables against Top Ten Percent:

1   De-select the chart if it is selected, click the **Name drop-down arrow**, click **Top_Ten-_Percent**, press and hold down the [**Ctrl**] key, click the **Name drop-down arrow**, click **Grad_Percent**, and then select **Male_Percent** and **Female_Percent** in the same way.

2   Click **Insert > Chart > As New Sheet**.

3   Verify that the Range box contains **=$B$1:$B$11,$E$11:$E$11,$G$1:$G$11,$I$1:$I$11** in the ChartWizard - Step 1 of 5 dialog box; then click **Next**.

4   Click **XY (Scatter)**, then click **Next**.

5   Click **1**, a scatterplot with no lines or grids, then click **Next**.

6   Make sure the **Columns option button** is selected, that the X Data spin box is set to **1**, that the Legend Text spin box is set to **1**; then click **Next**.

7   Click the **Chart Title box**, then type **Big Ten Graduation Percent**, press [**Tab**], type **Top Ten Percent** as the Category (X) axis title, press [**Tab**], type **Grad Percent** as the Value (Y) axis title, and then click **Finish**.

Excel creates a new sheet with the name Chart1 in your workbook. Your new chart should look like Figure 4-19.

Figure 4-19
Chart sheet with plot of Top Ten Percent vs. several y-variables

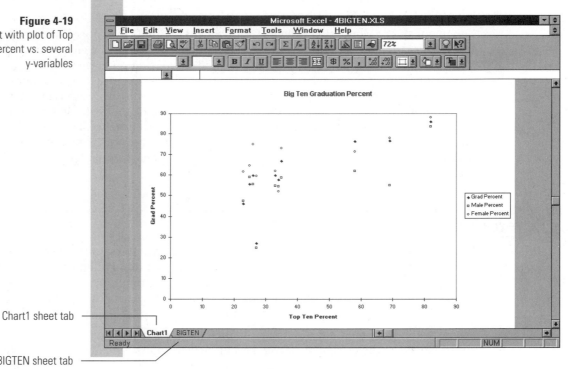

Chart1 sheet tab

BIGTEN sheet tab

To distinguish the three independent variables, Excel assigns three different data point symbols and colors. Your chart might look different depending on your monitor type.

Adding Data Labels

It would be useful to know which schools are associated with which data points. You can use the CTI Add-Ins to do this for each of the three data series.

To add data labels to each of the three data series, first select the series, then paste the labels:

1   Click one of the **data series** in the chart sheet, right-click to open the shortcut menu, then click **Insert ID Labels**.

2   With the dialog box still active, click the **BIGTEN sheet tab**.

3   Highlight the university names in cells **A1:A11** (you might have to scroll through the worksheet to see the range). The reference BIGTEN!$A$1:$A$11 appears in the Input box.

4   Verify that the **Selection Includes Header Row check box** is selected, then click **OK**.

5   Repeat steps 1 through 4 for the other two data series.

Your chart should now look like Figure 4-20.

Figure 4-20
Chart sheet with university name labels

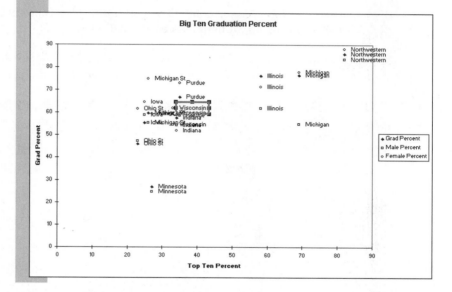

The chart sheet shows that the Female Percent data points are plotted highest up the y-axis for most schools, meaning that female athletes at most Big Ten schools have better graduation rates than male athletes and other students. The difference is most striking for Minnesota, where the female athletes are far ahead. Graduation rates for male athletes, on the other hand, are less than the university percentages at most schools. Finally, notice that the graduation percentage for female athletes at Ohio State is comparable to other Big Ten schools, whereas the graduation percentage for Ohio State male athletes is lower.

To give the chart sheet a more descriptive title and to save the workbook with the new chart:

1   Right-click the **Chart1 sheet tab**.

2   Click **Rename**.

3   Type **Graduation Percentage Plots** in the Name box, then click **OK**.

At this point you're done working with the Big Ten graduation data.

4   Click the **Save button** 🔲 to save your file as **4BIGTEN.XLS** (the drive containing your Student Disk should already be selected).

5   Click **File > Close**.

Plotting Two Variables with a Grouping Variable

When the data consist of several groups such as males and females, plots that show which values belong to one group (males) and which belong to the other (females) can be very helpful. This is different from the scatterplot you just created where you plotted a single x–variable against several y–variables. Now you are going to produce a plot that shows one x–variable against one y–variable, but the values from each group appear differently on the plot because they are distinguished by a third variable.

To illustrate this, let's use the JRCOL.XLS data set, which includes salary data for 81 faculty members at a junior college. It also includes variables such as number of years employed at the college and gender. The data were used in a legal action that was eventually settled out of court. Female professors sought help from statisticians to show that they were underpaid relative to the male professors.

To get started:

1 Open **JRCOL.XLS** (be sure you select the drive and directory containing your Student files).

2 Click **File > Save As**, select the drive containing your Student Disk; then save your workbook on your Student Disk as **4JRCOL.XLS**.

The JRCOL workbook appears in the Worksheet window. See Figure 4-21. The workbook contains predefined names and a freeze pane that keeps column headers always visible.

Figure 4-21
4JRCOL.XLS workbook

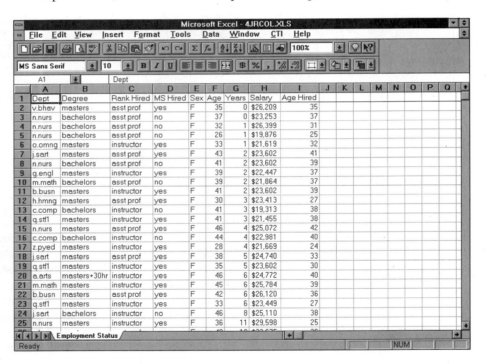

The salary data are included in the data series H1:H82 (with the range name Salary), the gender data are included in E1:E82 (Sex), and the number of years employed in G1:G82 (Years). To see the relationship between gender, salary, and years, you can plot salary against years, using colors (or, on a monochrome monitor, symbols) to differentiate males from females. First plot the data series showing salaries and years employed for just the females. Then select the data series for just the males, and add those data points to the same chart.

To create charts in Excel that differentiate among groups of data, you can use the AutoFilter techniques discussed in Chapter 2. You apply a filter to the data to show only those values that meet specified criteria; in this case, you want to display just the data series for the female employees by filtering the Sex variable to display only cells containing F for female.

To produce a chart that uses only the filtered data:

1. Click **Data > Filter > AutoFilter**. The filter drop-down arrow appears in each cell.

2. Click the **Sex drop-down arrow** in cell E1, then click **F**. Excel now displays only the female rows.

Now use the Chart Wizard to create the plot for the females, with salary as the y–variable and years as the x–variable.

To plot the female data:

1. Select the range **G2:H38**.

2. Click **Insert > Chart > As New Sheet** to create a new chart sheet.

3. Verify that the Range box contains the range **=$G$2:$H$38**, then click **Next**.

4. Click **XY (Scatter)**, then click **Next**.

5. Click **1**, a scatterplot with no lines or grids, then click **Next**.

6. Make sure the **Columns option button** is selected, and that the X Data spin box is set to **1**, because the first column in the range G2:H38 contains data for the x-axis. Verify that the Legend Text spin box is set to **0**, because the row containing the column label was *not* selected in the range G2:H38; then click **Next**.

7. Click the **Yes option button** to add a legend.

8. Click the **Chart Title box**, then type **Salary vs. Years Employed by Gender**, press [**Tab**], type **Years Employed** as the Category (X) axis title, press [**Tab**], type **Salary** as the Value (Y) axis title, and then click **Finish**.

Excel places the new chart in a chart sheet titled Chart1. See Figure 4-22.

Figure 4-22
New chart with Salary vs. Years Employed for female employees

series name is Series1

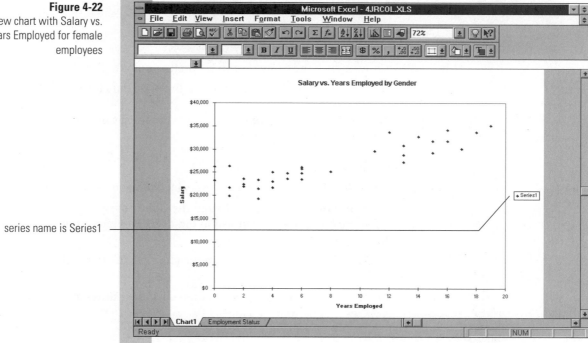

Excel names this data series Series1 because it is the first data series in this chart.

To give the series a more descriptive name:

1   Double-click one of the data points to select the series and open the Format Data Series dialog box.

2   Click the **Name and Values tab**.

3   Click the **Name box**, then type **Female Employees**. See Figure 4-23.

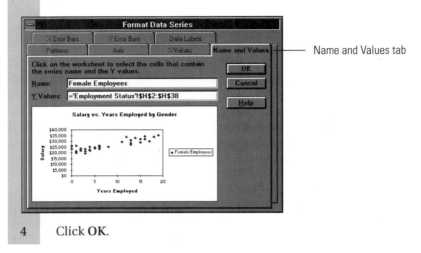

Name and Values tab

4   Click **OK**.

Excel changes the name of the data series to Female Employees. Now that the female data series is complete, return to the Employment Status worksheet and select the male employee data. By placing the male data on the same chart, you'll be able to compare the two data series.

To select the male employment data:

1   Click the **Employment Status sheet tab**.

2   Click the **Sex drop-down arrow** (in cell E1), then click **M**. Only the data for the male employees are visible.

3   Select cells **G39:H82**.

4   Click **Edit > Copy** or click the **Copy button** 📋 on the toolbar.

5   Click the **Chart1 sheet tab** so that you can paste the male data into the scatterplot.

Don't worry that the female data have disappeared. The data series is momentarily hidden because of the effects of the AutoFilter.

To paste the male data into the scatterplot:

1   Click **Edit > Paste Special** to open the Paste Special dialog box.

2   Click the **New Series option button**, verify that the **Columns option button** is selected and that the **Categories (X values) in First Column check box** is selected, then click **OK**.

The data series for the males now appears in the chart. Name this series in the same way you named the series for the females.

3   Double-click one of the data point symbols, then click the **Names and Values tab** if necessary.

4   Click the **Name box**, type **Male Employees**, then click **OK**.

With both the female and male data series now plotted on the chart, the only task remaining is to view them both at the same time.

5 Click the **Employment Status sheet tab**.

6 Click **Data > Filter > AutoFilter** to remove filtering from the data list, click the **Chart1 sheet tab** to return to the chart and view both data series, then click the chart to deselect the data series.

Chart1 now looks like Figure 4-24.

Figure 4-24
Plot of Salary vs. Years Employed with different symbols for gender

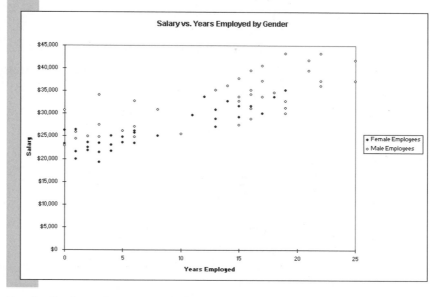

For the final touches, you could set a higher y-axis scale minimum so that the y-axis scale goes from $15,000 to $45,000 (see the discussion of modifying a chart's scale earlier in this chapter).

Figure 4-24 shows that the female employees are paid less. No matter how long the professors have been employed, the highest paid faculty are males. You can tell that the situation has not improved for those recently hired, because the highest paid workers with less than five years service are males and the lowest paid are females. On the plot, several males stand out as having much higher salaries than other recent hires. This suggests a bias toward higher salaries for males, but you can't reach a conclusion based only on this chart. Is it possible that the college administration can provide an explanation to show that females are treated fairly? Could there be some other factor that influences salaries and somehow explains the differences shown here? Could it be that the high-salaried males have more years of relevant experience? Or might they have important educational advantages? These are the types of questions that must be explored in order to fully explain the data points on our chart.

Name the chart sheet and save the workbook:

1 Right-click the **Chart1 sheet tab**, then click **Rename**.

2 Type **Salary vs. Years Employed**, then press [**Enter**].

3 Click the **Save button** 🖫 to save your file as **4JRCOL.XLS** (the drive containing your Student Disk should already be selected).

4 Click the **Employment Status sheet tab** to return to the worksheet.

Scatterplot Matrices

One of the other factors that is important here is Age Hired, the age when hired. There is no variable that explicitly indicates relevant experience previous to being hired, but Age Hired should give some idea of experience. It would be useful to construct a plot that shows how years employed and age hired influence salary.

One way to visualize data with more than two variables is the **scatterplot matrix**, also called a SPLOM. A **SPLOM** is a grid of several scatterplots showing the relationship between all the pairs of variables in the selection. Excel does not provide a SPLOM command, but the CTI Statistical Add-Ins provided with the Student Disk include a command to allow you to create a SPLOM.

Creating a Scatterplot Matrix

To create a SPLOM for Salary, Years, and Age Hired:

1   Click **CTI > Scatterplot Matrix**. The Create Scatter Plot Matrix dialog box opens.

2   Type **G1:I82** in the Input Columns box.

3   Verify that the **Selection Includes Header Row check box** is selected.

4   Click the **New Sheet option button** and type **Scatterplot Matrix** in the corresponding box. The dialog box should look like Figure 4-25.

Figure 4-25
Completed Create Scatter
Plot Matrix dialog box

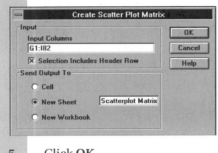

5   Click **OK**.

Excel adds a new worksheet to the workbook with the name Scatterplot Matrix, which displays the scatterplot matrix of Years vs. Salary vs. Age Hired, as shown in Figure 4-26.

Figure 4-26
SPLOM of Years Employed,
Salary, and Age Hired

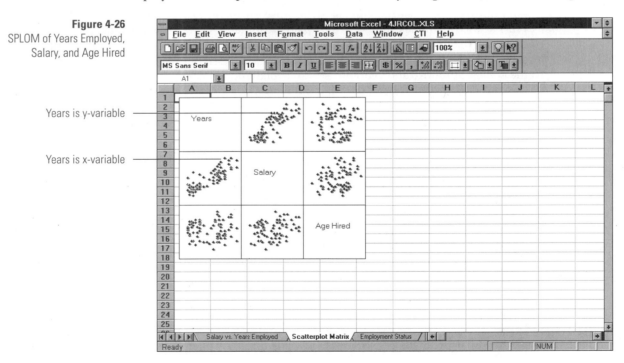

- Make sure that the CTI Statistical Add-Ins are installed.

- Click CTI > Scatterplot Matrix from the menu.

- Enter the range of data that you want to plot (the range need not be contiguous). Indicate whether the range includes a header row.

- Enter the output destination: cell, new worksheet, or new workbook, then click OK.

Interpreting a Scatterplot Matrix

In a scatterplot matrix, for each plot, the y variable is indicated by the variable name in the row of the matrix, whereas the x–variable is indicated by the variable name in the column. For example, the middle plot in the first column of the matrix is the scatterplot of Salary (the y–variable) vs. Years (the x–variable), which you examined with regard to gender earlier in this chapter. You can also see the relationship between Salary and Age Hired in the middle plot of the third column of the matrix.

Can you tell from these two plots whether Years or Age Hired is a better predictor of Salary? For good prediction, the points lie in a narrow vertical range for each value of the predictor. This occurs when the points lie in a narrow band from lower left to upper right on the plot or from the upper left to the lower right. If you know the predictor, then you know the y variable within a narrow range. For each value of Years, there is a fairly narrow vertical range of Salary, so knowing years permits a reasonably accurate prediction of salary. Later you will study this relationship in more depth with the help of regression analysis. On the other hand, for each value of Age Hired, there is a wide vertical range of Salary, so knowing a person's age when hired does not allow you to make a good prediction of salary.

The right plot in the top row shows Years vs. Age Hired. If there is any relationship between these two variables, it must be very weak, because it does not show up strongly here. For each value of Age Hired, there is a very wide range of Years, so you cannot use Age Hired to predict Years with any accuracy. Is this consistent with your intuition? Age Hired gives the time before hiring, and Years gives the time after hiring, so it is reasonable for there to be very little relationship between them. Suppose that you used instead two variables such as Age and Years. Would an older person be likely to have more years of experience? In other words, would you expect a relationship between Age and Years? You'll have an opportunity to examine these questions during the exercises at the end of the chapter.

Editing a Scatterplot Matrix

Because the SPLOM is made up of nine individual boxes that have been grouped together using Excel's grouping capabilities, you can edit each chart individually if you want. When you create a **group** in Excel, you combine individual objects into one object so that you can treat them as one unit. You can resize, move, copy, and paste a group, in this case, the SPLOM, like any other Excel worksheet object. However, when you want to work with just one of the group elements, such as a single chart in a SPLOM, you can ungroup the SPLOM and work with an individual chart. A SPLOM can be an excellent tool for exploring the relationships between many pairs of variables and then allowing you to focus on and modify only those charts that are of interest.

To work with the Salary vs. Age Hired chart:

1	Right-click the **SPLOM** (anywhere within the matrix) to open the shortcut menu.
2	Click **Ungroup**.
3	Click any cell outside the SPLOM to deselect the matrix.

The scatterplot matrix divides into nine individual charts, any of which you can select.

4   Double-click the **middle chart** in the third column (Age Hired vs. Salary).

5   Click the **Copy button** 📋 on the toolbar (or click **Edit > Copy**).

6   Click cell **G1** twice, then click the **Paste button** 📋 (or click **Edit > Paste**).

Excel pastes a duplicate of the chart of Salary vs. Age Hired in cell G1.

7   Drag the lower-right handle of the chart to cell **K15** to enlarge it. See Figure 4-27.

Figure 4-27
Duplicated and enlarged scatterplot of Salary vs. Age Hired

copy this chart to this location

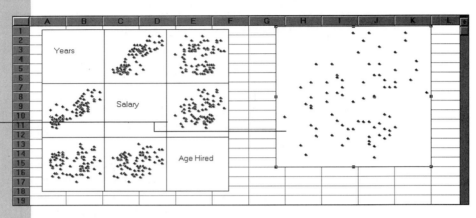

Now add axes to the plot you pasted.

8   Double-click the **chart** to activate it, then right-click anywhere within the chart to open the shortcut menu.

9   Click **Insert Axes** on the shortcut menu.

10   Click **Category (X) Axis** and **Value (Y) Axis** so that both check boxes are selected, then click **OK**.

The plot now contains axes, as shown in Figure 4-28.

Figure 4-28
Enlarged scatterplot matrix with axes

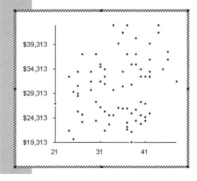

From this point, you can add labels and text or modify the axes using the methods described earlier in this chapter to create a finished chart.

To regroup your scatterplot matrix, save it, and exit Excel:

1   Double-click the **upper-left box** in the SPLOM (the one that contains the word "Years"). This deactivates the chart you were just working on.

2   Press and hold down the [**Shift**] key and click each of the eight remaining charts in the SPLOM to select them.

3   Right-click the **SPLOM** (anywhere within the borders of the nine charts) to open the shortcut menu.

4   Click **Group** on the shortcut menu.

5 Click the **Save button** 🖫 to save your file as **4JRCOL.XLS** (the drive containing your Student Disk should already be selected).

6 Click **File > Exit**.

E XER CISES

1. The JRCOL.XLS file contains a column variable named Age, which is the age hired plus number of years employed.

   a. Create a scatterplot of Years vs. Age.

   b. Are these two variables more strongly related than age hired and years employed? How do you explain this?

   c. Save your workbook as E4JRCOL.XLS.

2. The CALC.XLS data set includes grades (the variable range name is Calc) for the first semester of calculus, along with various predictor scores. (See Edge and Friedberg (1984).) The predictors include high school rank (HS Rank), American College Testing Mathematics test score (ACT Math), and an algebra placement test score (Alg Place) from the first week of class. Because admission decisions are often based on ACT math scores and high school rank, you might expect that they would predict freshman success accurately.

   a. Plot Calc against HS Rank, and also plot Calc against ACT Math (you might have to rescale the x- and y-axes in these charts).

   b. For each predictor value, do you find a narrow range of Calc on these plots? In other words, does it appear that these variables predict Calc accurately?

   c. Comment on and compare the use of ACT Math tests and high school rank in admissions and placement of students in classes.

3. For the CALC.XLS data described in Exercise 2, obtain a scatterplot matrix for Calc, HS Rank, ACT Math, and Alg Place.

   a. Among the three predictors, which seems to predict Calc best?

   b. Would you call it a good predictor?

   c. Save your workbook as E4CALC.XLS.

4. The ALUM.XLS data record the mass and volume of eight chunks of aluminum, as measured in a high school chemistry lab.

   a. Plot Mass against Volume (because the Chart Wizard uses the left-most column for the x-axis, you'll have to copy Volume so that it is to the left of Mass, as described earlier in this chapter).

   b. Is there an outlier? How could this occur, and how could you justify deleting the outlier?

   c. Do the other points seem to form a nearly straight line? The ratio of Mass to Volume is supposed to be a constant (the density of aluminum), so the points should fall on a line through the origin. Draw the line, and estimate the slope (the ratio of the vertical change to the horizontal change) along the line.

   d. How does your answer in Exercise 4c compare to the value 2.699 given in chemistry books as the density of aluminum?

   e. Save your workbook as E4ALUM.XLS.

5. The BIGTEN.XLS data set includes Computer, the ratio of the number of students to the number of computer outlets (terminals and desktop computers) on campus.

   a. Plot Grad Percent against Computer, and also plot Grad Percent against log(Computer). Note that you have to copy Computer so that it is to the left of Grad Percent, as described earlier in this chapter, because the Chart Wizard requires that the x–variable column appear to the left of the y–variable column.

   b. Which plot shows the strongest trend? Does a trend imply a causal relationship between two variables? If a university increases its computer availability, would an increase in graduation rate follow?

   c. Rather than a causal relationship, is it better to imagine that both variables are affected by a quality factor? In particular, the 1992 issue of *U.S. News and World Report's* "America's Best Colleges" rates Northwestern and Michigan in the top 25 universities. Illinois, Minnesota, and Wisconsin are in the next 26. The highly rated schools tend to have high graduation rates and lots of computers.

   d. Save your workbook as E4BIGTEN.XLS.

6. Plot Protein vs. Carbo for the WHEAT.XLS data (to make the Chart Wizard place Protein on the y-axis, copy the values to a new column to the right of the Carbo column), and use Brand to label the points.

   a. On the plot, the data seem to fall in two groups, with low Carbo on the left and high Carbo on the right. Which brand has the highest protein in the low Carbo group? Is this a surprise?

   b. Which has the lowest protein in the high Carbo group? This might not be so surprising if you read on the package that sugar and brown sugar are the second and third ingredients.

   c. Save your workbook as E4WHEAT.XLS.

7. The BASE.XLS data set includes salaries and statistics for major league baseball players at the start of the 1988 season. Plot Salary against Aver Career (career batting average).

   a. Create a new variable, LogSalary (the log of salary), instead of Salary, and see if this gives a smoother trend on the plot.

   b. Another important variable is Years, the number of years in the big leagues. Plot LNSalary against Years. Does it appear that Years is strongly related to LNSalary?

   c. Of Years and Aver Career, which do you consider to be the better predictor?

8. For the BASE.XLS data of Exercise 7, construct a scatterplot matrix of LNSalary, Years, and Aver Career.

   a. Save your workbook as E4BASE.XLS.

9. The LONGLEY.XLS data set has seven variables related to the economy of the United States from 1947 to 1962. The last variable is Total, total employment in thousands. Plot Total against Armforce, which is the total number in the armed forces, again in thousands.

   a. Use the Insert ID Labels command on the Chart shortcut menu (available through the CTI Statistical Add-Ins) to add ID labels to the points based on the Year.

   b. Use your knowledge of history to explain why four points on the lower left of the scatterplot stand out.

10. Refer again to the LONGLEY.XLS data set. Construct a scatterplot matrix of Total vs. Armforce and Unemploy, which is unemployment in thousands. How would you rate the ability of Armforce and Unemploy to predict total employment?

    a. Save your workbook as E4LONG.XLS.

CHAPTER 5

PROBABILITY DISTRIBUTIONS

O B J E C T I V E S

In this chapter you will learn to:

- Work with random variables and probability distributions

- Generate random, normally distributed data and obtain z-scores

- Create a normal probability plot

- Explore the distribution of the sample mean

- Apply the Central Limit Theorem

Up to now, you've used tools such as frequency tables, descriptive statistics, and scatterplots to describe and summarize the properties of your data list. Now you'll learn about probability, which provides the foundation for understanding and interpreting these statistics. You'll also be introduced to statistical inference, which uses such summary statistics to help you reach conclusions about the underlying nature of the data.

Random Variables

If you throw a die, there are six possible outcomes: 1, 2, 3, 4, 5, or 6. The outcome is a variable with six possible values. This is an example of a **random variable**, a variable whose value occurs at random. Another example of a random variable is this semester's text expenditures of a randomly chosen student. The resulting value for the random variable (such as throwing the die or asking a single student how much he or she paid for textbooks) is called an **observation**. Statisticians usually collect a set of observations and then study that set in order to learn about the random variable. For the textbook expenditure random variable, you would call a number of students, not just one. For example, if you call five students, the amounts each spent on textbooks might be $342, $287, $292, $331, and $310. This is an example of a **random sample**, a set of values obtained through independent realizations of a random variable.

If a random variable can take on only a finite number of values, it is called **discrete**. For example, the result of a die throw is a discrete random variable because it can take on only six values. On the other hand, if the possible values for a random variable cover a whole interval of real numbers, it is called **continuous**. For example, textbook expenditure for this semester is essentially a continuous random variable. You could argue that, strictly speaking, it is discrete because the value involves an integer number of cents, but for practical purposes the variable is continuous. As you'll see, there are many powerful statistical techniques that are associated with continuous random variables.

Probability Distributions

Although you can't predict perfectly the resulting values of random variables, you can still learn about them. An important concept in statistical inference is the concept of the probability distribution. The pattern of probabilities of a random variable is called its **probability distribution**. For example, the probability distribution for discrete random variables is the probability assigned to each of the possible values that the random variable can assume. When you toss a fair die, the number that appears follows a probability distribution with a probability of about 0.167 (16.7%) assigned to each of the six possible outcomes. The total of the probabilities must always equal 1.

For continuous random variables the situation is a bit more complex. With continuous random variables, you look at the probability of the random variable attaining a value within a range of possible values. Let's look at an example of this with the instructional template PDF.XLT.

To open PDF.XLT, be sure that Excel is started and the windows are maximized, and then:

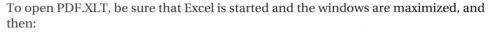

1   Open **PDF.XLT** (be sure you select the drive and directory containing your Student files).

The PDF Template

The template opens in your Worksheet window. See Figure 5-1.

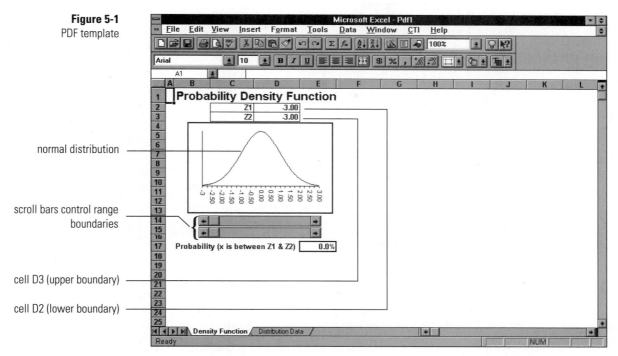

Figure 5-1
PDF template

normal distribution

scroll bars control range boundaries

cell D3 (upper boundary)

cell D2 (lower boundary)

The chart shows the classic bell-shaped curve that might already be familiar to you as the normal distribution, discussed later in the chapter. This chart is an example of a **probability density function**, because the curve shows where values are most likely to occur. In this example the most probable range of values is in the center of the curve. The probability associated with a range of values is equal to the area under the probability density curve between the lower and upper boundaries. Thus, a single value has zero probability because its area under the curve is zero. The PDF template provides an interactive way for you to determine the probability of a random variable having a value within a given range.

For example, suppose you are looking at the daily change in price per share for stocks on the New York Stock Exchange and you know that the distribution of the change in price follows the normal distribution shown in this template. What is the probability that a stock's change in price will be less than $1 per share on a given day? That is, if your stock is worth $54 today, what is the probability that tomorrow it will stay in the range $53–$55? This is equivalent to calculating the area under the probability density function curve between -1 and 1.

Notice the two scroll bars below the curve; you use these to control the range boundaries. You click the scroll bar arrows to set the upper and lower ranges (Z1 and Z2) to different values.

To find the area under the probability density function curve between -1 and 1:

1 Click the **right scroll arrow** of the top scroll bar until the value **-1.00** appears in cell D2.

2 Click the **right scroll arrow** of the bottom scroll bar until the value **1.00** appears in cell D3.

Cells D2 and D3 now display -1.00 and 1.00, respectively, the lower and upper ranges you are testing. See Figure 5-2. Cell E17 displays the probability of a random variable having a value between the boundaries you set. In the example, this tells you that the probability is equal to 68.3% (cell E17) that your stock, currently valued at $54, will be valued somewhere in the range of $53–$55 tomorrow. This assumes that the distribution of change in price follows the curve shown in Figure 5-2.

Figure 5-2
PDF template with range
from -1 to 1 selected

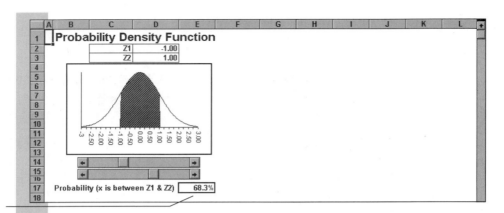

probability in cell E17 —

Try experimenting with other range boundaries until you get a good feel for the relationship between probabilities and areas under the curve.

3   Click the **left scroll arrow** of the top scroll bar until the value **-3.00** appears in cell D2.

4   Click the **right scroll arrow** of the bottom scroll bar until the value **3.00** appears in cell D3.

Note that when you move the lower boundary to –3 and the upper boundary to 3, the probability is 99.7%, not 100%. That is because this particular probability distribution extends from minus infinity to positive infinity; the area under the curve (and hence the probability) is never 1 for any finite range.

5   Click **File > Close** to close the PDF instructional template.

6   Click **No** when asked if you want to save the changes (because this is a template, you can always open it again and experiment further).

The template you just examined showed you one example of many possible probability density functions. Probability density functions all share the two following properties:

- The area under the probability density function curve is equal to one.

- The values of the probability density function are always greater than or equal to zero.

Once you know what a probability distribution looks like, how can you relate it to an actual random sample? The shape of the histogram for a random sample tends to follow the shape of the underlying probability density function.

Let's further examine these concepts using the instructional template TARGET.XLT.

To open TARGET.XLT:

1   Open **TARGET.XLT** (be sure you select the drive and directory containing your Student files).

The Target Template

One way to think about the relationship between the probability distribution and the distribution of sample values is to imagine a marksman engaging in target practice. Figure 5-3 shows the target. The bull's eye is at the center.

Every shot the marksman takes is random because the marksman cannot predict with perfect accuracy where the next shot will end up on the target. However, if you know how accurate the marksman is, you can estimate how closely the shots will surround the bull's eye. Let's use the template to create a series of random shots.

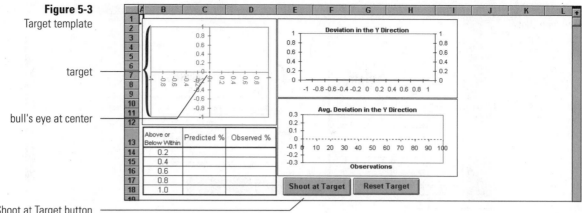

Figure 5-3
Target template

target

bull's eye at center

Shoot at Target button

To generate a set of random shots:

1 Click the **Shoot at Target button**.

The Accuracy dialog box opens, letting you choose from five degrees of accuracy: Highest, Good, Moderate, Poor, and Lowest. See Figure 5-4.

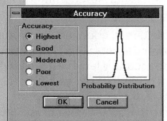

Figure 5-4
Accuracy dialog box

narrow distribution

A highly skilled marksman will be able to distribute shots within a narrow range of values (that is, closely clustered around the bull's eye) whereas a poor marksman's shots will be distributed over a wider range. The dialog box opens to a default of Highest, as shown in Figure 5-4. The values are closely clustered around the center of the curve. Now, see what happens when you select a marksman who has had only moderate practice.

2 Click the **Moderate option button**. Note that the corresponding probability distribution is more widely dispersed around the center, indicating that shots farther from the center are nearly as probable as a shot on the center. See Figure 5-5. In statistical terms, the shots of the highly skilled marksman are less variable.

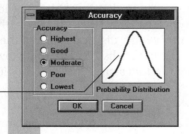

Figure 5-5
Accuracy dialog box with Moderate accuracy

wider distribution

3 Click **OK**.

The template generates a set of 100 shots at the target with moderate accuracy. See Figure 5-6.

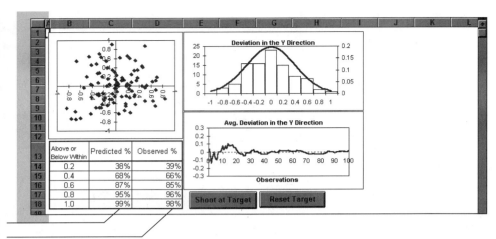

Figure 5-6
Template with shots marked on target

predicted values

observed values

The XY coordinate system on the target shows the bull's eye located at the origin (0,0). The distribution of the shots around the origin is described by a **bivariate** density function, because it involves two random variables, the vertical placement (or Y value of the shot) and the horizontal placement (or X value). For the purposes of this example, let's concentrate only on the vertical deviation of the shots from the bull's eye or origin and ignore the horizontal component. The vertical deviation is measured from the x-axis, so the values below the x-axis are negative.

Although many of the shots are near the target, about a third of them are farther than 0.4 vertical units away, either higher or lower. Because this is randomly simulated data, your values might be different. Based on the accuracy level you selected, a probability distribution showing the expected distribution of shots above or below the target is also generated in cells C14 through C18. In this example, the predicted proportion of shots within 0.4 of the target is 68% (cell C15), which is close to the observed value of 66% (cell D15). In other words, the distribution predicts that a marksman of moderate ability is able to hit the target within 0.4 of the bull's eye 68% of the time. This marksman came pretty close.

You can also examine the distribution of these shots by looking at the histogram of the vertical deviation of shots from the origin. For the purposes of this template, a shot below the target has a negative deviation, and a shot above the target has a positive deviation. The solid line in the histogram indicates the theoretical expected distribution of shots above or below the target, whereas the bars indicate the observed distribution of shots.

The 100 shots represent a random sample. They are drawn from a **population**—a collection of units (people, objects, or whatever) that you are interested in knowing about. In this case, the population of interest is not finite but is a theoretical, infinite population of all possible locations of the shots.

The shots themselves constitute continuous random variables. The locations of the shots above or below the bull's eye are the observations. How you expect the shots to be distributed above or below the target is the probability distribution, and the distribution of the observed shots is summarized by the frequency distribution or histogram (as discussed in Chapter 3). Try generating some more random samples using other levels of marksmanship.

To generate other random samples:

1 Click **Shoot at Target**, click **Lowest**, then click **OK** to generate a distribution where the values are widely scattered, and notice how the histogram changes.

2 Repeat step 1 using **Highest**.

The shape of the probability density function gives you some indication of how widely scattered the individual shots will be.

Now consider the converse situation. What if you didn't know how accurate the marksman was? You could estimate the accuracy by observing the distribution of the marksman's shots. If all the shots land within a few units of the origin, you would consider the marksman's accuracy high, with the corresponding probability density function tightly clustered around the center. In other words, you would use statistical inference to reach some conclusions about the underlying probability distribution.

Population Parameters

Probability distributions are characterized by their probability density functions. These functions are, in turn, determined by their parameters, known as **population parameters**. In the case of the normal distribution, two parameters determine the probability density function: the **population mean**, which is the value of the center of the curve, and the **population standard deviation**, which gives the spread of values around the center.

You've already seen the terms mean and standard deviation used in Chapter 3. Those values are estimators of the corresponding population parameters. Remember, the probability distribution is a theoretical construct—it is how you expect the sample of observed values to be distributed. You can also look at the distribution of the observed values to infer the shape of the true probability distribution that describes the entire population. Similarly, the population parameters have values that are unknown but can be estimated by calculating the values of statistics from the observed sample. To avoid confusion, statisticians designate the population mean by the Greek symbol μ (pronounced *mu*) and the population standard deviation by the Greek symbol σ (*sigma*). The sample mean, as you saw in Chapter 3, is designated by $\bar{x}$ (*xbar*) and the sample standard deviation by the letter s. The values $\bar{x}$ and s have a special and important property: they are not only estimators of μ and σ, they are **consistent estimators**, which means that as you increase the sample size, the values of $\bar{x}$ and s come closer and closer to the true population parameters μ and σ. With a large enough sample size, $\bar{x}$ and s will estimate the true values of μ and σ to whatever degree of precision you want.

The variability of the data influences how large a sample you need. Consider the target template from Figure 5-6. Included in the template is the average deviation above or below the target (the lower chart). The value the marksman is trying to hit has a deviation of zero (the origin), so the average deviation from the target should approach zero as the marksman tries more and more shots and the errors cancel each other out. Figure 5-7 shows the changing value of the average deviation above or below the target as the number of observations in the sample increases from 1 to 100 for two levels of accuracy, High and Low. (You don't have to generate this figure.)

For the distribution of shots corresponding to a marksman shooting with high accuracy, the sample mean quickly settles down to a value close to the origin—the value of the population parameter (remember that the curves you generate might be different, because these are random data). For the highly variable shots coming from a marksman with low accuracy, the sample mean oscillates back and forth between extreme values above or below the origin. The average value eventually will be as close to the value of the population mean (in this case zero) as you want, but it will take many more observations to do so. Just how many observations are required will be discussed later in this chapter.

This concludes your work with the target template. The remainder of this chapter investigates specific properties of the normal distribution.

Figure 5-7
Changing values of the
sample average

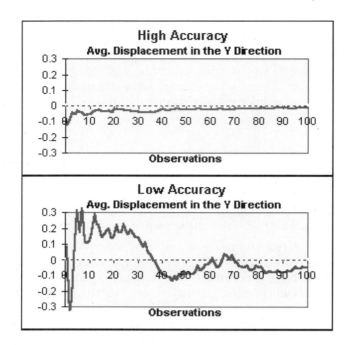

To finish with target practice:

1   Click **File** > **Close** to close the target instructional template.

2   Click **No** when asked if you want to save the changes (because this is a template, you can always open it again and experiment further).

The distributions of shots above or below the bull's eye you just generated all followed the basic bell-shaped curve known as the **normal distribution**. Variables like this or scores on an exam often follow this distribution shape. Statisticians have identified other distributions, though, that better fit other kinds of data. For example, the random flips of a coin do not follow a normal curve because the result can take on only one of two values: heads or tails. Data resulting from this kind of experiment have a discrete distribution, because they only take on discrete values. There are many other distributions that you will come across in your study of statistics, such as the *t* distribution (discussed in Chapter 6), and the chi-square distribution (discussed in Chapter 7).

Normal Data

The normal distribution is extremely important in statistics. There are many real-world examples of normally distributed data, and this distribution is frequently assumed in statistical practice. In fact, most of the remainder of this book assumes normally distributed data. But how can you determine whether the observed values originate from a normal distribution?

Figure 5-8 shows a histogram displaying the distribution of career batting averages for 263 major league ball players, from the BASE.XLS file, superimposed by the normal probability density function. (You don't have to generate this figure.)

The normal curve in Figure 5-8 uses the value of $\bar{x}$ (0.263) for the value of the sample mean μ and the sample standard deviation s (0.02338) for the value of the population standard deviation σ. The shape of a histogram for normal data should roughly follow the normal curve, as it does here. The normal curve drawn over the histogram is created using a CTI add-in, introduced later in this chapter.

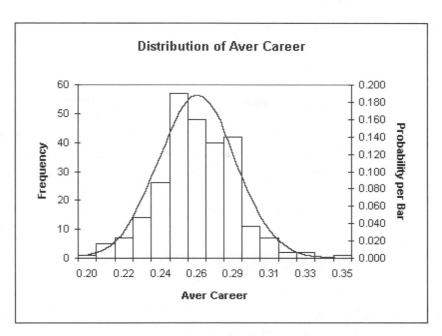

For normally distributed data, about 68.3% of the data should be within one standard deviation of the mean. For the batting average data, 68.3% of the data should be between 0.263 ± 0.02338, or the range 0.2398 to 0.2866. There are actually 181 out of 263 career batting averages in this interval, which is 68.82% (the observed percentage generally differs from the theoretical 68.3% even when the data are normal).

Figure 5-9 shows other percentages under the normal curve. About 95.45% of data should be within two standard deviations of the mean. For the batting average data, this is the range 0.216 to 0.310 (0.263 – 2 * 0.02338 to 0.263 + 2 * 0.02338) and encompasses 251 of 263 players, or 95.44%. If you go as far as three standard deviations each side of the mean, which is from 0.193 to 0.333 for the career batting averages, the interval should include 99.73% of the values. This means that very few observations should be more than three standard deviations from the mean for normal data. Out of the 263 major league players, 262, or 99.62%, lie within three standard deviations of the sample mean. (You don't have to generate this figure.)

Figure 5-9

Percentages under the normal curve

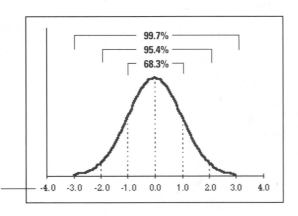

number of standard deviations away from the sample mean

To summarize what the normal curve predicts:

- The distribution is symmetric, with most of the probability concentrated near the center.
- About 2/3, or 68.27%, of the observations are within one standard deviation of the mean.
- About 95.45% are within two standard deviations of the mean.
- About 99.73% are within three standard deviations of the mean.

Although the career batting average data match the normal distribution percentages very well, does this constitute proof that the data are normal?

Using Excel to Generate Random Normals

What do normal data look like? Excel includes a function with the Analysis ToolPak Add-Ins that generates normal data; you can use it to create a sample of 100 normal observations that help answer this question.

To obtain 100 random normal values:

1   Click the **New Workbook button** 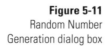 to open a new workbook.

2   Click cell **A1**, type **Standard Normal**, then press [**Enter**].

3   Click **Tools > Data Analysis**. If you do not see this option in the Tools menu, you need to load the Analysis ToolPak Add-Ins (see Chapter 3).

4   Click **Random Number Generation** in the Analysis Tools list box (you might have to scroll to see it) in the Data Analysis dialog box (Figure 5-10), then click **OK**.

Figure 5-10
Data Analysis dialog box

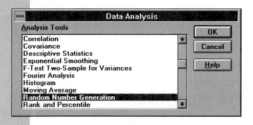

The Random Number Generation dialog box appears.

To specify how many and what type of random numbers you want:

1   Type **1** in the Number of Variables text box to indicate that you want one column of random numbers, then press [**Tab**].

2   Type **100** in the Number of Random Numbers text box to indicate that you want 100 rows of numbers in that one column.

3   Click **Normal** in the Distribution drop-down list.

4   Double-click the **Mean text box**, type **0**, press [**Tab**], then type **1** in the Standard Deviation text box.

5   Click the **Output Range option button**, click the corresponding text box, then type **A2** (you can also select this cell with your mouse to save typing). This tells Excel where to start the random number generation.

6   When the dialog box looks like Figure 5-11, click **OK**.

Figure 5-11
Random Number
Generation dialog box

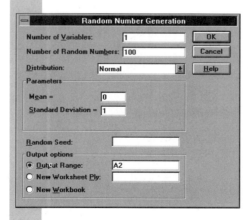

Excel generates 100 random normal values and places them into cells A2:A101 of the Sheet1 worksheet. See Figure 5-12. Because this is randomly generated data, your numbers might look different.

Figure 5-12
Random normal values in
column A

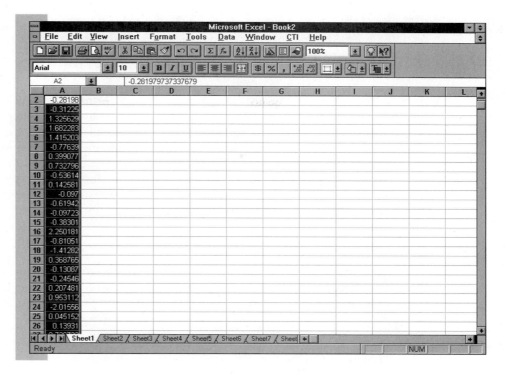

The Standard Normal

Because the data you just generated were derived from the normal distribution with mean 0 and standard deviation 1, about 68% of the numbers should be within one standard deviation of the mean (between –1 and 1), and about 95% should be within two standard deviations of the mean (between –2 and 2). There should be occasional numbers exceeding 2 in absolute value, but only about 1 in 20 should do so. If these numbers look familiar to you it's because the PDF template used earlier in this chapter was based on a normal distribution with a standard deviation of 1 and a mean of 0.

The data in column A come from the **standard normal population**, which has a mean of 0 and a standard deviation of 1. Probabilities from the standard normal distribution are often quoted in performing statistical analyses, as you'll see in later chapters. Generating pseudo-data in this fashion is called a **simulation**. The target template you used earlier in this chapter is another example of a simulation.

Try to simulate the batting average data. In Excel, you can simulate the data either by rerunning the Random Numbers Generation command with the values for batting average sample mean and standard deviation, or by using the standard normal data you generated in column A to create this random sample. Let's try the latter method. Any normal distribution is related to the standard normal in this way: multiply the standard normal (the values in column A) by the standard deviation (0.02338), and add the mean (0.263). The mean locates the center of the distribution, and the standard deviation determines how widely the data are spread about the mean.

To generate a normal random sample using the parameters from the batting average data:

1. Widen column **A** so that the column label in cell A1 (scroll up to see it if necessary) fits in the cell, then click cell **B1**.

2. Type **Batting Avg Sim** in cell **B1**, then press [**Enter**].

3. Highlight the range **B2:B101**.

4. Scroll up to view cell B2, then type **=A2*0.02338+0.263** and press [**Enter**].

5. Click **Edit > Fill > Down** (or press [**Ctrl**]+[**D**]).

6. Widen column **B** so that the column label fits in the cell.

Excel applies the formula you entered in cell B2 to all the cells. Column B now contains random data based on a normal distribution with a mean of 0.263 and a standard deviation of 0.02338.

Z-scores

You could reverse the algebra and create data that follow the standard normal by entering the formula =(B2–0.263)/0.02338 in column C2 and fill down as in step 5 above; you would obtain the same values in column A. You can obtain the standard normal values from any normal distribution by subtracting the mean and dividing by the standard deviation. These standard normal values are called **z-scores**. Subtracting the mean and dividing by the standard deviation to create z-scores is often done in statistics to express the observed values from a normal distribution in terms of the standard normal curve.

Creating Histograms Using the CTI Statistical Add-Ins

To obtain histograms for the random data in columns A and B, you could use the histogram chart offered by the Analysis ToolPak Add-Ins discussed in Chapter 3. For the purposes of this book, a second histogram tool is offered by the CTI Statistical Add-Ins; this histogram differs from the Analysis ToolPak histogram in the following ways:

- overlays the histogram with a smooth normal curve
- is dynamic so that if you change the data, the histogram and the normal curve update automatically
- assumes evenly spaced bins, and prompts the user on the total number of bins needed
- creates bin boundaries halfway between succeeding bin values, so that a histogram with bin values of 1, 3, and 5 has boundaries for these bins at (0, 2) for bin 1, (2, 4) for bin 3, and (4, 6) for bin 5

To create a histogram of the random data entered into column B of the current worksheet:

1   Verify that the CTI Statistical Add-Ins are installed (the CTI menu appears on the menu bar if they are). If the add-ins are not installed, install them using the methods described in Chapter 3.

2   Click **CTI > Histogram** to open the Create a Histogram dialog box.

3   Type **B1:B101** in the Input Columns text box (you can also select this range using the mouse).

4   Click the **Selection Includes Header Row check box**.

5   Accept the default bin number of **15**.

6   Click the **Cell option button**, then type **D1** in the corresponding text box.

7   When your dialog box looks like Figure 5-13, click **OK**.

Figure 5-13
Create a Histogram
dialog box

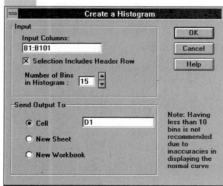

Figure 5-14 shows the histogram (you might need to scroll to see it). Because the data are randomly generated, your histogram may look different.

Figure 5-14
Histogram of simulated standard normal data

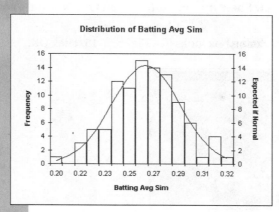

With a sample of 100 observations, there is much variability, although the histogram does follow the curve approximately. One message of Figure 5-14 is that even when you generate normal data randomly, the histogram and curve correspond only roughly. As you will see, if you significantly increase the number of samples, the histogram and curve correspond more closely. Relating this back to the batting average data, you would conclude that the distribution of the career batting averages seems to be very similar to the distribution of a randomly generated sample of normal data.

8   Click **File > Save As**, select the drive containing your Student Disk, then save your workbook on your Student Disk as **5NORMAL.XLS**.

9   Click **File > Close**.

The Normal Probability Plot (PPlot)

It is not always easy to tell from the histogram whether the data are normal. A **normal probability plot**, a graph of the observed values versus the expected z-scores, gives you another view of the data. If the data are normal or approximately normal, the points should nearly fall in a straight line. If this is not the case, you should hesitate to apply many of the standard statistical procedures that are discussed later in this book, although you might be able to straighten things out with a logarithm or square-root transformation. You saw an example of such a transformation in Chapter 3, with the employee variable in the women-owned business data set.

Creating a Normal Probability Plot

Excel does not include a probability plot, but one has been provided for you with the CTI Statistical Add-Ins.

To create a normal probability plot for the batting average data:

1   Open **BASE.XLS** (be sure you select the drive and directory containing your Student files).

2   Click **File > Save As**, select the drive containing your Student Disk, then save your workbook on your Student Disk as **5BASE.XLS**.

3   Click **CTI > Normal P-Plot**.

4   Type **aver_career** (be sure to include the underscore) in the Input Range text box (the range name for the range H1:H264).

5	Click the **Selection Includes Header Row check box** because this selection includes a label.
6	Click the **New Sheet option button**, then type **Normal Pplot** in the corresponding text box.
7	When your dialog box looks like Figure 5-15, click **OK**.

Figure 5-15
Create a Normal Probability
Plot dialog box

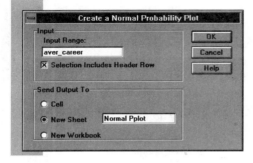

Interpreting a Probability Plot

Excel calculates 263 normal scores in column A. **Normal scores** are the expected set of values from a sample where each value comes from a standard normal distribution. Specifically, if you create a sample of 263 standard normal values, the expected value for the largest of these 263 will be approximately 2.8237, the second largest expected value will be approximately 2.5021, and so on. (Excel calculates these numbers using percentiles from the normal distribution; you can verify that these values are used in column A by sorting the column in descending order.) To express these normal scores in the scale of the batting average data, multiply the values in column A by the sample standard deviation (0.02338) and add the sample mean (0.263). For these data, this results in an expected maximum value (if the data come from a normal distribution) of 2.8237 * 0.02338 + 0.263 = 0.3290. The actual observed maximum value is a batting average of 0.3521, which is larger than you would expect.

The normal probability chart plots the expected normal scores (y-axis) against the observed data values (x-axis) so that the largest normal score is paired with the largest observed value, and so on. See Figure 5-16.

Figure 5-16
Output for the normal
probability plot

y-axis: expected normal
scores

x-axis: observed values

numbers indicate
duplicate values

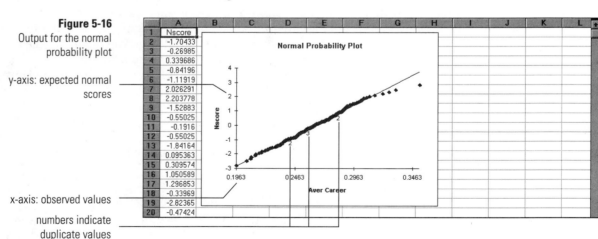

If the data are normal the points fall close to a straight line. You can judge the linearity of the plot by comparing the points to a line superimposed upon the plot (the CTI add-in superimposes a line automatically). The add-in labels duplicate values with a number showing the number of duplicates.

In this case, the batting average data follow a straight line very well. The only departures from the line occur for the five largest batting averages. Because the values fall below the line, this means that the expected batting averages are lower than those you observed. For example, the expected maximum batting average (0.3290) is less than the observed maximum batting average (0.3591).

Skewed data show a definite pattern on a normal probability plot. Positively skewed data fall below linearity at the left of the plot, whereas negatively skewed data tend to rise above the straight line on the right of a normal probability plot. See Figure 5-17. (You don't have to generate this figure.)

Figure 5-17
Normal probability plots for positively and negatively skewed data

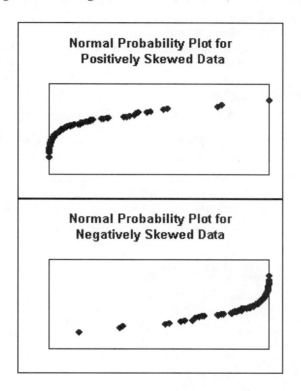

To close the batting average workbook and save the changes to file 5BASE.XLS:

1   Click **File > Close** to close the file.

2   Click **Yes** when asked if you want to save the changes to 5BASE.XLS.

The Distribution of the Sample Mean

Of all the statistics that we use in our lives, the mean is the most important. The newspapers are full of information presented in the form of averages, and sports statistics are dominated by averages. Why do you think we compute the average of many observations instead of just using a single observation? For example, wouldn't it be simpler to study student textbook expenditures by asking just one student rather than conducting a survey? Intuitively, you might believe that the average is more accurate, and that more observations will increase the accuracy. After all, why collect more data if they will not improve the accuracy?

The Average of Normal Data

Earlier in the chapter, you learned that the sample mean is a consistent estimator of the population mean μ because, given enough observations, it estimates μ to whatever degree of precision you want. Just as each sample value is a random variable from a probability distribution, the sample mean is a random variable too, and follows its own probability distribution. The distribution of a sample statistic such as the mean is called the **sampling distribution**. If you know the sampling distribution of the mean, you can use it to quantify how many samples you need in order to estimate μ with a certain precision.

For a sample of 9 observations from a normal population centered at 0 with standard deviation 1 (in other words, from a standard normal distribution), can you determine the sampling distribution of the mean? To find out, let's use a simulation to generate 100 samples, each with 9 observations. Then you can compute the mean for each sample and examine the distribution of the 100 means.

To create 100 samples with 9 observations each:

1 Click the **New Workbook button** 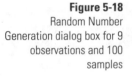 to create a new workbook.

2 Click **Tools > Data Analysis**.

3 Click **Random Number Generation** in the Analysis Tools list box, then click **OK**.

4 Type **9** in the Number of variables (observations) text box, **100** in the Number of Random Numbers text box, choose **Normal** for the Distribution, **0** for the Mean, **1** for the Standard Deviation, and **$A$1** for the Output Range.

5 When the Random Number Generation dialog box looks like Figure 5-18, click **OK**.

Figure 5-18
Random Number
Generation dialog box for 9
observations and 100
samples

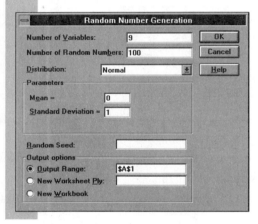

Depending upon your computer's memory and processor speed, this might take a while. When Excel has generated the 900 random numbers, create a 10th column in your worksheet that calculates the average of the first 9.

To calculate the averages of the random numbers:

1 Click **Edit > Go To**.

2 Type **J1:J100** in the Reference text box, then click **OK**.

3 Type **=AVERAGE(A1:I1)** in cell **J1**, then press **[Enter]**.

4 Click **Edit > Fill > Down** (or press **[Ctrl]+[D]**).

The new worksheet now contains 10 columns and 100 rows, and the last number in each row is the average of the first 9. Notice that the average is much less variable than the other values. When you compute the average, it smooths out the highs and lows. Next create a histogram, with the normal curve smoothed over it for comparison.

To create the histogram:

1   Click **CTI > Histogram**.

2   Type **J1:J100** in the Input Range box (verify that the Selection Includes Header Row check box is not selected).

3   Click the **New Sheet option button**, then type **Avg Simulation** in the corresponding text box.

4   Click **OK**.

The new sheet Avg Simulation shows the histogram table and histogram of the average of 9 observations with 100 samples. See Figure 5-19. Yours may look different.

Figure 5-19
Histogram of simulated
averages

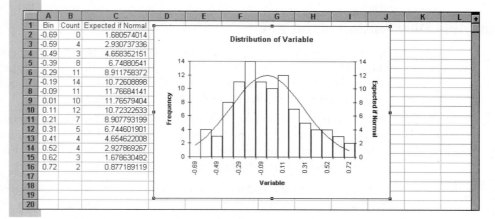

The distribution of the sample mean looks normal and, in fact, is normal. The distribution is centered at 0 as you would expect for a mean based on samples of observations from the standard normal distribution, but it differs in one respect. The sampling distribution of the mean is much narrower than the distribution for a single observation. For the standard normal distribution, most of the values are between –2 and 2 (as indicated in Figure 5-9); here most of the values are between –0.7 and 0.7. If there is any justice in the world, the sample mean should be more accurate in estimating μ than would a single observation. To verify this, calculate the statistics for the simulated mean values.

To obtain statistics on the averages of the simulated samples:

1   Click the **Sheet1 sheet tab** to return to the simulation data worksheet.

2   Click **Tools > Data Analysis**.

3   Click **Descriptive Statistics** in the Analysis Tools list box, then click **OK**.

4   Type **J1:J100** in the Input Range text box, make sure that the data are grouped by **Columns**, click the **New Worksheet Ply option button**, then type **Avg Stats** in the corresponding text box.

5   Verify that **Summary Statistics** is checked, then click **OK**.

The descriptive statistics appear as shown in Figure 5-20 (you can reformat your columns to display all the labels and click any cell to deselect the columns).

Figure 5-20
Descriptive Statistics for
average of simulation

	A	B	C	D	E	F	G	H	I	J
1	Column1									
2										
3	Mean	-0.02153								
4	Standard Error	0.029357								
5	Median	-0.01037								
6	Mode	#N/A								
7	Standard Deviation	0.293569								
8	Sample Variance	0.086183								
9	Kurtosis	-0.39564								
10	Skewness	0.188269								
11	Range	1.383233								
12	Minimum	-0.63444								
13	Maximum	0.748793								
14	Sum	-2.15335								
15	Count	100								
16	Confidence Level(95.000%)	0.057538								
17										

The Sampling Distribution of the Mean

The standard deviation of the sample means shown in cell B7 of Figure 5-20 is about 0.29 (you might have slightly different values for your randomly generated sample means), or about one-third of the standard deviation of the standard normal distribution. The average of sample means is close to 0 (cell B3), as is the median (cell B5).

The standard deviation here illustrates an important principle of statistics. If a sample is composed of n independent observations from a population having a normal distribution with mean μ and standard deviation σ, then the sampling distribution of the average ($\bar{x}$) is the normal distribution with mean μ and standard deviation $\sigma/\sqrt{n}$.

The estimated standard deviation of $\bar{x}$ is more commonly known as the standard error of the mean. If you know the population standard deviation for the individual observations, you can easily estimate what the standard error should be. In this example the standard deviation for the individual observations is 1 because each observation comes from the standard normal, so the standard deviation of the mean equals $\frac{1}{3}$:

$$\text{standard deviation } (\bar{x}) = \sigma/\sqrt{n} = 1/\sqrt{9} = \frac{1}{3}$$

Estimating the Precision of the Sample Mean

You can use the distribution of the sample mean to tell how accurately the sample mean estimates the true population mean. For example, suppose you have 100 observations that come from a standard normal distribution so that the value of the population mean μ is 0, and population standard deviation σ is 1. You've just learned that the sampling distribution of the mean is a normal distribution with mean 0 and standard deviation 0.1 (because $1/\sqrt{100} = 0.1$). Let's apply this knowledge to what you already know about the normal distribution, namely that about 95% of the values fall within 2 standard deviations of the mean. Because the sample mean is a random variable that comes from a normal distribution with a standard deviation of 0.1, there is a 95% probability that its value will fall within 0.2 units of the mean.

You can also use this information to draw conclusions about the population mean. If you have a sample size of 100 and you know that σ is equal to 1, there is a 95% probability that the sample mean lies within 0.2 units of the true mean. Thus if the sample mean is 5.3, you can be fairly confident that the true mean is between 5.1 and 5.5. To be even more precise, you can increase the sample size. For example, if you want $\bar{x}$ to fall within ± 0.02 of the true mean value 95% of the time, you need a sample size of 10,000, because $1/\sqrt{10,000} = 0.01$. So if the sample mean is 5.30 with a sample size of 10,000, you can be fairly confident that the true mean is between 5.28 and 5.32.

To close the current workbook:

1   Click **File > Clos**e.

2   Click **No** when asked if you want to save the changes (because this is simulated data, you won't need it again).

The Central Limit Theorem

What happens if the data come from a distribution that is not normal? Can you say anything about the sampling distribution of the mean in that case? The Central Limit Theorem helps answer this question. To see how it works, open the instructional template CENTRAL.XLT.

To open CENTRAL.XLT:

1   Open **CENTRAL.XLT** (be sure you select the drive and directory containing your Student files).

Excel displays the workbook based on the template CENTRAL.XLT, as shown in Figure 5-21.

Figure 5-21
Central workbook

distribution parameters

summary statistics

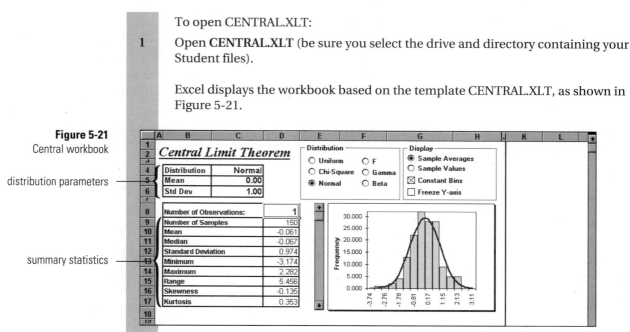

The template lets you create 150 random samples from one of six common probability distributions—uniform, chi-square, normal, *F*, gamma, and beta—with up to 9 observations per sample. (These are standard distributions used in statistics that you haven't worked with yet; some you'll see in subsequent chapters.) The template also calculates the sample means for each of the 150 samples, so you can view how the distribution of these means changes as the number of observations in each sample increases from 1 to 9. Let's begin using the template by again examining the distribution of the sample means for the normal distribution.

Generating Sample Means from a Normal Distribution

The template shown in Figure 5-21 opens with the standard normal probability distribution selected; cells C5 and C6 display the parameter values. Descriptive statistics for the simulated data appear in cells D8 to D17. Currently, you're looking at the distribution of the average from 150 samples of standard normal data with 1 observation per sample. The average of the sample means is –0.061, and the standard deviation of the sample means is 0.974. The chart shows a frequency distribution of the sample averages with the normal probability density function curve superimposed upon the bars. Because there is only 1 observation per sample, this is equivalent to 150 individual observations from a standard normal distribution. Now increase the number of observations in each sample to 9.

To increase the number of observations:

1   Drag the scroll bar down until you see the number **9** in cell **D8**.

Depending upon the speed of your computer, the template might take a few seconds to recreate the plot and recalculate the sample statistics, shown in Figure 5-22.

Figure 5-22
Sample averages from samples with 9 observations

9 observations

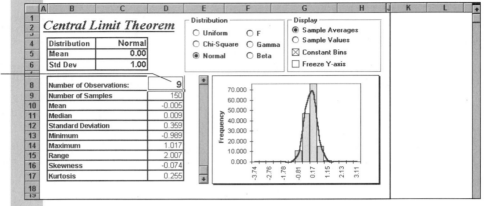

The average of the sample means is now –0.005 and the standard deviation of the sample means is 0.359, or about $\frac{1}{3}$ of the standard deviation for the sample means with a sample size of just 1. See Figure 5-22. This is what you would expect, because $1/\sqrt{9} = \frac{1}{3}$. Note as well that the normal probability density function and the histogram are distributed much more tightly around the mean.

Note: If you want to view the histogram with different bin values for different histograms, you can deselect the Constant Bins check box. When the bin values are the same, you can compare the spread of the data from one histogram to another because the same bin boundaries are used for all histograms. Deselecting the check box fits the bin values to the data and gives you more detail, but it's more difficult to compare histograms when you have different bin boundaries. You can also "freeze" the y-axis to better compare one chart to another.

2   Click the **Constant Bins check box** to deselect it so that you can see more detail and produce the histogram shown in Figure 5-23.

Figure 5-23
Histogram with different bin values

Constant Bins deselected

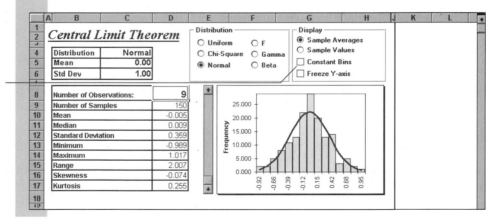

The distribution of the sample means is approximately normal, as you would expect. The histogram is the same, but the bin ranges are based only on the data used to generate the sample. If you want to view the histogram and normal curve for the distribution of the sample means for other sample sizes, just drag the scroll box to the sample size you want.

To return the template to its original state:

1   Drag the scroll box back up to the top of the scroll bar so that the Number of Observations is 1.

2	Click the **Constant Bins check box** to select it.
3	Verify that the **Sample Averages button** is selected.

Generating Means from a Uniform Distribution

Another commonly used probability distribution is the uniform distribution. The target template you used earlier in Figure 5-3 showed an example of the uniform distribution. Shots from the marksman are distributed in a circular fashion around the origin, with an error in one direction no more likely than an error in another direction. The angles of each shot as measured from the x-axis follow the uniform distribution from 0° to 360° because probability is evenly distributed over this range. Figure 5-24 shows the probability density function for the uniform distribution.

Figure 5-24
Probability density function
for the uniform distribution

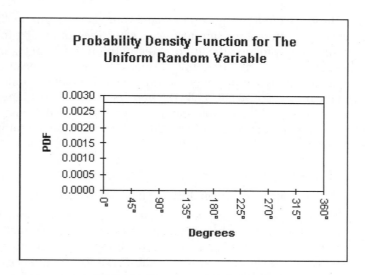

As you can see, unlike in the normal distribution, probability values in the uniform distribution are spread evenly over a range. And unlike the normal distribution, the uniform distribution covers a finite range between a lower and upper boundary. The population mean μ and population standard deviation σ for the uniform distribution are:

$$\mu = \frac{\text{upper boundary + lower boundary}}{2}$$

$$\sigma = \frac{\text{upper boundary} - \text{lower boundary}}{\sqrt{12}}$$

So for data that follow a uniform distribution with a lower boundary of 0 and upper boundary of 100, $\mu = (100 + 0)/2 = 50$ and $\sigma = (100 - 0)/\sqrt{12} = 28.87$

Although the uniform distribution differs from the normal distribution, can you still say anything about the sampling distribution of its mean?

To find out, generate 150 samples of uniformly distributed data:

1	Click the **Uniform option button** on the Central Limit Theorem template. The Uniform Distribution Parameters dialog box opens.
2	Type **0** in the Minimum text box (if necessary) and **100** in the Maximum text box, then click **OK**.

The template generates 150 random samples from uniformly distributed data. See Figure 5-25.

Figure 5-25

Output of sample means for
the uniform distribution

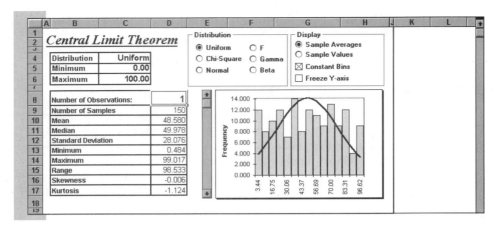

Because these are randomly generated data, your values might be different. Figure 5-25 shows the distribution of the sample means when the sample size for each sample is 1. The distribution roughly follows the uniform distribution shown in Figure 5-24 and does not resemble at all the superimposed normal curve. Remember, because there is 1 observation per sample, this sample represents 150 individual observations following a uniform distribution. The mean value of 48.580 (cell D10; your value might be different) is close to the population mean of 50, and the sample standard deviation is 28.076, just a little less than the population standard deviation value of 28.87. Now increase the sample size for each of the 150 samples to 9.

To increase the sample size:

1 Drag the scroll bar down until the number of observations equals **9**.

The average of the sample means for sample sizes of 9 is 49.542 (much closer to the population mean value of 50), and the standard deviation 5.233 is about $\frac{1}{3}$ of the previous value. See Figure 5-26. More importantly, the histogram now reveals a distribution that looks close to normal. This is a consequence of the Central Limit Theorem, which states that the sampling distribution of the sample means is close to normal.

Figure 5-26

Output for the uniform
distribution where the
sample size = 9

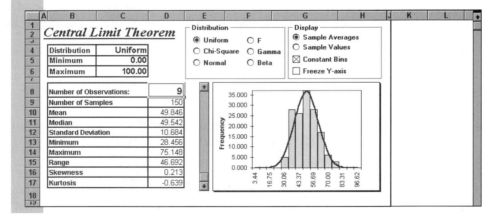

Specifically, the Central Limit Theorem states that the sampling distribution of the mean from *any* probability distribution with a mean μ and standard deviation σ approximately follows a normal distribution with mean μ and standard deviation $(\sigma/\sqrt{n}$ (where n is the sample size) if *n* is big enough. In the case of the uniform distribution (μ = 50 and σ = 28.87) the distribution of the average of 9 uniformly distributed values approximately follows a normal distribution with mean 50 and standard deviation 9.623 $(28.87/\sqrt{9})$, in accordance with the Central Limit Theorem. This is fairly close to what you saw in the simulation.

You should consider a few points. First of all, the Central Limit Theorem applies to any probability distribution if the population mean and standard deviation exist. Secondly, the degree to which the sampling distribution of the mean follows the normal distribution is governed by the sample size and the original probability distribution. For large sample sizes, the approximation can be very good, whereas for small sample sizes the approximation is less accurate. If the original probability distribution is extremely skewed, you would need a larger sample size before the normal distribution approximates the sampling distribution of the mean. How large is large? Generally, for a nice distribution that isn't too skewed, a sample size of 20 to 30 is sufficient.

The Central Limit Theorem is arguably the most important theorem in statistics. With it, you can make statistical inferences about the sample mean even in situations where you don't know the shape of the population's probability distribution.

Experiment with some of the other distributions. The *F*, gamma, and chi-square distributions can be very skewed depending upon the parameters you specify. Examine how the distribution of the sample means changes as you increase the number of observations.

You will learn more about the *F* and chi-square distribution as well as the *t* distribution in the next few chapters.

To end this session:

1. Click **File > Exit**, then click **No** when asked if you want to save changes to the workbook created by the template. You can always open a new copy of the template another time.

E X E R C I S E S

1. Define the term "random variable."

   a. How is a random variable different from an observation?

   b. What is the distinction between a population mean and a sample mean?

2. Does the sample of the top 50 women-owned businesses in Wisconsin discussed in Chapter 3 constitute a random sample of women-owned businesses? Explain your reasoning. Can you make inferences about any women-owned businesses based on this sample?

3. The PDF template is based on the standard normal distribution. Open the PDF template.

   a. Calculate the probability of a standard normal random variable having a value between –1.5 and 2.5.

   b. Calculate the probability of a value falling between 0 and 2.

4. Excel includes a function NORMSDIST(X), which calculates the probability of a standard normal random variable having a value ≤*x*.

   a. Use the function to find the probability of a standard normal random variable having a value of 0.5 or less.

   b. What is the probability of a standard normal random variable having a value exactly equal to 0.5?

   c. Excel includes another function, NORMDIST(X,MEAN,STDEV,CUM), which you can use to find the probability of a random variable having a value ≤*x* from a normal distribution with mean MEAN and standard deviation STDEV. If the value of CUM is TRUE, the function calculates the probability of the range variable being ≤*x* (if the value is FALSE, it calculates the value of the probability density function at *x*). Using this function, find the probability of a random variable from a normal probability distribution with mean 10 and standard deviation 2 having a value of less than 5.

5. Open the BASE.XLS file. The mean career batting average is 0.263, and the standard deviation is 0.02338.

   a. Using Excel's NORMDIST(X) function from Exercise 4 above, assume that the career batting averages are normally distributed, and find the probability of a player batting 0.300 or better (that is, 1 – probability of batting less than 0.300).

   b. Using Excel's AutoFilter capability, find out how many players' batting averages were 0.300 or more. Compare this to the expected number.

6. The BASE.XLS file also contains salaries for 263 major league players at the start of the 1988 season.

   a. Using the CTI Statistical Add-Ins, create a histogram with a normal curve of the salary data.

   b. Create a normal probability plot of the salary data. Do the data appear normal?

   c. Calculate the expected maximum salary based on the normal scores (you will need to calculate the sample mean and the sample standard deviation). How does this compare to the observed maximum salary?

   d. Is the distribution of the salary positively skewed, negatively skewed, or symmetric?

   e. Save your file as E5BASE.XLS.

7. The file 3WBUS.XLS you created in Chapter 3 contains a variable showing the logarithm of the number of employees for each company. You calculated this variable in Chapter 3 to reduce skewness in the employee data. Did the transformation also make the distribution look more normal?

   a. Create a histogram with a normal curve of the log(employees) data using the CTI Statistical Add-Ins. Does the histogram follow the normal curve well?

   b. Create a normal probability plot of the log(employees) data using the CTI Statistical Add-Ins.

   c. Using the results of parts 7a and 7b above, do you think that the log(employee) data follows the normal distribution? Defend your answer.

8. The vertical deviations of shots in the target instructional template follow a normal distribution with mean 0. The standard deviations at each level of accuracy are as follows:

Accuracy	Standard Deviation
Highest	0.1
Good	0.2
Moderate	0.4
Poor	0.6
Lowest	1.0

   Open the TARGET.XLT file and create a distribution of shots around the target with Good accuracy.

   a. Explain why the predicted percentages have the values that they have.

   b. For a marksman with the lowest accuracy, how many shots would the marksman have to take before the marksman could assume with 95% probability that the average vertical deviation from the target was less than 0.2?

   c. How many shots would a marksman with the highest accuracy have to take before achieving similar confidence in the average of the shots?

   d. *True or false.* As the number of shots taken by the marksman increases, the vertical deviation of each shot from the origin decreases. Is this statement true or false for the target simulation? Is this statement true or false for the real world? Discuss how the target simulation can differ from the real world.

9. Using the Analysis ToolPak, generate 4 columns with 100 rows of standard normal variables, then create a 5th column of the 100 averages from those 4 columns.

   a. Does the average have a standard deviation that is less than the standard deviation of the 4 original variables?

   b. Is the standard deviation reduced approximately in accordance with theory?

10. Use the Analysis ToolPak to create 9 columns with 100 rows of uniform variables with a minimum value of 0 and a maximum value of 50, then create a 10th column of the 100 averages from those 9 columns.

    a. What are the standard deviation and mean of the first column? Of the column of averages?

    b. What are the population mean and standard deviation for the first column and the column of averages?

    c. What is the approximate probability distribution of the column of averages?

11. *True or false.* According to the Central Limit Theorem, for any probability distribution, as you increase the sample size of values based on that distribution, the distribution of the observations will be approximately normal. Defend your answer.

12. You want to generate a random sample of observations that follow the uniform distribution with an unknown mean but with a known standard deviation of 10.

    a. How many observations will you need in order to estimate the mean value within 2 units about 95% of the time?

    b. How many observations will you need in order to estimate the mean value within 1 unit about 68% of the time?

    c. If the sample size is 25 and the population mean is 50, what is the probability that the sample mean will have a value of 48 or less?

CHAPTER 6

STATISTICAL INFERENCE

O B J E C T I V E S

In this chapter you will learn to:

- Assess the accuracy of the sample mean using confidence intervals

- Reach conclusions about data using hypothesis testing

- Use the t distribution when σ is unknown

- Perform a one-sample t-test

- Perform a two-sample t-test

- Analyze data using nonparametric approaches

The probability distributions you learned about in the last chapter are needed for the two main tools of statistical inference: confidence intervals and hypothesis testing. You use confidence intervals to calculate a range of values within which the true population mean might lie. Obtaining a confidence interval for a sample mean value gives you some idea of how far off you may expect the true population mean to be. With hypothesis testing, you state two mutually exclusive hypotheses and choose between them by assuming one of the hypotheses is true and determining whether such an assumption is consistent with the observed values.

Confidence Intervals

In Chapter 5 you learned that approximately 95% of observations from a normal distribution fall within two standard deviations of the mean. Consequently, about 95% of sample means fall within $2 * \sigma/\sqrt{n}$ units of the population mean. You can use this information to estimate the **confidence interval**, which is the range of values around the sample mean that will cover the true mean value about 95% of the time. The range roughly takes the form:

$$\left(\bar{x} - 2\frac{\sigma}{\sqrt{n}}, \bar{x} + 2\frac{\sigma}{\sqrt{n}} \right)$$

The more general form of the confidence interval expression is:

$$\left(\bar{x} - Z_{1-\alpha/2}\frac{\sigma}{\sqrt{n}}, \bar{x} + Z_{1-\alpha/2}\frac{\sigma}{\sqrt{n}} \right)$$

where this range, centered at $\bar{x}$, is expected to cover the true mean $100(1 - \alpha)\%$ of the time. See Figure 6-1.

Figure 6-1
Normal curve with values $\pm Z_{1-\alpha/2}$ unknown location of population mean μ

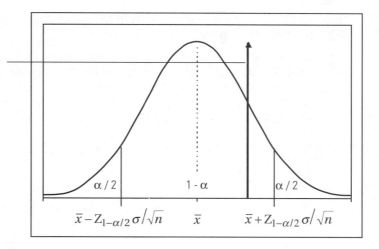

The value of $Z_{1-\alpha/2}$ is based on the standard normal distribution, where for a random variable x following a standard normal distribution, the probability that $x \leq Z_{1-\alpha/2}$ is equal to $1 - \alpha/2$. Equivalently, $100(1 - \alpha/2)\%$ of the standard normal distribution lies to the left of the point $Z_{1-\alpha/2}$. For a 95% confidence interval, $\alpha = 0.05$, $1 - \alpha/2 = 0.975$, and $Z_{0.975} = 1.96$ (which is why you use the value 2 for a rough approximation of the 95% confidence interval).

Confidence intervals are extremely important in statistics, because whenever you report a sample mean, you need to be able to gauge how precisely it estimates the population mean. If you know that the confidence interval for the mean return on a sample of different investments is very large, you might be cautious about future speculation because the true mean return value could be much greater, but it also could be much less than your initial estimate.

Calculating a Confidence Interval

When you have a random sample and have calculated the sample mean, you can use Excel to calculate confidence intervals for that mean. For example, suppose you are conducting a survey of the cost of a medical procedure as part of research on health reform. The cost of the procedure varies following the normal distribution and you know that the population standard deviation of the cost σ is $1,000. After sampling the cost of the procedure at 50 hospitals, you calculate the sample mean to be $5,500. What is the 90% confidence interval for the mean cost? (That is, how far above and below $5,500 must you go to be able to comfortably say, "I'm 90% confident that the population mean cost of this procedure lies in this range"?)

To calculate the 90% confidence interval for the population mean:

1 Start Excel, open a new workbook if necessary, then maximize both windows.

2 Type **Average** in cell A1, then press [**Tab**].

3 Type **Std Err** in cell B1, then press [**Tab**].

4 Type **Lower** in cell C1, press [**Tab**], type **Upper** in cell D1, then press [**Enter**].

5 Click cell **A2**, then type **5500** (the sample mean).

6 Click cell **B2**, type **=1000/SQRT(50)**, then press [**Enter**].

This formula calculates the standard deviation of the sample mean, $\sigma/\sqrt{n}$, a value more commonly known as the **standard error of the mean**. You now have the numbers you need to calculate the confidence interval.

Excel includes a function, NORMSINV(x), which calculates the value of a $Z_{1-\alpha/2}$. For a 90% confidence interval, $\alpha = 0.1$, $1 - \alpha/2 = 0.95$ and $Z_{0.95}$ equals approximately 1.645.

7 Click cell **C2**, type **=A2–B2*NORMSINV(0.95)**, press [**Tab**], type **=A2+B2*NORMSINV(0.95)**, then press [**Enter**].

Excel returns a 90% confidence interval in cells C2 and D2 that covers the range **$5267.38 to $5732.62**. If you were trying to estimate the cost of government programs that cover this procedure for any hospital, you would state that you were 90% confident that the true mean cost of the procedure was not going to be less than $5,267 or greater than $5,732. See Figure 6-2.

Figure 6-2
90% confidence interval

	A	B	C	D	E	F	G	H	I	J	K	L
1	Average	Std Err	Lower	Upper								
2	5500	141.4214	5267.383	5732.617								
3												
4												

8 Click **File > Close**, then click **No** when asked if you want to save the changes.

Interpreting the Confidence Interval

It's important that you understand what is meant by statistical confidence. When you discovered that the 90% confidence interval for the true mean cost of the procedure was between $5,267 and $5,732, you were not saying that the probability is 90% that the true mean falls between $5,267 and $5,732. This would incorrectly imply that the range you've calculated or the population mean are random variables. But after drawing a specific sample and from that sample calculating a specific confidence interval, you are no longer working with random variables but with observed values. These observed values are not random, and the population parameter is, as always, not random. The population mean either will be within the confidence range or it will not. The 90% confidence refers to the method of calculating the range. It means that 90% of the time the procedure results in a range that includes the true population mean.

Calculating a Confidence Interval When σ is Known

- For a 100(1 – α)% confidence interval, find the value of $Z_{1-\alpha/2}$ (Excel function NORMSINV(x) where x = 1 – α/2).

- The standard error of the mean will be equal to $\sigma/\sqrt{n}$ (σ is a known constant divided by the Excel function SQRT(n) where n is the sample size).

- The sample mean will be equal to $\bar{x}$ (Excel function AVERAGE(range) where range is the range containing the data).

- The 100(1 – α)% confidence interval = AVERAGE(range) ± NORMSINV(x) * σ/SQRT(n).

The Confidence Interval Template

To get a visual picture of the confidence interval, you can use the Confidence Interval instructional template, CONF.XLT.

To open the Confidence Interval instructional template:

1 Open **CONF.XLT** (be sure you select the drive and directory containing your Student files).

The template opens showing 100 simulated confidence intervals from 100 different samples. See Figure 6-3.

Figure 6-3
The CONF.XLT template

confidence intervals that don't include the sample means

true mean line

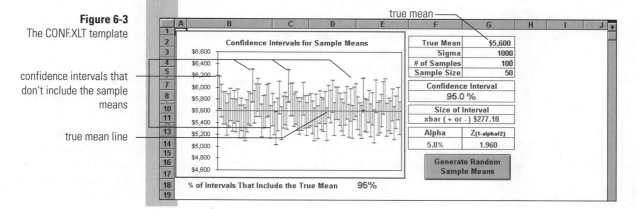

This template uses the values specified for the previous problem. The value of σ is equal to $1,000, but this time the template assumes that the population mean μ is known and equals $5,600. The template has generated 100 samples from this population with 50 observations per sample, and calculated the sample means and the 95% confidence intervals for each of the samples. These are plotted as vertical lines on the chart; confidence intervals that miss the population mean ($5,600) are shown in red (if you have a color monitor, dark gray if you have a monochrome monitor). You would expect that about 95 of the confidence intervals will include the true mean value ($5,600) and that around 5 will not. In this particular example, 95% of the randomly generated samples do indeed include the true mean within their confidence intervals, and 5 of the 100 samples do not. The boundaries of each of the 100 confidence intervals are $\bar{x} \pm 1.96 * \$141.42 = \bar{x} \pm \$277.18$.

Using this simulation you see that the sample means estimate the true mean cost of the procedure within $277 dollars 95% of the time. You can change the size of the confidence interval by changing the confidence value in cell F8.

To reduce the width of the confidence interval:

1 Click cell **F8**, type **75** as the confidence level percentage, then press [**Enter**].

The boundaries of each confidence interval (Size of Interval in cell F10) are reduced to a value of $\bar{x} \pm \$162.68$. Now the true mean is captured within the confidence interval only 66% of the time in this simulation. Even though the probability of capture is 75%, usually you will not have exactly 75% of the intervals capturing the true mean. See Figure 6-4.

Figure 6-4
75% confidence interval

confidence interval is 75%

true mean captured 66% of the time

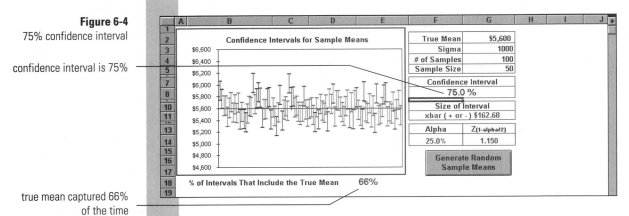

Now increase the probability of capturing the true mean by increasing the confidence interval to 99.9%:

2 Click cell **F8**, type **99.9**, then press [**Enter**].

All of the samples include the true mean cost in their confidence intervals and the size of the confidence interval has increased to $\bar{x} \pm \$465.34$. As you can see, there is a trade-off in determining the appropriate size of the confidence interval. Making the confidence interval too narrow increases the probability that the true mean value will not be captured, whereas making the confidence interval too wide will almost certainly capture the true mean value, but will leave you with a range of values that might be too broad to be useful. Statisticians have generally favored the 95% confidence interval as a compromise between these two positions.

An important lesson to learn from this simulation is not to take a single sample mean at face value. Confidence intervals help you quantify how closely that sample mean estimates the true mean. The next time you hear in the news that a study has shown that a drug causes a mean decrease in blood pressure or that a government program has caused an average increase in family income, you should ask, "What is the confidence interval?"

If you want to create a new set of random samples, you can click the Generate Random Sample Means button. Continue working with the confidence interval instructional template until you understand the relationship between the confidence interval, the sample mean value, and the population mean.

To finish working with the template:

3 Click **File > Close**, then click **No** when asked if you want to save the changes.

Hypothesis Testing

Closely related to confidence intervals is hypothesis testing. **Hypothesis testing** is a decision process that compares two hypotheses—a null hypothesis, usually denoted by the symbol H_0, and an alternative hypothesis, indicated by H_a—and chooses between the two. The null hypothesis often represents the status quo, whereas the alternative hypothesis might represent a new theory or approach.

You can use what you've learned so far to perform a hypothesis test. For example, assume you are studying the number of defective resistors found in a production batch. Previous quality control studies have indicated that the number of defective resistors in each batch follows (approximately) a normal distribution with $\mu = 50$ and $\sigma = 15$. Hoping that a new manufacturing process will reduce the number of defects, the quality assurance manager might test 25 batches. Let's say the average number of defective resistors from the 25 batches is 45. Is this a significant decrease in the mean number of defects?

In the language of hypothesis testing, you state and then decide between two hypotheses:

H_0: The new process has not changed the mean number of defective resistors.
or
H_a: The new process has changed the mean number of defective resistors.
This is the same as saying:
H_0: The population mean of the new process is equal to 50.
H_a: The population mean of the new process is not equal to 50.

The Hypothesis Test Template

Your Student files include a template that lets you see how hypothesis testing works.

To experiment with the instructional template HYPOTH.XLT:

1   Open **HYPOTH.XLT** (be sure you select the drive and directory containing your Student files).

The template opens to a visual depiction of the hypothesis test. See Figure 6-5.

Figure 6-5
The HYPOTH.XLT template
standard error
observed sample mean
acceptance region

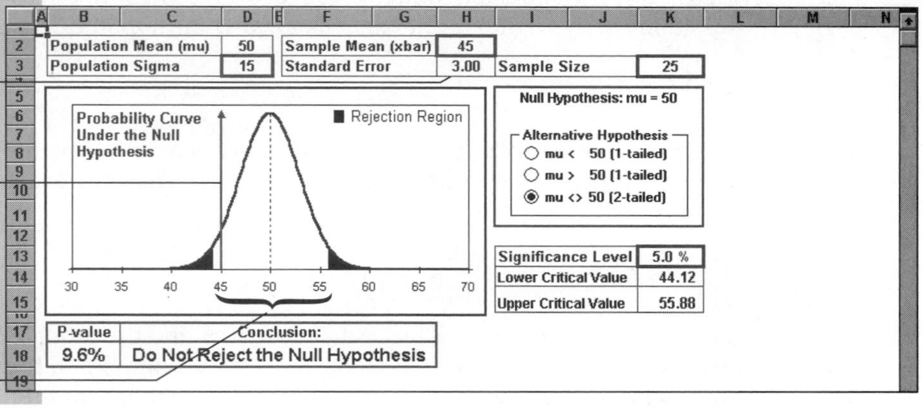

With hypothesis testing, you accept the null hypothesis unless you have compelling reasons to choose otherwise. Hypothesis testing focuses on the probability of making an error if you choose the alternative hypothesis instead of the null hypothesis.

The chart in the template shows the sampling distribution of the mean if the null hypothesis is true. Under the null hypothesis you are assuming that the sample mean follows a normal distribution with mean 50 and standard deviation $15 / \sqrt{25} = 3$. Recall that this value is also known as the standard error and is labeled as such in cell H3. The observed sample mean is 45 (cell H2) and is indicated by the vertical red line on the chart. You want to determine the probability that a sample mean of 45 or less could come from a normal distribution with $\mu = 50$ and $\sigma = 3$.

Cell K13 displays the **significance level**, which is the probability of error that you would accept in erroneously rejecting the null hypothesis. The significance level is identified by the Greek symbol α and often has a value of 5%. In this case, the probability of rejecting the null hypothesis is 5% when the null hypothesis is true.

The Acceptance and Rejection Regions

You use α to calculate a **rejection region**, a range of values for which you would reject the null hypothesis. This template displays the area under the probability curve corresponding to the rejection region with solid black. If your sample mean lies within the rejection region, you would reject the null hypothesis.

The complement of the rejection region is the **acceptance region**, which includes only those values that would support accepting the null hypothesis. Because $100\alpha\%$ is the probability of the sample mean lying in the rejection region, the probability of the sample mean lying in the acceptance region is $100(1 - \alpha)\%$. So to find the acceptance region you calculate the boundary of a range centered at μ whose area under the normal curve is $100(1 - \alpha)\%$. If you think you've seen this before, it's because you have. The equation for the $100(1 - \alpha)\%$ acceptance region around μ is in the same form as the $100(1 - \alpha)\%$ confidence interval around $\bar{x}$ discussed earlier in this chapter. Using the same reasoning, the boundary of the acceptance region for the null hypothesis is:

$$\left(\mu - Z_{1-\alpha/2}\frac{\sigma}{\sqrt{n}}, \mu + Z_{1-\alpha/2}\frac{\sigma}{\sqrt{n}} \right)$$

These boundary values, known as **critical values**, are shown in K14 and K15 of the template and are equal to 44.12 and 55.88, so the width of the acceptance region is 11.76 (55.88 – 44.12). Because 45 lies within the acceptance region, you do not reject the null hypothesis, and you decide there is not enough evidence to claim that the mean number of defects in the new process is less than the mean number of defects in the old process.

Notice that the rejection region extends in both directions. What you are trying to decide is whether 45 is an unusual sample average for this distribution. In doing so you include the possibility of departures either above or below the hypothesis mean.

P-values

You could have also calculated the probability of a sample mean being as extreme as the observed value (45) in its departure from μ. To do so, you calculate the z-score of the sample mean where:

$$Z = \frac{\bar{x} - \mu}{\sigma/\sqrt{n}} = \frac{45 - 50}{15/\sqrt{25}} = \frac{-5}{15/5} = -\frac{5}{3}$$

Z is a standard normal variable and you want to find the probability that an observation will be beyond Z. In other words, for a positive Z you want to find the probability of a standard normal random variable being greater than or equal to Z. In this case, Z is negative, so you want to find the probability of a standard normal random variable being less than or equal to Z. The probability of a standard normal random variable being less than $-\frac{5}{3}$ is 0.0478 (you can verify this using Excel's NORMSDIST function as discussed in Exercise 4 of Chapter 5). Because you're looking at extreme values of this magnitude in either direction, double this value to arrive at a *p*-value of 0.0956, or 9.6%—the value shown in cell B18 of the template. Because the *p*-value is greater than 5% you do not reject the null hypothesis at the 0.05 level.

Two-Tailed and One-Tailed Tests

What if there was something about the new process that absolutely ruled out the possibility of an increase in the number of defects? Hypothesis tests that assume a departure from the null hypothesis can occur only in one direction are called **one-tailed tests** (as opposed to the **two-tailed tests**, which assume the difference can go in either direction). In this example you might state a one-tailed alternative hypothesis as:

> H_0: The population mean of the new process = 50.
> H_a: The population mean of the new process is < 50.

To generate the one-tailed hypothesis test:

1   Click the **mu < 50 (1-tailed) option button**.

The boundary of the lower rejection region changes to the value 45.07 whereas the upper rejection region disappears entirely. See Figure 6-6.

Figure 6-6
The one-tailed test

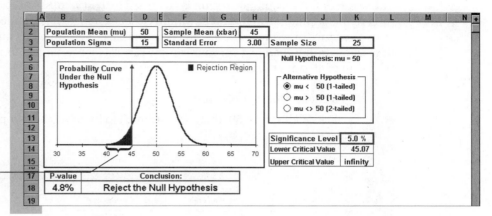

rejection region on left ──────

This test assumes that the average number of defects cannot increase with the new process. The $(1 - \alpha)\%$ acceptance region for the null hypothesis now covers the range:

$$\left(\mu - Z_{1-\alpha} \frac{\sigma}{\sqrt{n}} , \infty \right)$$

The *p*-value for the test is 4.8%, because you are only interested in extreme values in the lower tail. Because this is less than 5%, you reject the null hypothesis at the 5% level and accept the alternative that the average number of defects with the new process is less than 50.

You should use one-tailed tests only when you are absolutely sure that you can ignore the other tail or in situations where you are interested only in a change in one direction. One-tailed tests are generally frowned on because they make it easier to erroneously reject a true null hypothesis. To counteract this problem, you should state your alternative hypothesis *before* doing your statistical analysis (rather than deciding whether to do a one-tailed test after seeing the resulting *p*-values). Return the template to a two-tailed test.

To calculate the two-tailed hypothesis test:

1   Click the **mu < > 50 (2-tailed) option button**.

....

Z-test: Calculating a p-value

When σ is known under a null hypothesis that the population mean $\mu = x$:

- Calculate the value of the Z score so that Z = (AVERAGE(range) – x) / (σ / SQRT(n)) where range is the range reference of the data and n is the sample size. The values x and σ are known.

- For a one-tailed test where Z is negative, calculate the probability to the left of Z with the Excel function =NORMSDIST(Z).

- For a one-tailed test where Z is positive, calculate the probability to the right of Z with the Excel function =1 – NORMSDIST(Z).

- For a two-tailed test where Z is negative, calculate the probability of an extreme value of Z with the Excel function =2*NORMSDIST(Z).

- For a two-tailed test where Z is positive, calculate the probability of an extreme value of Z with the Excel function =2*(1 – NORMSDIST(Z)).

....

Other Factors that Influence Hypothesis Tests

The Hypothesis Testing instructional template allows you to vary these other factors: population standard deviation, sample size, sample mean, and significance level. Let's see what happens as you change some of these other factors.

The width of the acceptance region is governed by two values: the size of the sample and the underlying population standard deviation σ. Let's change these values to see how they affect the acceptance or rejection of the null hypothesis. For example, if you examined 50 batches of resistors rather than just 25, your sample size is twice as large. For now, let's say the sample mean of 45 stays the same.

To change the sample size:

1 Click cell **K3**, type **50**, then press [**Enter**] to increase the sample size from 25 to 50. See Figure 6-7.

Figure 6-7

Increasing the sample size

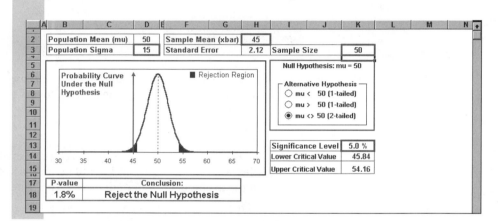

The width of the acceptance region shrinks from 11.76 (55.88 – 44.12) with a sample size of 25 to 8.32 (54.16 – 45.84) under a sample size of 50, as shown in Figure 6-7. The observed sample mean, 45, lies within the rejection region, and thus you reject the null hypothesis with a *p*-value of 1.8%. Increasing the sample size allows you to estimate the mean with more precision. With a large enough sample size, you can make the acceptance region as small as you want. Taken to the extreme, with enough observations you can reject any value for the null hypothesis mean other than the value of the observed sample mean. As the sample size increases, the sample encompasses more and more of the population. In effect, the sample becomes indistinguishable from the population.

Now try changing the value of the population standard deviation σ, which measures the variability in the number of defects per batch.

To change the value of σ:

1 Click cell **D3**, type **20**, then press [**Enter**] to increase the value of the population standard deviation from 15 to 20.

Because the population standard deviation has increased, the value of the standard error has increased from 2.12 to 2.83. See Figure 6-8.

The lower critical value is now 44.46, so you do not reject the null hypothesis. With this much variability, you cannot say that the new process affects the mean number of resistors. The variability of the data is one of the most important factors in hypothesis testing; much of statistical analysis is concerned with reducing or explaining variability.

You can also modify the value of the sample mean in cell H2 or the proposed significance level in cell K13. Try various combinations of these factors to see how they affect the acceptance or rejection of the null hypothesis.

Figure 6-8

Increasing σ

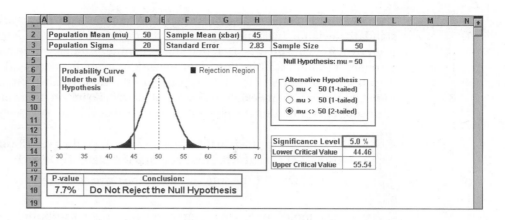

An Important Point

When you accepted the null hypothesis you were not stating that the null hypothesis was true. Rather, you were stating that there was insufficient reason to reject it and to accept the alternative. The distinction is subtle but important. To state that the mean number of defects equals 50 excludes the possibility that the mean number of defects under the new process might equal 49 or 49.9 or 49.99. But you didn't test any of these possibilities. What you *did* test was whether the data are compatible with the null hypothesis. You found in some cases that they were not.

1

To finish working with the template:

Click **File > Close**, then click **No** when asked if you want to save the changes.

Comparing Confidence Intervals and Hypothesis Testing

You've looked at two approaches to statistical inference: the confidence interval and the hypothesis test. For a particular value of α, the width of the confidence interval around the value of $\bar{x}$ is equal to the width of a two-sided acceptance region around μ. This means that each of these two statements implies the other:

- The value μ lies outside the $100(1 - \alpha)\%$ confidence interval around $\bar{x}$.

- Reject the null hypothesis that the population mean equals μ at the $100\alpha\%$ significance level.

The *t* Distribution

Up to now, you've been assuming that the value of σ is known. What if you did not know the value of σ? In the 19th century, it was common to use the value of the sample standard deviation as a substitute for σ in calculating the *p*-value.

However, in the early years of this century, William Gossett became worried about the error caused by using the wrong standard deviation. He was working at the Guinness brewery in Ireland on small experiments, and he thought that the error might be especially bad if the sample size (*n*) was small. What Gossett found is that the distribution is not standard normal when the sample standard deviation is used. Instead, the values follow a probability distribution called the *t* distribution. That is, when you use the sample standard deviation (*s*) instead of the population standard deviation (σ), the ratio

$$t = \frac{\bar{x} - \mu}{s/\sqrt{n}}$$

is a random variable that follows the *t* distribution and not the standard normal.

The *t* distribution is characterized by a parameter that represents the accuracy of the sample standard deviation in estimating σ called the **degrees of freedom**. For a sample of size *n*, the degrees of freedom are ($n-1$). If that seems counterintuitive, remember that the sample standard deviation is a function of the deviations from the mean and that the sum of the deviations is always equal to zero. If you know the values of the first ($n-1$) deviations, the value of the remaining deviation can be calculated. So in fact, there are only ($n-1$) degrees of freedom in estimating the population standard deviation.

To help you appreciate the difference between the *t* distribution and the standard normal, you can use the instructional template, TPDF.XLT:

1   Open **TPDF.XLT** (be sure you select the drive and directory containing your Student files).

The template opens to a chart comparing the *t* distribution to the standard normal. See Figure 6-9.

Figure 6-9
The TPDF.XLT file

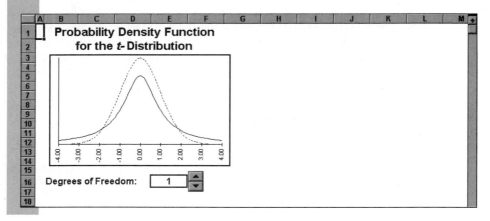

The template shows the distribution of the *t* with one degree of freedom (corresponding to a sample size of two), plotted with the standard normal curve (dotted line). The *t* distribution is symmetric like the normal, but it is more likely to have extreme observations because of the randomness of *s* in the denominator of the *t* ratio.

Experiment with the template to view how the *t* distribution changes as you increase the degrees of freedom.

To increase the degrees of freedom:

1   Click the **up scroll arrow** and watch as the *t* moves closer to the normal. Continue to increase the degrees of freedom.

As the degrees of freedom increase, the *t* distribution better approximates the standard normal. This is because the value of the sample standard deviation estimates the value of σ more precisely.

2   Click **File > Close**, then click **No** when asked if you want to save the changes.

One-Sample *t*-Tests

Let's apply the *t* distribution to a problem involving textbook expenditures. The college administration has stated in promotional literature that textbook expenditures are $200 on the average. Twenty-five students from the student telephone directory are randomly selected and asked about the total cost of their textbooks for the current semester. If the sample mean of these 25 students is $230 and the sample standard deviation is $50, does this indicate that the average reported by the college administration is wrong?

First construct a hypothesis test with hypotheses:

H_0 : The average textbook expenditure = \$200.
H_a : The average textbook expenditure >\$200.

Because you are using the t distribution to compare the mean of one sample to a hypothesized value, this type of test is called a **one-sample t-test**. The hypothesis test is one-tailed, because you are not interested in whether textbooks cost less than the administration's estimate, only more. You want to find the probability under the null hypothesis of the average textbook expenditure from a sample of 25 students being \$230 or more.

To test the hypothesis calculate the t ratio (also called the t statistic):

$$t = \frac{\bar{x} - \mu}{s/\sqrt{n}} = \frac{230 - 200}{50/\sqrt{25}} = \frac{30}{10} = 3$$

Here $s = 50$ represents the estimated standard deviation for individual students, and $s/\sqrt{n} = 50/\sqrt{25} = 10$ is the standard error of the sample mean.

You can use Excel to calculate the probabilities for the t distribution to see if the t statistic is statistically significant. Use the statistical function TDIST(t, df, tails), where t is the value of the statistic, df is the degrees of freedom, and tails indicates whether the test is one- or two-tailed.

To test the hypothesis that the mean textbook expenditure is different from \$200, find the one-tailed probability of a t-ratio of 3 with 24 degrees of freedom:

1   Click the **New Workbook button** .

2   Type =**TDIST(3,24,1)** in cell A1 and press [**Enter**].

The TDIST function returns a p-value of 0.003 (rounded to three decimal places). The probability of the t statistic being equal to 3 or more is 0.003. Because 0.003 is less than 1%, you reject the null hypothesis ($\mu = \$200$) at the 5% and the 1% significance levels. This means that a sample mean as big as \$230 from a sample of 25 is very unlikely to occur if the population mean is really \$200. Perhaps prices have gone up since the college arrived at its figure.

....

t-Test: Calculating a p-Value

When σ is unknown under a null hypothesis that the population mean $\mu = x$:

- Calculate the value of the t-test statistic so that t = (AVERAGE(range) – x) / (STDEV(range) / SQRT(n)), where range is the range reference of the data, n is the sample size, and x is a known value.

- For a one-tailed test, calculate the probability with the Excel function =TDIST(t, n–1,1).

- For a two-tailed test, calculate the probability with the Excel function =TDIST(t, n–1,2).

....

Two-Sample t-Tests

In the one-sample test, you compared the sample mean to a fixed value expressed in the null hypothesis. In a two-sample test you compare two independent means. Think in terms of two distributions—one for each sample. The two samples might be the salaries of male and female faculty at a junior college or weight loss under two different diet plans. Because you want to compare the population means of the two distributions to see if they are definitely different, your t statistic, called the **two-sample t**, needs to involve the difference of sample means. If $\bar{x}_1$ and $\bar{x}_2$ are the two sample means, n_1 and n_2 are the two sample sizes, and μ_1 and μ_2 are the two population means, then:

$$t = \frac{(\bar{x}_1 - \bar{x}_2) - (\mu_1 - \mu_2)}{s\sqrt{\dfrac{1}{n_1} + \dfrac{1}{n_2}}}$$

The s in the denominator is an estimate of the standard deviation that comes from pooling the standard deviations of the two groups. If s_1 is the first standard deviation and s_2 is the second standard deviation, then:

$$s = \sqrt{\frac{(n_1 - 1)s_1^2 + (n_2 - 1)s_2^2}{n_1 + n_2 - 2}}$$

This is a type of average of the two standard deviations, and s is always somewhere between s_1 and s_2. If n_1 and n_2 are the same, then the part inside the radical is just a simple average of the variances (squares of the standard deviations). Otherwise, it is a weighted average based on the degrees of freedom, $(n_1 - 1)$ and $(n_2 - 1)$. The formula for t might look complicated, but it follows the same pattern as the t for the one-sample test except that here you start with the difference $\bar{x}_1 - \bar{x}_2$. Then you subtract the corresponding population mean $(\mu_1 - \mu_2)$, and divide by the standard error. This pattern, subtracting the population mean and dividing by the standard error, often leads to a ratio with a t distribution. If the data are normal and both populations have the same population standard deviation, then the t ratio has the t distribution with $(n_1 + n_2 - 2)$ degrees of freedom—$(n_1 - 1)$ from the first group plus $(n_2 - 1)$ from the second group.

For example, consider two groups—one that has learned to write with a word processor on a Macintosh computer and one that has learned to write with a word processor on an IBM PC. There are 25 students in each group. At the end of the word-processing unit, each student writes an essay that is graded on a 100-point scale. The average grade for students using the Macintosh is 75, with a standard deviation of 8, and the average for those using the IBM PC is 80, with a standard deviation of 6. Could the difference of sample means, 75 vs. 80, be easily attributed to chance, or is it statistically significant? If the population mean score for the Macintosh users is μ_1 and the population mean score for the PC users is μ_2, the hypotheses are:

$$H_0 : \mu_1 - \mu_2 = 0$$
$$H_a : \mu_1 - \mu_2 \neq 0$$

Calculate the t statistic assuming the null hypothesis that the population means are equal, and check to see how extreme the t statistic is. If the t is unusually large—which says that the difference of means is great compared to its estimated standard deviation—the difference is statistically significant.

To compute the t, you first need s, which is:

$$s = \sqrt{\frac{(n_1 - 1)s_1^2 + (n_2 - 1)s_2^2}{n_1 + n_2 - 2}} = \sqrt{\frac{(25 - 1)8^2 + (25 - 1)6^2}{25 + 25 - 2}} = \sqrt{50} = 7.07$$

The t is:

$$t = \frac{(\bar{x}_1 - \bar{x}_2) - (\mu_1 - \mu_2)}{s\sqrt{\dfrac{1}{n_1} + \dfrac{1}{n_2}}} = \frac{(75 - 80) - (0)}{7.07\sqrt{\dfrac{1}{25} + \dfrac{1}{25}}} = \frac{-5}{2} = -2.5$$

Is a t of -2.5 enough for statistical significance?

1 To calculate the probability to the left of -2.5 and to the right of 2.5:

Click cell A2, type =TDIST(2.5,48,2) and press [**Enter**].

Note that you do not type –2.5 because Excel works with the absolute value to calculate the *p*-value. The degrees of freedom are $(25 - 1) + (25 - 1) = 48$, and because you don't assume a direction in the difference between the scores, use the two-tailed probability value. Excel returns a probability value of about 0.016. Because this is less than 0.05, you reject the null hypothesis that the difference between the two population means equals zero at the 5% significance level, and conclude that students write better on the IBM PC.

This example is artificial, but it is based on research by Dr. Marcia Peoples Halio, of the University of Delaware (*Academic Computing*, January, 1990, pp. 16-19, 45). She found that when students had a choice of computer, the students choosing the PC had much better writing scores than the students choosing the Macintosh. Note that the students chose which computer to use. Because the Macintosh has been advertised as the computer for those who know nothing about computers, you might wonder if the more confident students tended to choose the PC.

To finish with this example:

1   Click **File > Close**, then click **No** when asked if you want to save the changes.

The Robustness of *t*

So far, all your work with the *t* distribution has assumed that the data are normal. What happens if this assumption is violated? The *t* distribution has a property called **robustness**, which means that even if the assumption of normality is violated moderately, the *p*-values returned by the *t*-test will not be too far wrong.

The remainder of this chapter involves analyzing two data sets. You'll use some of the tools you've been learning about (confidence intervals, hypothesis testing, and the *t* statistic) plus a few new ones in order to reach conclusions about the data.

Paired Comparisons

The first example explores the concept of **paired data**, where pairs of observations are dependent. For example, a clinical researcher might investigate the efficacy of a new drug by measuring the response of the same sample of volunteers to a traditional drug and to the experimental drug. Because each patient has two sets of responses, one under the traditional treatment and one under the experimental approach, those two values are said to be paired, or dependent. Ideally, any differences between the observations will be due to the treatment alone. Thus, an analysis of change in response due to the effect of the new drug would be called a **dependent**, or **paired**, **comparison**.

As another example, you could have paired data involving just one treatment, with measurements taken before and after the treatment to see if it has any effect. For example, the treatment could be a law that restricts pollution emissions, with air quality measured before and after the law goes into effect.

In Chapter 3 you looked at the distribution of personal computer price data in the PCINFO.XLS workbook. Part of the workbook includes a processor score that measures the speed of each computer. Many manufacturers in the list produce several computers that are identical in all components except for the type of central processing unit (called a **processor** or **CPU**) installed in the computer. See Figure 6-10 (you don't have to open PCINFO.XLS; this figure helps you remember it).

Because each pair of the manufacturers' computers shown in Figure 6-10 shares the same characteristics except for one variable, CPU, this is a good example of paired data. Notice that the Aberdeen Mini EISA VLB computer listed first sells with either a DX/33 or a DX2/66 processor. Everything else being equal, any difference in either price or performance between these models will be due to the difference in CPU.

Figure 6-10
PCINFO data

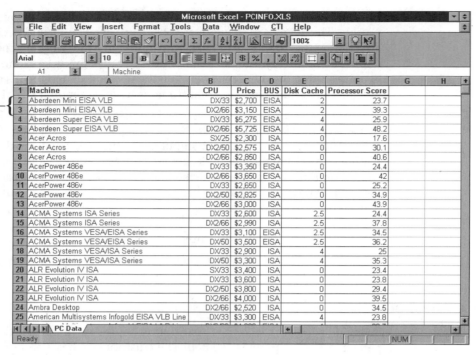

same computer with
different processor

With paired data, it's useful to present the pairs in two columns. The PCPAIR.XLS data contain 56 pairs of DX/33 and DX2/66 machines. Continue with the computer-purchase investigation by taking processor speed into consideration.

To open PCPAIR.XLS, be sure that Excel is started and the windows are maximized, and then:

1. Open **PCPAIR.XLS** (be sure you select the drive and directory containing your Student files). See Figure 6-11.

Figure 6-11
PCPAIR.XLS

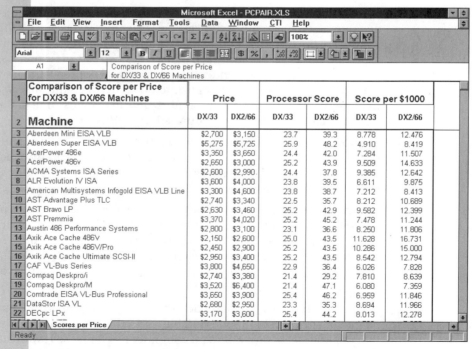

2. Click **File > Save As**, select the drive containing your Student Disk, then save your workbook on your Student Disk as **6PCPAIR.XLS**.

This file, as shown in Figure 6-11, contains the price, processor score, and score per $1,000 (score/price x 1000) with a different column for each processor. Columns B and C show the cost of a PC with the DX/33 and DX2/66 processor, respectively. Columns D and E compare the processor score of the two CPUs, and columns F and G indicate the processor score per $1,000.

The DX2/66 chip employs "clock-doubling" technology to increase the speed of the processor. Theoretically the speed should be twice as great as the DX/33 chip, though this is rarely the case. The processor score for the DX2/66 chip should always be faster than the score for the DX/33 chip and also will be more expensive, so these comparisons are not very interesting. But what about the processor score per dollar? Is the increase of price justified by a similar increase in processor score? To determine the answer, let's calculate the t statistic for the paired difference of processor score per $1,000.

A Paired Difference

Label the true mean processor score per $1,000$ for the DX/33 chips as μ_1, and the true mean score per $1,000$ for the DX2/66 chip as μ_2. Define $\mu_d = \mu_1 - \mu_2$ to be the difference between the two means. You want to choose between two hypotheses:

$$H_0 : \mu_d = 0$$
$$H_a : \mu_d \neq 0$$

Because these are paired data, you can create a single column of the differences between the score per $1,000$ for each model.

To calculate the difference in processor score per $1,000$:

1 Click cell **H2** (you might have to scroll to see it), type **Difference**, and press **[Enter]**.

2 Click **Edit > Go To**, type **H3:H58** in the Reference text box, and click **OK**.

3 Type **=G3–F3** in cell H3 to calculate the difference in score per $1,000$ for row 3, and press **[Enter]**.

4 Click **Edit > Fill > Down** (or press **[Ctrl]+[D]**) to fill in the formula for the rest of the selection, then click any cell to remove the highlighting. See Figure 6-12.

Figure 6-12
Differences between score per $1,000 for DX/33 and DX2/66 processors

	B	C	D	E	F	G	H	I	J	K	L
1	Price		Processor Score		Score per $1000						
2	DX/33	DX2/66	DX/33	DX2/66	DX/33	DX2/66	Difference				
3	$2,700	$3,150	23.7	39.3	8.778	12.476	3.698				
4	$5,275	$5,725	25.9	48.2	4.910	8.419	3.509				
5	$3,350	$3,650	24.4	42.0	7.284	11.507	4.223				
6	$2,650	$3,000	25.2	43.9	9.509	14.633	5.124				
7	$2,600	$2,990	24.4	37.8	9.385	12.642	3.258				
8	$3,600	$4,000	23.8	39.5	6.611	9.875	3.264				
9	$3,300	$4,600	23.8	38.7	7.212	8.413	1.201				
10	$2,740	$3,340	22.5	35.7	8.212	10.689	2.477				
11	$2,630	$3,460	25.2	42.9	9.582	12.399	2.817				
12	$3,370	$4,020	25.2	45.2	7.478	11.244	3.766				
13	$2,800	$3,100	23.1	36.6	8.250	11.806	3.556				
14	$2,150	$2,600	25.0	43.5	11.628	16.731	5.103				
15	$2,450	$2,900	25.2	43.5	10.286	15.000	4.714				
16	$2,950	$3,400	25.2	43.5	8.542	12.794	4.252				
17	$3,800	$4,650	22.9	36.4	6.026	7.828	1.802				
18	$2,740	$3,380	21.4	29.2	7.810	8.639	0.829				
19	$3,520	$6,400	21.4	47.1	6.080	7.359	1.280				
20	$3,650	$3,900	25.4	46.2	6.959	11.846	4.887				
21	$2,680	$2,950	23.3	35.3	8.694	11.966	3.272				
22	$3,170	$3,600	25.4	44.2	8.013	12.278	4.265				

Scores per Price

Ready

Checking for Normality

Before using Excel to calculate the *t* statistic, you should look at the distribution of the difference in processor score per $1,000 to see if it is normal. After all, the *t* statistic is designed for normal data, although it is robust to deviations from normality.

To create a histogram of the paired differences:

1   Click **CTI > Histogram** (see Chapter 3 if you need to install the CTI Statistical Add-Ins).

2   Type **H2:H58** in the Input Columns text box, then click the **Selection Includes Header Row check box.**

3   Click **New Sheet**, type **Difference Histogram**, then click **OK**.

Excel creates a new sheet entitled Difference Histogram. See Figure 6-13.

Figure 6-13
Difference histogram

second peak —
first peak —

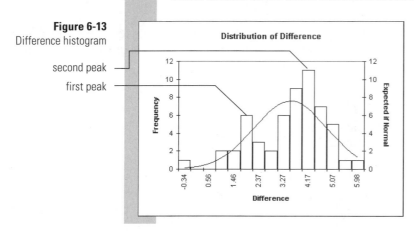

The histogram does not look very good on the left, where a negative difference stands out. There also appear to be two peaks to the histogram, rather than one as in a bell-shaped curve. Recall from Chapter 5 that when you have reason to doubt whether your data are normally distributed, the normal probability plot helps you with that assessment. If the points fall nearly in a straight line, you have evidence that your data are nearly normal.

To see how your paired difference data look on a normal probability plot:

1   Click the **Scores per Price sheet tab** to bring the worksheet containing the paired data to the front.

2   Click **CTI > Normal P-Plot**.

3   Type **H2:H58** in the Input Range text box, then click the **Selection Includes Header Row check box.**

4   Click **New Sheet**, type **Difference Pplot**, then click **OK**.

The Difference Pplot sheet shows the normal probability plot for the paired differences between the DX/33 CPU and the DX2/66 CPU. See Figure 6-14.

The large negative paired difference is more extreme than expected for normal data. Aside from that one point, the rest of the curve seems generally linear. The normal probability plot is supposed to be nearly straight for normal data, and normal data are assumed with the *t* statistic, so there is some concern about the validity of the *t* statistic. However, the robustness of the *t* should be helpful here.

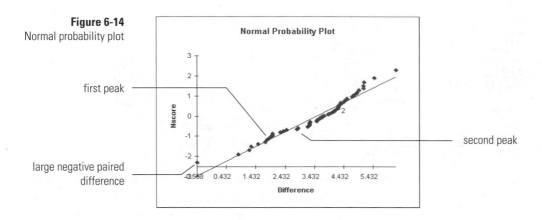

Figure 6-14
Normal probability plot

first peak

large negative paired
difference

second peak

Testing Paired Data with the Analysis ToolPak

Let's perform the *t*-test using the Analysis ToolPak Add-In.

To apply the dependent (or paired) *t*-test to these data:

1   Click the **Scores per Price sheet tab**.

2   Click **Tools > Data Analysis**.

3   Click *t*-**Test: Paired Two Sample for Means** in the Analysis Tools list box, then click **OK**.

In the *t*-Test Paired Two Sample dialog box you only need to specify the two columns of paired values. (Although you calculated the differences between the columns to get some insight into the distribution of the data, you don't have to do so to perform the test using the Analysis ToolPak.)

4   Type **F2:F58** in the Variable 1 Range text box, then press [**Tab**].

5   Type **G2:G58** in the Variable 2 Range text box, then press [**Tab**].

6   Type **0** in the Hypothesized Mean Difference text box because under the null hypothesis you assume that the difference is zero.

7   Click the **Labels check box** because you've included the header row for these columns.

8   Verify that **0.05** is entered for the Alpha value, otherwise known as the significance level.

9   Click the **New Worksheet Ply option button**, click the corresponding text box, then type **Difference Paired t-test**. The t-Test: Paired Two Sample for Means dialog box should look like Figure 6-15.

Figure 6-15
t-Test: Paired Two Sample
for Means dialog box

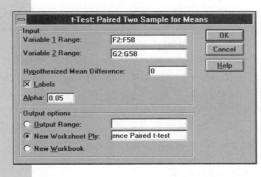

10   Click **OK**.

Excel places the results of the paired *t*-test into the Difference Paired t-test worksheet. See Figure 6-16, shown with the first column widened.

Figure 6-16

Paired *t*-test results

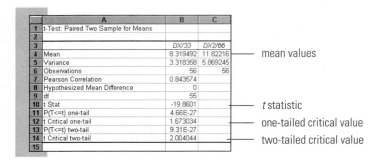

	A	B	C
1	t-Test: Paired Two Sample for Means		
2			
3		DX/33	DX2/66
4	Mean	8.319492	11.82216
5	Variance	3.318358	5.869245
6	Observations	56	56
7	Pearson Correlation	0.843574	
8	Hypothesized Mean Difference	0	
9	df	55	
10	t Stat	-19.8601	
11	P(T<=t) one-tail	4.66E-27	
12	t Critical one-tail	1.673034	
13	P(T<=t) two-tail	9.31E-27	
14	t Critical two-tail	2.004044	
15			

mean values — (rows 4)

t statistic — (row 10)

one-tailed critical value — (row 12)

two-tailed critical value — (row 14)

The mean processor score per $1,000 for the DX/33 chip is 8.319, whereas it is 11.822 for the DX2/66 chip. The mean difference between the processors' points per $1,000 is therefore 11.822 – 8.319 = 3.503. There is a substantial average increase in the processor score per dollar when switching from the DX/33 chip to the DX2/66.

Using the Analysis ToolPak to Calculate the Paired t-Test

- Click t-Test: Paired Two Sample for Means in the Analysis Tools list box.

- Type the range of the first sample in the Variable 1 Range text box.

- Type the range of the second sample in the Variable 2 Range text box.

- Indicate the appropriate significance level, direction for output, and whether the ranges specified include labels.

Evaluating the *t* Statistic

But is the difference *statistically* significant? The *t* statistic given in Figure 6-16 is –19.86, computed under the assumption of the null hypothesis that changing the processor does not result in a change in the processor score per dollar.

To determine whether to accept or reject the null hypothesis, compare the absolute value of the *t* statistic to the critical value. If the absolute value of the *t* statistic is larger than the critical value, you reject the null hypothesis; if it is smaller, you accept the null hypothesis. Excel provides you with two critical values, shown in Figure 6-16 (cells B12 and B14).

The alpha value is 0.05, and the alternative hypothesis is two-sided. Because the critical value for the two-tailed *t* is 2.004 (cell B14), which is less than the *t* statistic absolute value of 19.86, you reject the null hypothesis of no difference in favor of the alternative hypothesis that the processor scores per dollar are different between the two CPUs.

A Nonparametric Approach

You might still be concerned about whether the data follow the normal distribution and hence whether you should use the *t* statistic for hypothesis testing. The *t* -test is an example of a **parametric** test because it assumes that the data follow a specific distribution whose form you can quantify with the population parameters. When this assumption is violated or you are unsure of its validity, you can use a nonparametric method that makes fewer (and simpler) assumptions about the distribution of the data. **Nonparametric** methods are based on counts or ranks of the data and not the actual values, and thus do not assume a specific distribution.

Nevertheless, nonparametric methods also make some assumptions about the data. The one-sample Wilcoxon Sign Rank test, the nonparametric counterpart to the paired *t*-test, only assumes that the distribution of values is symmetric about the median. For the Wilcoxon test, your two hypotheses are:

H_0 : The median difference in processor score per $1,000 is zero.

H_a : The median difference in processor score per $1,000 is not zero.

This test first ranks the absolute values of the 56 differences, assigning a rank from 1 to 56 to each difference, multiplies the rank by the sign of the difference (either – 1 or 1) and calculates the total sum of these signed ranks. If most of the differences were positive, this would be a large positive number; if most of the differences were negative, this would be a large negative number. If there is a nearly equal number of positive and negative values, the total of the signed ranks would be close to 0.

The nice thing about using ranks is that the hypothesis test is much less sensitive to extreme values. If you take the maximum value and make it much more extreme, it has no effect on the ranks, but it would have a big effect on the mean and standard deviation, and thus a big effect on the *t* statistic. Because of this, the Wilcoxon statistic is said to be more robust with respect to outliers.

Excel does not provide a Wilcoxon Sign Rank test, nor does the Analysis ToolPak Add-In. The ability to perform a Wilcoxon Sign Rank test has been provided for you with the CTI Statistical Add-Ins.

To calculate the Wilcoxon Sign Rank test for the processor score data:

1 Click the **Scores per Price sheet tab**.

2 Click **CTI > 1 Sample Wilcoxon**.

3 Type **H2:H58** in the Input Range text box (this column contains the paired differences).

4 Click the **Selection Includes Header Row check box**.

5 Click the **New Sheet option button**, then type **Difference Wilcoxon Test** in the New Sheet text box.

Your dialog box should look like Figure 6-17.

Figure 6-17
Calculate 1-Sample Sign Rank Statistics dialog box

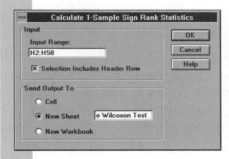

6 Click **OK**.

Figure 6-18
Output of Wilcoxon test

increases

decrease

The output in Figure 6-18 shows that there were 55 increases (cell B4) and only one decrease (cell B2) in processor score per $1,000 when switching from the DX/33 chip to the DX2/66. The nonparametric test thus results in an approximate two-tailed *p*-value of 8.01 x 10^{-11}, which says that the difference is extremely significant. An approximation is used in larger sample sizes (> 16) where calculating the exact Wilcoxon test statistic *p*-value would be difficult and time-consuming.

The very small p-value indicates that the null hypothesis, which states that the median difference is 0, is incompatible with the data observed.

Confidence Intervals Using the t Statistic

Having concluded that the processor score per $1,000 for the DX2/66 chip is different from the DX/33 chip, calculate a confidence interval around the sample mean difference. Figure 6-16 showed that, between the DX/33 and the DX2/66 chips, the mean processor score per $1,000 increased from 8.319 to 11.822 —a difference of 3.503. A confidence interval for this sample mean difference would have to be based on the t statistic because the value of the population standard deviation σ is not known, only estimated by s. The Analysis ToolPak Add-In doesn't produce the t confidence interval as part of the output, but you can compute it manually using the expression that follows.

The $100(1 - \alpha)\%$ t confidence interval has a form similar to the normal confidence interval discussed earlier in this chapter and is the range:

$$\left(\bar{x} - t_{1-\alpha/2,n-1} \frac{s}{\sqrt{n}}, \bar{x} + t_{1-\alpha/2,n-1} \frac{s}{\sqrt{n}} \right)$$

which includes the population mean about $100(1 - \alpha)\%$ of the time.

How is $t_{1-\alpha/2, n-1}$ defined? For a t-distributed random variable with $n - 1$ degrees of freedom, the probability to the left of $t_{1-\alpha/2, n-1}$ is $1 - \alpha / 2$. For a 95% confidence interval, $1 - \alpha / 2 = 0.975$, and $t_{0.975, n-1}$ will have a value close to two.

In Excel you calculate this value with the TINV function. The TINV function calculates the critical value for the two-tailed t-test, given the significance level. Specifically, to calculate the value of $t_{1-\alpha/2, n-1}$ with the TINV function, you enter the function as TINV(α, $n - 1$).

To calculate the 95% t confidence interval for the difference data in column H:

1. Click the **Scores per Price sheet tab**.

2. Click cell I2, type **Lower** in cell I2, press **[Tab]**, then type **Upper** in cell J2.

3. Click cell I3, type **=AVERAGE(H2:H58)–TINV(0.05,55)*STDEV(H2:H58)/SQRT(55)** (this is the lower boundary of the 95% confidence interval), then press **[Tab]**.

4. Click cell J3, type **=AVERAGE(H2:H58)+TINV(0.05,55)*STDEV(H2:H58)/SQRT(55)** (this is the upper boundary of the 95% confidence interval), then press **[Enter]**.

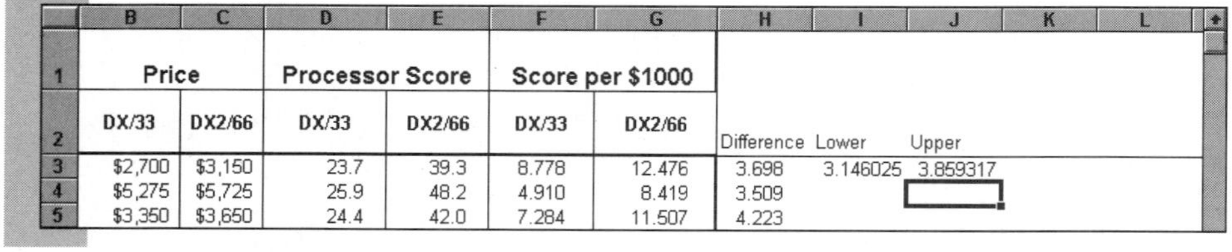

Figure 6-19
Cells I2:J3 in the workbook

	B	C	D	E	F	G	H	I	J	K	L
1	Price		Processor Score		Score per $1000						
2	DX/33	DX2/66	DX/33	DX2/66	DX/33	DX2/66	Difference	Lower	Upper		
3	$2,700	$3,150	23.7	39.3	8.778	12.476	3.698	3.146025	3.859317		
4	$5,275	$5,725	25.9	48.2	4.910	8.419	3.509				
5	$3,350	$3,650	24.4	42.0	7.284	11.507	4.223				

As shown in Figure 6-19, the 95% confidence interval for the sample mean difference covers the range 3.146 to 3.859. Because the confidence interval includes only positive values, you have clear evidence that replacing the DX/33 chip with the DX2/66 chip provides faster processor speed per $1,000.

To save and close the 6PCPAIR workbook:

1. Click the **Save button** ▣ to save your file as **6PCPAIR.XLS** (the drive containing your Student Disk should already be selected).

2. Click **File > Close**.

Two Sample Comparisons

A second type of comparison looks for differences between two separate groups of individuals who don't share the same characteristics. This is called a **two-sample**, or **independent**, **comparison**, because the two groups are presumably independent, like males and females. You could also assign subjects randomly to two groups, give the first group treatment 1, give the second group treatment 2, and compare the two groups. The random assignment is important. For example, if a self-selected group takes a commercial course to prepare for the GMAT (graduate management admissions test), and you compare their scores to a group that didn't receive the training, how do you know that their performance is due to the training? Because the study group was self-selected and not randomly selected, they might do better just because they were more motivated in the first place.

Two-Sample *t* for Independent Comparisons

In a NASA-funded study, seven men and eight women spent 24 days in seclusion to study the effects of gravity on circulation. Without gravity, there is a loss of blood from the legs to the upper part of the body. The study started with a nine-day control period in which the subjects were allowed to walk around. Then followed a 10-day bed-rest period in which the subjects' feet were somewhat elevated to simulate weightlessness in space. The study ended with a five-day recovery period in which the subjects again were allowed to walk around. Every few days, the researchers measured the electrical resistance at the calf, which increases when there is blood loss. The electrical resistance gives an indirect measure of the blood loss and indicates how the individuals respond to the procedure. The data from the test are found in the SPACE.XLS file.

To open SPACE.XLS:

1. Open **SPACE.XLS** (be sure you select the drive and directory containing your Student files).

2. Click **File** > **Save As**, select the drive containing your Student Disk, then save your workbook on your Student Disk as **6SPACE.XLS**. Figure 6-20 shows the first few rows of SPACE.XLS.

Figure 6-20
First few rows of SPACE.XLS

	A	B	C	D	E	F	G	H	I	J	K	L	M
1	Subject ID		Day1		Day5		Day9		Day10		Day13		D
2	Males	Females	Males	Females	Males	Females	Males	Females	Males	Females	Males	Females	Males
3	70	87	165.9	212.1	169.5	217.2	154.3	230.4	181.4	254.9	193.1	245.5	172.4
4	71	89	210.3	203.5	188.3	210.3	191	202.8	220.6	263.1	205.8	249	215.6
5	72	90	166.8	210.3	163.6	241.5	163.4	202.8	181.6	279.8	159.1	245	191.6
6	76	94	182.3	228.4	173.2	218.9	168.6	216.8	187.3	259.6	204	243.9	198.3
7	77	97	182.1	206.2	180.8	190.5	187	192.9	197.1	245.5	196.2	231	212.5
8	78	99	218	203.2	221.4	205.1	200.4	194.4	244.9	225.1	223.1	204.8	236.2
9	80	101	170.1	224.9	162.3	214.3	162.5	211.7	185.9	256.5	188.1	244.3	188.1
10		102		202.6		190.4		178.3		251.2		245.3	
11													

Let's use the two-sample *t*-test to compare men and women on the 19th day (columns O and P), at the end of the bed-rest period.

3. Scroll to the right to see the electrical resistance data for day 19.

The best way to begin analyzing data from two separate samples is to look at the distributions of the groups. You could create histograms (using the CTI or Analysis ToolPak command) for both males and females, making sure that the scales of the x and y axes cover the same range for ease of comparison. Although this is not difficult with only two groups, it can be time-consuming in cases when the number of groups increases to three or more. To save you time and effort, the CTI Statistical Add-Ins include a command that creates multiple histograms following a common format.

To create multiple histograms for the male and female electrical resistance:

1   Click **CTI > Multiple Histograms**.

2   Type **O2:P10** in the Input Range text box to select the Day 19 space leg data. The fact that the column lengths are unequal will not affect the charts.

3   Click the **Selection Includes Header Row check box**.

4   Click the **New Sheet option button** to send the output to a new worksheet, then type **Day 19 Histograms** in the New Sheet text box.

The Create Histograms for Multiple Columns dialog box should look like Figure 6-21.

Figure 6-21
Completed Create
Histograms for Multiple
Columns dialog box

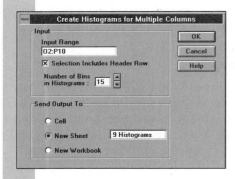

5   Click **OK**.

Excel places the histograms and histogram table into a new worksheet, Day 19 Histograms. See Figure 6-22.

Figure 6-22
Histograms

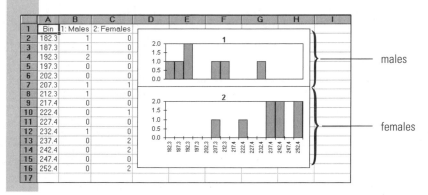

You can easily compare the distributions because they are stacked one above the other. Histogram 1 corresponds to the males, and histogram 2 corresponds to the females. Both histograms use a common range for the x and y axes. It appears that the females have higher resistance, but there is substantial overlap. There really are not enough data to be able to tell whether the observations are normal, but the data look consistent with the normal distribution. In addition to normality, the two-sample t-test assumes equal variances in the two groups. In Figure 6-22, the two groups do have about the same variability.

Is the apparent gender difference shown in Figure 6-22 statistically significant? To answer that question, do a two-sample t-test. You want to test whether the mean resistance for females (μ_1) is equal to the mean resistance for males (μ_2). Or in terms of a hypothesis test:

$$H_0: \mu_1 - \mu_2 = 0$$
$$H_a: \mu_1 - \mu_2 \neq 0$$

To use Excel's Analysis ToolPak to perform such a two-sample t-test:

1   Click the **Space Legs sheet tab**.

2   Click **Tools > Data Analysis**.

3   Click **t-Test: Two-Sample Assuming Equal Variances** (you are assuming that the variability of data around the mean in both groups is the same), then click **OK**.

4   Type **O2:O9** in the Variable 1 Range text box, press [**Tab**], then type **P2:P10** in the Variable 2 Range text box.

5   Press [**Tab**], then type **0** in the Hypothesized Mean Difference text box.

6   Click the **Labels check box** because the ranges include labels.

7   Verify that **0.05** is entered in the Alpha text box.

8   Click **New Worksheet Ply option button** and type **Day 19 t-Test** in the New Worksheet Ply box.

Your dialog box should look like Figure 6-23.

Figure 6-23
Completed Two-Sample
t-test dialog box

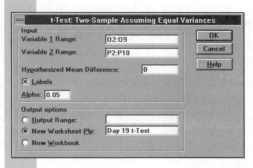

9   Click **OK**.

The Day 19 t-Test worksheet contains the two-sample *t* statistics. See Figure 6-24. You will have to widen the columns to see all the row titles.

Figure 6-24
Two-sample *t*-test with
equal variances output

	A	B	C
1	t-Test: Two-Sample Assuming Equal Variances		
2			
3		*Males*	*Females*
4	Mean	200.4143	236.75
5	Variance	330.4048	238.6686
6	Observations	7	8
7	Pooled Variance	281.0084	
8	Hypothesized Mean Difference	0	
9	df	13	
10	t Stat	-4.18815	
11	P(T<=t) one-tail	0.000531	
12	t Critical one-tail	1.770932	
13	P(T<=t) two-tail	0.001063	
14	t Critical two-tail	2.160368	
15			

The statistics in Figure 6-24 agree with the histogram plots shown in Figure 6-22. On the average, the females have higher resistance (236.75 as compared with 200.41 in cells B4 and C4). The variances in the two samples are not exactly the same (330.40 for the males versus 238.67 for the females in cells B5 and C5), but there is some robustness with the two-sample *t*-test, so the population variances need not be exactly the same. Of course, the two observed variances will not be exactly the same, even if the populations do have exactly the same population standard deviations.

The value of the *t* statistic is –4.19. The two-tailed *p*-value is given as 0.001 (cell B13), which is less than 0.05, so the difference is significant at the 5% level. Because the probability is so small, the null hypothesis of no difference between males and females seems incompatible with the data, and you should be inclined to reject it and state that the electrical resistance is different for males and females.

Before you do that, you might want to calculate a *t* statistic allowing unequal variances. The robustness of the *t* statistic breaks down if the variances in the two groups are extremely different. The worst situation occurs when the two samples are of very different sizes, and the small sample has much larger standard deviation. Then the resulting *t* statistic does not follow the *t* distribution, and some insignificant differences might be called significant.

Because of this problem, Excel's Analysis ToolPak also provides a command to calculate a *t* statistic allowing a different variance for each group. The generated statistic has an approximate *t* distribution. If there are equal numbers of observations in the two groups, the degrees of freedom are adjusted downward if one group has a higher variance. The degrees of freedom are never more than (n – 2), the number for the usual two-sample *t*.

To calculate the *t* statistic allowing unequal variances:

1   Click the **Space Legs sheet tab**.

2   Click **Tools > Data Analysis**.

3   Click **t-Test: Two Samples Assuming Unequal Variances**, then click **OK**.

4   Type **O2:O9** in the Variable 1 Range text box, press [**Tab**], type **P2:P10** in the Variable 2 Range text box, press [**Tab**], then type **0** in the Hypothesized Mean Difference text box.

5   Click the **Labels check box** because the ranges include labels, then verify that **0.05** appears in the Alpha box.

6   Click the **New Worksheet Ply option button**, type **Day 19 t-Test (Unequal)** in the New Worksheet text box, then click **OK**.

The resulting output looks like Figure 6-25. Once again you will have to resize the columns to see all the row titles.

Figure 6-25
Two-sample *t*-test with unequal variances output

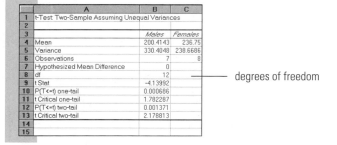

	A	B	C
1	t-Test Two-Sample Assuming Unequal Variances		
2			
3		Males	Females
4	Mean	200.4143	236.75
5	Variance	330.4048	238.6686
6	Observations	7	8
7	Hypothesized Mean Difference	0	
8	df	12	
9	t Stat	-4.13992	
10	P(T<=t) one-tail	0.000686	
11	t Critical one-tail	1.782287	
12	P(T<=t) two-tail	0.001371	
13	t Critical two-tail	2.178813	
14			
15			

—— degrees of freedom

It is nice to see that the *t*-test allowing different variances gives almost the same answers as the *t*-test requiring equal variances. This is generally true as long as the two standard deviations and the sample sizes are close. Note that the effect of assuming different variances has reduced the degrees of freedom from 13 to 12 (actually the approximate degrees of freedom are close to 11.9, which Excel has rounded up). If the two approaches differ by much, then allowing different variances between the groups is probably better, because the resulting *p*-value is more accurate when the variances differ greatly. On the other hand, you will have less explaining to do if you use the usual pooled *t* when it gives nearly the same answer, because it is more widely known.

Using the Analysis ToolPak to Calculate the Paired t-Test

- Click t-Test: Two-Sample Assuming Equal Variances or t-Test: Two-Sample Assuming Unequal Variances in the Analysis ToolPak list box.

- Type the range of the first sample in the Variable 1 Range text box.

- Type the range of the second sample in the Variable 2 Range text box.

- Indicate the appropriate significance level, direction for output, and whether the ranges specified include labels.

Two-Sample t Confidence Interval

The output in the Day 19 t-Test worksheet gives you enough information to compute a 95% confidence interval for the true (population) difference of means. This requires first calculating the standard error of the difference of sample means. You can get this from the value of the t statistic that is given, because:

$$t = \frac{\text{mean(males)} - \text{mean(females)}}{\text{Std. Error}}$$

Solving for the standard error gives:

$$\text{Std.Error} = \frac{\text{mean(males)} - \text{mean(females)}}{t}$$

To calculate the standard error and then the confidence interval:

1   Click the **Day 19 t-Test sheet tab**.

2   Click cell **A15**, type **Difference**, then press [**Tab**].

3   Type **=B4–C4** in cell B15 to calculate the difference in means between the two groups, then press [**Enter**].

4   Click cell **A16**, type **Std Error (Difference)**, then press [**Tab**].

5   Type **=B15/B10** in cell B17, the estimate of the standard error of the difference, then press [**Enter**].

6   Click cell **A17**, type **95% Confidence Level**, then press [**Tab**].

The 95% confidence level is calculated by multiplying the critical value (cell B14) by the standard error, which you just calculated in cell B16.

7   Type **=B14*B16** in cell B17, then press [**Enter**]. Your worksheet should now resemble Figure 6-26.

Figure 6-26
Obtaining the two-sample
t confidence interval

	A	B	C
1	t-Test: Two-Sample Assuming Equal Variances		
2			
3		Males	Females
4	Mean	200.4143	236.75
5	Variance	330.4048	238.6686
6	Observations	7	8
7	Pooled Variance	281.0084	
8	Hypothesized Mean Difference	0	
9	df	13	
10	t Stat	-4.18815	
11	P(T<=t) one-tail	0.000531	
12	t Critical one-tail	1.770932	
13	P(T<=t) two-tail	0.001063	
14	t Critical two-tail	2.160368	
15	Difference	-36.3357	
16	Std Error (Difference)	8.675834	
17	95% Confidence Level	18.743	
18			
19			

Thus the 95% confidence interval for the mean difference between the male and female groups is approximately: -36.3 ± 18.7 or the range $(-55, -17.6)$. A clear difference exists between genders in how their calf muscles respond to prolonged weightlessness. You are 95% confident that the true mean difference in the electrical resistance between the genders is not less than -55 and not greater than -17.6. This could mean that men and women who spend extended periods of time in space will need to undergo different rehabilitation programs when they return to Earth.

To save and close the 6SPACE workbook and exit Excel:

1   Click the **Save button** ▣ to save your file as **6SPACE.XLS** (the drive containing your Student Disk should already be selected).

2   Click **File > Exit**.

E X E R C I S E S

1. Explain why this statement is false: "A 95% confidence interval that covers the range (–5, 5) tells you that the population mean μ has a probability 95% of being between –5 to 5."

2. For a sample of normally distributed data with mean of 50 from a sample size of 25:

   a. Calculate the 95% confidence interval if σ = 20.

   b. Calculate the 95% confidence interval if σ is unknown and s equals 20 (also explain why you'll have to use the t distribution).

3. The nationwide average price for a three-year-old Honda Civic is $8,500, with a known standard deviation of $600. Nine three-year-old Civics in San Francisco have an average price of $9,000. You wonder whether the price of Civics in San Francisco differs from the national average.

   a. State the question in terms of a null and an alternative hypothesis. What are you assuming about the distribution of the Civic prices?

   b. Will the alternative be one- or two-sided? Defend your answer.

   c. If the significance level is 5%, is the difference in price between the San Francisco Civics and the national average statistically significant?

   d. Redo the hypothesis test in part 3c, but with the change that the San Francisco Civic information is based on a sample of 10 Civics and that $600 is the sample standard deviation.

4. In tests of stereo speakers, 10 American-made speakers had an average performance rating of 90 with standard deviation 5, and five imported speakers had an average rating of 85 with standard deviation 4.

   a. Use the formula for the two-sample t-test to determine whether the difference is significant at the 5% level.

   b. What about the 10% level?

5. Explain why this statement is false (or at least incomplete): "The example comparing the processor score per $1,000 for the DX/33 and the DX2/66 chip from this chapter shows that a 486 computer equipped with a DX2/66 chip will usually show greater performance per dollar than another 486 with a DX/33 chip."

6. Open the workbook BIGTEN.XLS.

   a. Construct a histogram of the difference between male and female athlete graduation rates.

   b. Perform a paired t-test on the graduation rates for male and female athletes. State your hypothesis, significance level, and conclusion.

   c. Construct a 95% confidence interval for the difference in graduation rate.

   d. Can you apply your results to universities in general? Defend your answer.

   e. Why is this an example of paired data?

   f. Save your workbook as E6BIGTEN.XLS.

7. Open the workbook POLU.XLS. Create a new column for the average number of unhealthy days between 1985 and 1989.

   a. Perform a paired t-test comparing the number of unhealthy days in 1980 to the average for the period 1985 through 1989.

   b. Construct a 95% confidence interval for the difference in the number of unhealthy days.

   c. Create a normal probability plot of the paired difference. Do you see any violations of the assumption of normally distributed data?

d. Using the CTI Statistical Add-Ins, perform a one-sample Wilcoxon test. How do your results compare to the paired *t*-test values? What might account for this?

e. Save your workbook as E6POLU.XLS.

8. Create two new columns in POLU.XLS of the logarithm of the number of unhealthy days in 1980 and the logarithm of the average number of unhealthy days for the period 1985 through 1989.

a. Perform a paired *t*-test of the difference in the logarithm of the counts between 1980 and the average for the period 1985 through 1989.

b. Is the paired *t*-test significant this time? What would cause this?

c. Create a normal probability plot for the difference in part 8a. Compare the shape of the normal probability plot of the log counts with the counts from Exercise 7.

d. Save your workbook as E6POLU.XLS.

9. Open the workbook PCUNPAIR.XLS. The workbook shows the PRICE and PROCESSOR SCORE for 486 machines with the DX2/66 chip using either the EISA or the ISA BUS.

a. Create multiple histograms of the price variable for the EISA and the ISA BUS.

b. Perform a two-sample *t*-test on the price information assuming equal variances for the EISA BUS and the ISA BUS.

c. Does the assumption of equal variances appear justified? Redo the analysis, allowing the variability in the two groups to be different.

d. Write a report summarizing your conclusions. What is your null hypothesis? What is the alternative hypothesis? Is this a one- or two-tailed test, and why?

e. Save your workbook as E6PCUNPAIR.XLS.

10. Redo the analysis in Exercise 9 for the processor score in the EISA and the ISA groups.

11. Open the PCUNPAIR.XLS workbook, then create a new variable for the processor score per $1,000. Using the approach of Exercise 9 and 10, assess whether this value differs depending upon BUS type.

12. Open the workbook CALCFM.XLS. The workbook shows the first semester calculus scores for male and female students.

a. Perform a two-sample *t*-test on the final score assuming equal variance. Is there a difference between men and women?

b. Construct a 95% confidence interval for the difference in the score.

c. In using the *t* distribution for this data set, you've assumed that the scores are normally distributed. What might make this assumption untrue for test score data? Hint: Test scores are constrained to lie between 0 and 100. What property of the t-test might allow you to use the *t* anyway?

d. Create normal probability plots for the male and female groups.

e. In Exercise 6, the comparison between males and females was a paired comparison. In this exercise you performed a two sample *t*-test. Discuss why these data are not paired, as opposed to the data in Exercise 6.

f. Save your workbook as E6CALCFM.XLS.

13. Open the workbook MATH.XLS.

This file contains data from a study analyzing methods of teaching mathematics. Students were randomly assigned to two groups: a control group and an experimental group. The control group was taught in the usual way, which meant that homework was usually assigned, but not collected or graded. In the experimental group the students were regularly assigned homework, which was then collected. Frequent quizzes were given. Students in the experimental group were allowed to raise their grades by retaking exams, although the retakes were not the exact same tests. The score on the final exam was entered into the MATH.XLS workbook.

a. Perform a two sample *t*-test on the final exam score between the control group and the experimental group. Write down your initial conclusion regarding the effect of the experimental method on the final exam score.

b. Create multiple histograms of the final exam score for the two groups. What do the histograms tell you about the distribution of the test scores? Hint: Are the final exam scores distributed evenly across the range of scores?

c. Copy the scores of students who scored more than 10 points on the final from each group into two new columns. The groups represent those students who did not have low scores on the final exam. Perform a two-sample *t*-test comparing these two groups.

d. How do the results of the two-sample *t*-test with the group of mid- to high-scoring students compare with the two-sample *t*-test of the complete samples? Calculate the proportion of students who had scores of less than 10 under the experimental and control methods. What effect did the experimental method of teaching have on the final exam scores?

e. Does the experimental method raise the final exam score of the typical math student? Why?

f. Save your workbook as E6MATH.XLS.

TABLES

O BJECTIVES

In this chapter you will learn to:

- Create pivot tables of a single categorical variable

- Create pie charts and bar charts

- Relate two categorical variables with a two-way table

- Apply the chi-square test to a two-way table

- Compute expected values of a two-way table

- Combine or eliminate small categories to get valid tests

- Test for association between ordered categorical variables

- Create a custom sort order for your workbook

Variable Types: Continuous and Categorical

In previous chapters, data were usually continuous. **Continuous** variables can take on many values, whereas **categorical** variables can have only certain values, each of which represents a different category.

This chapter deals with categorical variables, like brand of cola or year of college, that take on just a few values. Think of such a variable as splitting the cases into categories. For example, year of college splits undergraduates into the categories of freshman, sophomore, junior, and senior.

Ordinal and Nominal Variables

Within categorical variables there is the special class called **ordinal** (or ordered) variables, in which categories have a natural order. For example, the year of college variable would be called ordinal because there is a natural order to the four levels.

If there is no natural order to the categories, such as with cola brands, then a categorical variable is called **nominal**, where one category doesn't necessarily come before the other.

Single Categorical Variables

Categorical variables are common in surveys, which help determine the count, or frequency (that is, how many), of each value in a category. How many people prefer Pepsi? How many went to college? Excel's pivot table procedure can create summary tables of frequency counts for a categorical variable. Pie charts and bar charts show the frequencies graphically.

To illustrate how to work with categorical variables, let's look at data from a survey of professors who teach statistics courses. The SURVEY.XLS data set includes 392 responses to questions about whether the course requires calculus, whether statistical software is used, how the software is obtained by students, what kind of computer is used, and so on.

> To open SURVEY.XLS, be sure that Excel is started and the windows are maximized, and then:
>
> 1   Open **SURVEY.XLS** (be sure you select the drive and directory containing your Student files).
>
> 2   Click **File > Save As**, select the drive containing your Student Disk, then save your workbook on your Student Disk as **7SURVEY.XLS**.
>
> The worksheet looks like Figure 7-1.

What computers are being used most often in statistics instruction? The column labeled "Computer" contains responses to the question, "What kind of computer is used in your classes?" The answers fall into four categories: two kinds of microcomputers (Macintoshes and PCs), minicomputers, and mainframes. (It makes sense to put the IBM PC and all of its clones in one category, because they all use the same type of software.) You can get a count for each response to the question about the kinds of computers used in statistics classes.

Pivot Tables

In Excel, you obtain category counts by generating a **pivot table**, a worksheet table that summarizes data from the source data list (in this case the Survey data). Excel pivot tables are **interactive**: you can update them whenever you change the source data list, and you can switch row and column headings to view the data in different ways (hence the term "pivot").

Figure 7-1
7SURVEY.XLS data

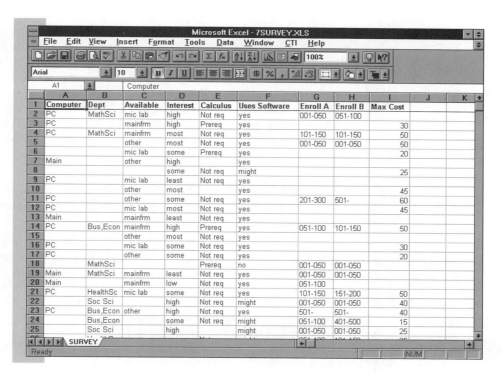

Figure 7-1 shows a Microsoft Excel window displaying 7SURVEY.XLS data. The spreadsheet contains the following data:

	A	B	C	D	E	F	G	H	I	J	K
1	Computer	Dept	Available	Interest	Calculus	Uses Software	Enroll A	Enroll B	Max Cost		
2	PC	MathSci	mic lab	high	Not req	yes	001-050	051-100			
3	PC		mainfrm	high	Prereq	yes			30		
4	PC	MathSci	mainfrm	most	Not req	yes	101-150	101-150	50		
5			other	most	Not req	yes	001-050	001-050	50		
6			mic lab	some	Prereq	yes			20		
7	Main		other	high		yes					
8				some	Not req	might			25		
9	PC		mic lab	least	Not req	yes					
10			other	most		yes			45		
11	PC		other	some	Not req	yes	201-300	501-	60		
12	PC		mic lab	most	Not req	yes			45		
13	Main		mainfrm	least	Not req	yes					
14	PC	Bus,Econ	mainfrm	high	Prereq	yes	051-100	101-150	50		
15			other	most	Not req	yes					
16	PC		mic lab	some	Not req	yes			30		
17	PC		other	some	Not req	yes			20		
18		MathSci			Prereq	no	001-050	001-050			
19	Main	MathSci	mainfrm	least	Not req	yes	001-050	001-050			
20	Main		mainfrm	low	Not req	yes	051-100				
21	PC	HealthSc	mic lab	some	Not req	yes	101-150	151-200	50		
22		Soc Sci		high	Not req	might	001-050	001-050	40		
23	PC	Bus,Econ	other	high	Not req	yes	501-	501-	40		
24		Bus,Econ		some	Not req	might	051-100	401-500	15		
25		Soc Sci		high		might	001-050	001-050	25		

Try creating a pivot table that summarizes the computer usage data. Excel provides a **PivotTable Wizard**, similar to the Chart Wizard you've used to make charts, that makes creating a pivot table easy.

To create a pivot table that shows the frequency of computer type in statistics classes:

1. Click **Data > PivotTable** to open the PivotTable Wizard - Step 1 of 4 dialog box.

2. Verify that the **Microsoft Excel List or Database option button** is selected, then click **Next**.

   Now the PivotTable Wizard requests the worksheet range that contains the data you want to use. If the data in your worksheet are arranged as a list, this range appears in the Range box.

3. Verify that the range **$A$1:$I$393** is entered in the Range box, then click **Next**.

   The PivotTable Wizard - Step 3 of 4 dialog box appears. See Figure 7-2.

Figure 7-2
PivotTable
Wizard — Step 3 of 4
dialog box

PivotTable Wizard - Step 3 of 4 dialog box showing:
Drag Field Buttons to the following areas to layout your PivotTable
- ROW — To show items in the field as row labels.
- COLUMN — To show items in the field as column labels.
- DATA — To summarize values in the body of the table.
- PAGE — To show data for one item at a time in the table.

PAGE, COLUMN, ROW, DATA areas. Field buttons: Computer, Enroll A, Dept, Enroll B, Available, Max Cost, Interest, Calculus, Uses Soft

field buttons correspond to column labels

ROW area

DATA area

You can double click field buttons to customize fields.

Help   Cancel   < Back   Next >   Finish

The Step 3 dialog box contains two important parts: a graphical layout of an empty pivot table, and field buttons that correspond to each of the fields in your worksheet. You specify the structure of the table by dragging fields into any of the four areas: PAGE, ROW, COLUMN, or DATA. For example, if you drag Computer into the ROW area on the Step 3 dialog box, Excel uses all values that occur in the Computer column as row labels in the pivot table. PAGE refers to different levels of tables that share row and column variables (for example, PAGE could refer to year if this survey were repeated in successive years). DATA refers to the variables that you want to analyze (in this case, you want to analyze which computers were used in statistics courses).

4　Drag the **Computer field button** to the ROW area of the pivot table. This tells Excel to use the different computer types (PC, Main, Mini, and Mac) as the row labels in the pivot table.

5　Drag the **Computer field button** to the DATA area of the pivot table. The label changes to "Count of Computer," indicating that the pivot table will show the counts of each computer within each of the computer categories in the ROW area. Your Step 3 dialog box should look like Figure 7-3.

Figure 7-3
PivotTable
Wizard — Step 3 of 4

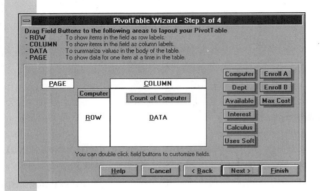

6　Click **Next**.

7　Click the **Grand Totals for Rows check box** to deselect it, because the pivot table you are creating has only one column, so you don't need to sum it.

8　Verify that the rest of the PivotTable Options are selected, as in Figure 7-4.

Figure 7-4
PivotTable
Wizard — Step 4 of 4

9　Click **Finish** in the Step 4 dialog box to place the pivot table on a new sheet.

Excel displays the pivot table in A1:B8 of Sheet1. See Figure 7-5.

Figure 7-5
Pivot table for Count
of Computer

DATA field button

missing values

counts of each category

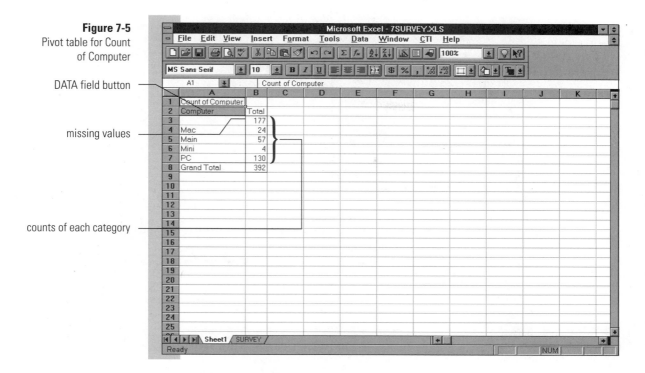

Creating a Pivot Table

- Click Data > PivotTable to start the PivotTable Wizard.

- Enter the range containing the source data (the range must be contiguous).

- In Step 3 of the PivotTable Wizard, drag the field buttons representing the columns that you want to be in the pivot table and drop them into the areas where you want them to appear (ROW, COLUMN, PAGE, or DATA).

- Indicate whether Excel should create Grand Totals for rows and columns, then click Finish.

The first column of the pivot table contains the four categories in the Computer column (Mac, Main, Mini, and PC), plus a blank cell for missing values (cell A3). The second column shows the number of statistics professors who use computers of each type in their classes. (Excel might also display a Pivot Table toolbar. You will not use the toolbar in this chapter, but you can learn more about it using Excel's on-line Help.) You discover that 130 of the 392 professors used only PCs, 57 used only mainframes, 24 only Macintoshes, and 4 only minicomputers. The number 177 is the count of missing values. It is large because many classes used more than one computer type, and nothing was entered for those classes in the Computer column. More explicit variables not included in this data set show that if you were to examine how many times the Mac or mainframe was mentioned, you would still find the PC far in front. More classes use the PC than all other types of computers put together.

Once you create a pivot table, you can refine it to show different aspects of the data. For example, you can include only the non-missing rows by hiding the row containing missing or blank entries for computer type.

To hide the blank row:

1   Double-click the **DATA field button** in cell **A2** to open the PivotTable Field dialog box, which lets you customize the rows in the pivot table.

2   Click the **blank entry** in the Hide Items list box to hide the blank row. See Figure 7-6.

Figure 7-6
PivotTable Field dialog
box with blank entry
selected in the Hide
Items list box

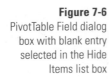

PivotTable Field

Name: Computer   OK

Orientation   Cancel
● Row   ○ Column   ○ Page
                                  Delete
Subtotals
● Automatic   Sum   Help
○ Custom   Count
               Average
○ None   Max
               Min
               Product

Hide Items:

click blank entry

Mac
Main
Mini
PC

3   Click **OK**.

Excel hides the missing values and recalculates the total. See Figure 7-7.

Figure 7-7
Pivot table with blank row
removed

	A	B	C	D	E	F	G	H	I	J	K	
1	Count of Computer											
2	Computer	Total										
3	Mac	24										
4	Main	57										
5	Mini	4										
6	PC	130										
7	Grand Total	215										
8												

You can now see that 215 professors who responded to the survey listed only one computer. Of these, 130 listed the PC. What percentage of professors used the PC to the exclusion of all others? You can modify the pivot table to express this value as a percentage rather than as a count.

To create a percentage pivot table:

1   Right-click **any of the count values in column B**, then click **PivotTable Field** in the shortcut menu.

2   Click the **Options button** in the PivotTable Field dialog box to display the Show Data as list box.

3   Click **% of column** in the Show Data as list box, then click **OK**.

The pivot table is modified to show values as percentages of the total rather than counts. See Figure 7-8.

Figure 7-8
Pivot table of percentages

	A	B	C	D	E	F	G	H	I	J	K	
1	Count of Computer											
2	Computer	Total										
3	Mac	11.16%										
4	Main	26.51%										
5	Mini	1.86%										
6	PC	60.47%										
7	Grand Total	100.00%										
8												

You can see at a glance that 60.47% of professors who listed only one computer use the PC, followed by 26.51% who use the mainframe, 11.16% who use the Macintosh, and less than 2% who use the minicomputer.

To view the counts of the responses again:

1   Right-click **any of the percentages in column B**, then click **PivotTable Field** in the shortcut menu.

2   If necessary, click the **Options button** to display the Show Data as list box.

3   Click **Normal** in the Show Data as list box, then click **OK**.

Pie Charts

You can show the proportion of computers in each category graphically with a pie chart. A **pie chart** displays a circle (like a pie), with each pie slice representing a different category.

To create a pie chart for computer usage:

1 Highlight the range **A3:B6**, click the **ChartWizard button** ⬛ and drag the crosshair pointer + over the range C1:F10.

2 Verify that the Range box contains =$A$3:$B$6 in the Step 1 dialog box, then click **Next**.

3 Double-click the **Pie** chart type in the Step 2 dialog box.

4 Double-click **7** (a pie chart with % and labels).

5 Verify that the **Columns data series option button** is selected, that the Use First Column(s) spin box contains **1**, and that the Use First Row(s) spin box contains **0**; then click **Next**.

6 Click **Finish** in the Step 5 dialog box to create the pie chart (you don't need to add a legend or titles).

The pie chart shown in Figure 7-9 appears.

Figure 7-9
Pie chart of computer usage

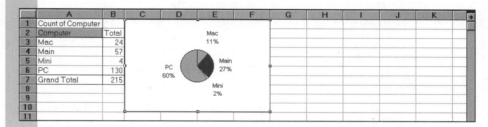

Notice that the categories are in alphabetical order: Mac, Main, Mini, and PC. Beginning at the top of the pie and running clockwise, Excel allots each category a pie sector proportional to its count frequency.

To save the 7SURVEY workbook:

1 Click the **Save button** ⬛ to save your file as 7SURVEY.XLS (the drive containing your Student Disk should already be selected).

Bar Charts

Bar charts offer another graphical view of categorical data. In the pie chart, the size of each slice of the pie is proportional to the number of observations in a category. In a bar chart, the length of the bar is proportional to the number of observations. You already created a histogram, a specialized form of a bar chart, in Chapter 3.

To create a bar chart for computer usage:

1 Highlight the range **A3:B6**, click the **ChartWizard button** ⬛ and drag the crosshair pointer + over the range C11:F20.

2 Verify that the Range box contains =$A$3:$B$6 in the Step 1 dialog box, then click **Next**.

3 Double-click the **Column** chart type in the Step 2 dialog box.

4 Double-click **1** in the Step 3 dialog box.

5   Verify that the **Columns option button** is selected, that the Use First Column(s) spin box contains **1**, that the Use First Row(s) spin box contains **0**, then click **Next**.

6   Click **No** to omit a legend, then click **Finish**.

A bar chart appears in the range C11:F20. See Figure 7-10.

Figure 7-10
Bar chart of computer
usage data

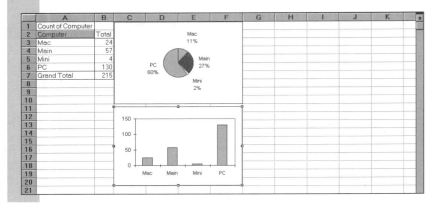

The height of each bar shows the number of classes that use one of the four categories of computer. Most statistical graphics specialists believe that the bar chart shows the comparative sizes of the groups more clearly than the pie chart, because it is easier to compare parallel lines in a bar chart than angles in a pie chart. The pie chart is common in business graphics, but it is used less often in statistics.

Give the worksheet a descriptive name:

1   Right-click the **Sheet1 sheet tab** to open the shortcut menu, then click **Rename**.

2   Type **Computer Use** in the Name box, then click **OK**.

3   Click the **SURVEY sheet tab** to return to the survey data.

Two-Way Tables

What about the relationship between two categorical variables? You might want to see how computer use varies by department. Do different departments tend to choose different types of computers? Does one department tend to use Macintoshes while another tends to use PCs? Excel's pivot table feature can compile a table of counts for computer use by department. Departments fall into four categories:

- Bus,Econ for business and economics
- HealthSc for health sciences
- Math Sci for mathematics, statistics, physical sciences, and engineering
- Soc Sci for psychology and other social sciences

To create a pivot table for computer by department:

1   Click **Data > PivotTable**.

2   Verify that the **Microsoft Excel List or Database option button** is selected, then click **Next**.

3   Verify that the Range box contains **$A$1:$I$393**, then click **Next**.

4   Drag the **Computer field button** to the ROW area, then drag the **Dept field button** to the COLUMN area.

5 Drag the **Computer field button** to the DATA area (you could also drag the Dept field button to the DATA area; the results would be the same because this pivot table will contain counts and not values of these variables).

The PivotTable Wizard - Step 3 of 4 dialog box should look like Figure 7-11.

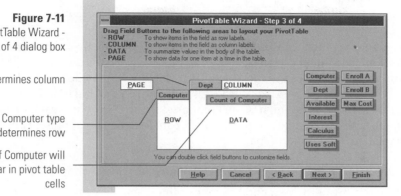

Figure 7-11
PivotTable Wizard - Step 3 of 4 dialog box

Dept determines column

Computer type determines row

Counts of Computer will appear in pivot table cells

When you created the first pivot table, you hid the missing values from view after you created the table. You can also hide them before you create the table by modifying each field's properties.

To hide the missing values:

1 Double-click the **Computer field button** in the ROW area, click the **blank item** in the Hide Items list box, then click **OK**.

2 Double-click the **Dept field button** in the COLUMN area, click the **blank item** in the Hide Items list box, then click **OK**.

3 Click **Next** in the PivotTable Wizard—Step 3 of 4 dialog box.

4 Accept the defaults in the Step 4 dialog box, then click **Finish**.

Excel creates the pivot table shown in Figure 7-12.

Figure 7-12
Two-way table of Computer versus Dept

	A	B	C	D	E	F	G	H	I	J	K
1	Count of Computer	Dept									
2	Computer	Bus.Econ	HealthSc	MathSci	Soc Sci	Grand Total					
3	Mac	4	0	12	6	22					
4	Main	12	3	22	13	50					
5	Mini	1	0	2	1	4					
6	PC	49	8	29	28	114					
7	Grand Total	66	11	65	48	190					
8											

The table in Figure 7-12 shows frequencies of different combinations of department and computer used. For example, cell E3, the intersection of Soc Sci and Mac, displays a 6, indicating that six professors in the Soc Sci departments reported the Macintosh as the choice for their classes. There are a total of 190 unique responses. Let's modify the pivot table to show the column percentages.

To show column percentages:

1 Right-click **any of the count values** in the table, then click **PivotTable Field** in the shortcut menu.

2 Click the **Options button** to display the Show Data as list box if it is not visible.

3 Click **% of column** in the Show Data as list box, then click **OK**. See Figure 7-13.

Figure 7-13

Column percentages for the
Computer by Dept table

	A	B	C	D	E	F	G	H	I	J	K
1	Count of Computer	Dept									
2	Computer	Bus,Econ	HealthSc	MathSci	Soc Sci	Grand Total					
3	Mac	6.06%	0.00%	18.46%	12.50%	11.58%					
4	Main	18.18%	27.27%	33.85%	27.08%	26.32%					
5	Mini	1.52%	0.00%	3.08%	2.08%	2.11%					
6	PC	74.24%	72.73%	44.62%	58.33%	60.00%					
7	Grand Total	100.00%	100.00%	100.00%	100.00%	100.00%					
8											

The PC percentages are high for Bus,Econ and HealthSc, and, relatively speaking, the Mac percentage is high for MathSci. That is, 74.24% of Bus,Econ classes use the PC (cell B6), but only 6.06% of Bus,Econ classes use the Macintosh (cell B3). For MathSci, only 44.62% use the PC (cell D6), but 18.46% use the Macintosh (cell D3). You are looking for *relative* differences here: the MathSci 18.46% is not high when compared to the PC percentage, but is when compared to the 6.06% Macintosh usage for Bus,Econ. In Excel, pivot tables can express the percentages relative to computer type.

To calculate the percentage of each cell relative to the number of PCs reported:

1 Right-click **any of the percentages** in the table, then click **PivotTable Field** in the shortcut menu.

2 If necessary, click the **Options button** to display the Show Data as list box.

3 Click **% Of** in the Show Data as list box (the third option in the list).

4 Click **Computer** in the Base Field list box if it's not already selected.

5 Click **PC** in the Base Item list box. See Figure 7-14.

Figure 7-14

PivotTable Field dialog
box with % Of option
selected

6 Click **OK**.

Figure 7-15 shows the pivot table with the % Of option selected; in this case, each Mac, Main, and Mini count is expressed as a percentage of the PC count.

Figure 7-15

Table with the % of PC
values

	A	B	C	D	E	F	G	H	I	J	K
1	Count of Computer	Dept									
2	Computer	Bus,Econ	HealthSc	MathSci	Soc Sci	Grand Total					
3	Mac	8.16%	0.00%	41.38%	21.43%	19.30%					
4	Main	24.49%	37.50%	75.86%	46.43%	43.86%					
5	Mini	2.04%	0.00%	6.90%	3.57%	3.51%					
6	PC	100.00%	100.00%	100.00%	100.00%	100.00%					
7	Grand Total										
8											

The Bus,Econ departments used Macintoshes at a rate of 8.16% of the PC usage rate (cell B3), while the MathSci departments used Macs at a rate of 41.38% of the rate of PC usage (cell D3). Overall, the Macs were used 19.30% as often as the PC. Perhaps you anticipated that business and economics departments would be inclined to use the PC because of its prevalence in industry. It's also probably not a surprise that the Macintosh is popular with mathematicians.

To reformat the table to show counts again:

1. Right-click **any of the percentages** in the pivot table, then click **PivotTable Field** in the shortcut menu.

2. If necessary, click the **Options button** to display the Show Data as list box.

3. Click **Normal** in the Show Data as list box, then click **OK**.

4. Click the **Save button** 🖫 to save your file as 7SURVEY.XLS (the drive containing your Student Disk should already be selected).

If there were no relationship between computer and department, the column percentages would be about the same for each department category, computer usage would be about the same for each department category, and you could say that computer usage is independent of department category. The column percentages given in the Grand Total column (Figure 7-13) would be applicable to each of the columns. However, you've seen that there might be a difference in usage of Macs between departments.

How can you test the hypothesis that there is no relationship between computer and department, that is, that computer usage is independent of department? To do so you can refer to a probability distribution called the chi-square distribution.

The Chi-Square Distribution

Statistical inference in this chapter will rely mostly on the chi-square (also written χ^2) distribution. The **chi-square distribution** is used in situations where you analyze proportions and counts. An example of the chi-square distribution is shown in the instructional template CPDF.XLT.

To open CPDF.XLT:

1. Open **CPDF.XLT** (be sure you select the drive and directory containing your Student files).

The template opens to a chart showing the probability density function of the chi-square distribution. See Figure 7-16.

Figure 7-16
The CPDF.XLT template

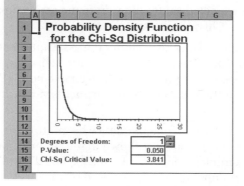

Like the t distribution, the chi-square distribution involves a single parameter—the degrees of freedom. Unlike the normal and t distributions, the chi-square can take on only positive values. The distribution is highly skewed when the degrees of freedom are low, but is less so as the degrees of freedom increase. As the degrees of freedom increase, higher values become more probable. To better understand the shape of the chi-square distribution, the CPDF template allows you to vary the degrees of freedom by clicking the degrees of freedom scroll arrows.

Experiment with the template to view how the distribution of the chi-square changes as you increase the degrees of freedom.

To increase the Degrees of Freedom:

1 Click the **Degrees of Freedom up scroll arrow** and watch as the areas of highest probability in the distribution shift to the right.

2 Continue to increase the degrees of freedom.

In this chapter, hypothesis tests based on the chi-square distribution will use the area under the upper tail of the distribution to determine the *p*-value. This template allows you to specify a particular *p*-value of interest in cell E15. The corresponding critical value appears in cell E16, and a vertical line in the chart indicates the location of the critical value.

To change the *p*-value:

1 Click cell **E15**, type **0.10**, then press [**Enter**]. This gives you the location of the critical value for the test at the 10% significance level.

Notice that the critical value shifts to the left, telling you that values of the chi-square statistic to the right of this point occur 10% of the time. Continue working with the chi-square template, trying different parameter values, until you feel clear about the nature of the chi-square distribution.

To close the template and return to the 7SURVEY workbook:

1 Click **File > Close**.

2 Click **No**, because you don't need to save your work with the template.

You should return to the 7SURVEY workbook; if you don't, click Window > 7SURVEY.XLS.

Computing Expected Values

When working with tables, such as the table that compares computer usage to department, you are trying to test whether the row and column variables are independent. The statistician Karl Pearson in 1900 devised such a test based on the idea of **expected values**, computing what is expected for each frequency under the assumption of independence, and then calculating how much the expected values differ from the actual observed frequencies. The test is known as the Pearson Chi-Square Test.

How do you compute the expected values? Under the assumption of independence, the percentages in Figure 7-13 for the different departments for each computer type should be about the same as the corresponding Grand Total given in column F. That is, the Grand Total for Macs is 11.58%, but the values in the departments vary from 0.00% (HealthSc) up to 18.46% (MathSci).

You can calculate the expected value for a frequency by multiplying the expected column percentage (that is, the overall percentages shown in the Grand Total column in Figure 7-13) by the column total. For example, in Figure 7-13, the expected percentage of Macs used is 11.58% (cell F3). The total number of computers used by the Bus,Econ departments is 66, as shown in Figure 7-12 (cell B7). Therefore, the expected number of Macs used in Bus,Econ is .1158(66), or 7.64, which is 11.58% of 66. Note that the actual observed value is 4 (cell B3 in Figure 7-12). Notice that the 11.58% is obtained from the ratio of 22, the total number of Macs used, to the grand total count of 190, the total number of computers reported. Therefore, you can obtain the expected value with this formula:

$$\left(\frac{22}{190}\right) \bullet 66 = 7.64$$

You can calculate the expected values for each of the 16 cells in the table by entering this formula in one of the cells and then using Excel's fill feature to enter it into the rest. The formula you enter in step 3 below contains a mix of absolute and relative cell references to accomplish this (recall from Chapter 2 that a relative reference, written without a $, identifies a cell location based on its position relative to other cells, whereas an absolute reference, written with a $, always points to the same cell).

To calculate the expected values:

1 Click cell **A9**, type **Expected Value**, then press [**Enter**].

2 Highlight the range **B9:E12**.

3 Type the formula **=$F3*B$7/$F$7** in cell **B9**, then press [**Enter**].

4 Click **Edit > Fill > Right** (or press [**Ctrl**]+**R**) to enter the formula in the highlighted cells on the right.

5 Click **Edit > Fill > Down** (or press [**Ctrl**]+**D**) to enter the formula in the rest of the highlighted cells.

Because of the combination of absolute and mixed cell references, Excel entered the expected values into all 16 cells. With absolute cell references, the reference does not change when you extend the formula down a column or across a row. Because you want to divide by the overall grand total in each of the sixteen equations of the expected value table,assign the cell that contains the grand total ($F$7) an absolute reference. On the other hand, you want each cell to take its row total from column F (an absolute column reference), but which row total the cell uses depends upon its row, so you use the mixed address $F3. Similarly, the column totals are always found in row 7, but which row total is used depends upon the column of the cell, so you use the mixed address B$7. Now view the formulas you entered into cells B9:E12.

To view the formulas:

1 Click **Tools > Options**.

2 Click the **View tab** in the Options dialog box if it isn't already selected.

3 Click the **Formulas check box** within the Window Options group box, click **OK**, then click any cell to remove the highlighting.

Excel now shows the formulas rather than the values. See Figure 7-17.

Figure 7-17
Formulas entered into cells B9:E12

	A	B	C	D	E	
1	Count of Computer	Dept				
2	Computer	Bus,Econ	HealthSc	MathSci	Soc Sci	G
3	Mac	4	0	12	6	2
4	Main	12	3	22	13	5
5	Mini	1	0	2	1	4
6	PC	49	8	29	28	1
7	Grand Total	66	11	65	48	1
8						
9	Expected Value	=$F3*B$7/$F$7	=$F3*C$7/$F$7	=$F3*D$7/$F$7	=$F3*E$7/$F$7	
10		=$F4*B$7/$F$7	=$F4*C$7/$F$7	=$F4*D$7/$F$7	=$F4*E$7/$F$7	
11		=$F5*B$7/$F$7	=$F5*C$7/$F$7	=$F5*D$7/$F$7	=$F5*E$7/$F$7	
12		=$F6*B$7/$F$7	=$F6*C$7/$F$7	=$F6*D$7/$F$7	=$F6*E$7/$F$7	
13						

Excel has entered the proper formula into each cell, and you've saved yourself a lot of typing by using a combination of mixed and absolute cell references.

To view values rather than formulas:

1 Click **Tools > Options**.

2 Click the **View tab** if it isn't already selected.

3 Deselect the **Formulas check box** within the Window Options group box, then click **OK**.

Pearson Chi-Square Statistic

The **Pearson chi-square statistic** provides a way of measuring the departure of the observed frequencies from the expected values. It is the summation of the squared differences between the expected and observed values for each cell in your pivot table (in this example, there are 16 cells: 4 computer types and 4 departments), divided by the expected value. Using f for frequency (that is, the observed count), e for expected (the values entered in B9:E12), and Σ for summation, the Pearson chi-square formula is:

$$\text{Pearson chi–square} = \sum \frac{(f-e)^2}{e}$$

For the data in Figure 7-12, this formula expands to:

$$\frac{(4-7.64)^2}{7.64} + \frac{(0-1.27)^2}{1.27} + \frac{(12-7.53)^2}{7.53} + \frac{(6-5.56)^2}{5.56} + ... + \frac{(28-28.8)^2}{28.8} = 14.525$$

If the frequencies all agreed with their expected values, this total would be 0. In general, small values of the Pearson chi-square indicate agreement between observed frequency and the hypothesis of independence.

If the Pearson chi-square is small, you can accept the idea that each department category uses the same proportions of PCs, Macs, mainframes, and minicomputers. Larger values of the Pearson chi-square show that the variables in the table are not independent, which means that computer preferences are not the same in all departments.

How do you decide whether the Pearson chi-square is small or large? When the sample size is big enough, the Pearson chi-square has approximately the chi-square distribution. The degrees of freedom are $(r-1)(c-1)$, where r is the number of rows and c is the number of columns. Here $(r-1)(c-1) = (4-1)(4-1) = 9$.

Even though the formula might seem strange, there is some logic to the degrees of freedom. Look at cells B3:E12 on your screen and notice that the differences between observed and expected values in each row and each column sum to 0. For example, in the first row, $(4-7.64) + (0-1.27) + (12-7.53) + (6-5.56) = 0$. Because this sum must be 0, the last difference is determined by the previous three. Because this pattern applies to each row and column, when the nine differences for the first three rows and the first three columns are specified, all of the others are determined. The last difference in each row and column has no freedom, because the total must be 0. Even though the Pearson chi-square statistic is the sum of 16 squares, there are only 9 degrees of freedom.

How then can you tell if the computed value of 14.525 is significant? The Excel function CHITEST calculates the p-value of the Pearson chi-square statistic (you provide the observed and expected frequencies).

To calculate the p-value for the observed data:

1 Click cell **A14**, type **Probability**, then press [**Tab**].

2 Type **=CHITEST(B3:E6,B9:E12)** in cell **B14**, then press [**Enter**] (the first range contains the observed cell counts; the second range contains the expected values you just calculated).

Excel returns the value 0.104827 in cell B14. Because this probability is not less than 0.05, there is not a significant departure from independence at the 5% level, and you accept the null hypothesis that the computer used and the department are independent of each other.

There are other statistics besides the Pearson chi-square that you can use to measure the association between rows and columns in a pivot table. You can generate a listing of these statistics (called table statistics) using the CTI Statistical Add-Ins.

To generate table statistics for the computer and department table:

1 Click **CTI > Table Statistics**. If you do not see this option, the CTI Statistical Add-Ins have not been loaded. See Chapter 3 for information on how to install the CTI Statistical Add-Ins.

2 Type **B3:E6** in the Input Range box to specify the range containing the frequency counts.

3 Click the **New Sheet option button** and type **Comp by Dept Stats** in the New Sheet box so that the Calculate Table Statistics dialog box looks like Figure 7-18.

Figure 7-18
Calculate Table
Statistics dialog box

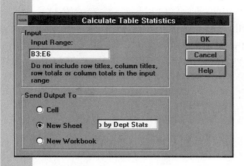

4 Click **OK**.

Excel produces the results shown in Figure 7-19.

Figure 7-19
Table statistics

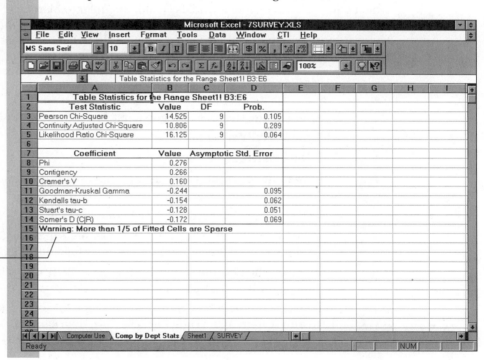

warning about small
frequencies

Calculating Table Statistics

- Click CTI > Table Statistics.

- Enter the cell range containing the table counts (exclude row and column titles, and row and column totals).

- Select an output destination (cell, worksheet, or workbook) for the table statistics.

- Click OK.

Validity of Chi-Square Test with Small Frequencies

The first thing to notice in Figure 7-19 is the warning printed after the table of statistics: "Warning: More than 1/5 of Fitted Cells are Sparse." Sparse means that there are cells with low counts. Before getting any deeper into the interpretation of results shown in Figure 7-19, you should consider this warning and the validity of the Pearson chi-square test you just used. The test requires large samples, and this means that small frequencies can be a problem. The rules for validity are usually given in terms of the expected values. Generally speaking, the expected values should be 5 or more, but this condition is conservative, so not every expected value must be at least 5.

In the warning, "fitted cells" means expected values. The warning suggests that you might get by with as many as one-fifth of the expected values under 5, meaning that the Pearson chi-square would still have approximately the chi-square distribution. However, more than one-fifth of the expected values are under 5 here. Looking at Figure 7-12, notice that six of the 16 expected values are under 5, and $6/16 = 0.375$, which is much greater than one-fifth. There is a problem.

Eliminating Sparse Data

What can you do to create a valid test? When there are low counts in a row or column, you can pool the values in that row or column with those in another row or column, or you can eliminate the row or column altogether. In the case of HealthSc, it would be hard to decide which other column to pool it with. Is HealthSc similar to MathSci? Soc Sci? It seems safer simply to eliminate the column. On the other hand, in the case of the minicomputers, you could argue for combining them with the mainframes. Let's recreate the table reducing the number of sparse cells by removing the HealthSc category and pooling the minicomputers with the mainframes (we could also restructure the present table instead of recreating it, but this way we'll have both tables in the workbook).

	To eliminate the HealthSc category, first create the pivot table as you did before:
1	Click the **SURVEY sheet tab**.
2	Click **Data > PivotTable**.
3	Verify that the **Microsoft Excel List or Database option button** is selected, then click **Next**.
4	Verify that the Range box contains **$A$1:$I$393**, then click **Next**.
5	Drag the **Computer field button** to the ROW area, drag the **Dept field button** to the COLUMN area, and drag a second **Computer field button** to the DATA area.
6	Double-click the **Computer field button** in the ROW area, click the **blank item** in the Hide Items list box, then click **OK**.
7	Double-click the **Dept field button** in the COLUMN area, click the **blank item** and the **HealthSc item** in the Hide Items list box, then click **OK**.
8	Click **Finish** to create the resulting pivot table, as shown in Figure 7-20.

Figure 7-20

Two-way table of Computer versus Dept with HealthSc removed

	A	B	C	D	E	F	G	H	I	J	K
1	Count of Computer	Dept									
2	Computer	Bus,Econ	MathSci	Soc Sci	Grand Total						
3	Mac	4	12	6	22						
4	Main	12	22	13	47						
5	Mini	1	2	1	4						
6	PC	49	29	28	106						
7	Grand Total	66	65	48	179						
8											

Of the 12 cells, 4, or 33%, have counts of less than 5, while 5 of 12, or about 42%, have values of 6 or less. There is still a problem with sparse cells. To compensate, group the mainframes and minicomputers together.

Combining Categories

You can use the PivotTable Wizard to combine various categories in the table. Combining categories does not affect the underlying data.

To combine the mainframe and minicomputer categories:

1   Select cells **A4:A5** (the row labels Main and Mini).

2   Right-click **the selection** to open the shortcut menu, then click **Group and Outline > Group**.

The row labels in column A are shifted to column B, and Excel adds a new row level in column A in the pivot table (Computer2), calling the grouping of Mains and Minis "Group 1." Let's give these more descriptive names.

3   Click cell **A2**, then type **Computer Groups** and press [**Enter**].

4   Click cell **A4**, then type **Mains/Minis** and press [**Enter**].

Now remove the ungrouped row fields from the table.

5   Click cell **B2**, the Computer field button, drag it off the table, and when the pointer appears as ✖, release the mouse button. Excel removes the ungrouped Computer field from the pivot table. See Figure 7-21.

Figure 7-21
Restructured Computer versus Dept table

	A	B	C	D	E	F	G	H	I	J	K
1	Count of Computer	Dept									
2	Computer Groups	Bus.Econ	MathSci	Soc Sci	Grand Total						
3	Mac	4	12	6	22						
4	Mains/Minis	13	24	14	51						
5	PC	49	29	28	106						
6	Grand Total	66	65	48	179						
7											
8											

Removing Sparse Cells

Hiding Categories:

- Double-click the row (or column) field button on the pivot table to display the PivotTable Field dialog box.
- Select the items to be hidden in the Hide Items list box.

Combining Categories:

- Select the row labels on the pivot table representing the categories you want to combine.
- Right-click the selection and choose Group and Outline > Group from the shortcut menu.
- Remove the ungrouped row fields from the pivot table by dragging the row (or column) field button representing the original grouping of the categories off the pivot table.

Notice that only one of the nine cells (11%) has a count below 5. Two of the nine cells (22%) have values of 6 or less. Restructuring the table has reduced the sparseness problem. What has it done to the probability value of the Pearson chi-square test?

To check the table statistics on the revised table:

1   Click **CTI > Table Statistics**.

2   Type **B3:D5** in the Input Range box.

3   Click the **Cell option button**, type **G1** in the Cell box, then click **OK**. Excel produces the table statistics shown in Figure 7-22 (scroll to see the output).

The Pearson chi-square test statistic *p*-value has changed greatly, from about 0.105 to 0.015. Because the *p*-value 0.015 is less than 0.05, the Pearson statistic is significant at the 5% level. To put it another way, at the 5% level you reject the null hypothesis, which states that there is no relationship between department and computer used.

Pearson *p*-value

Figure 7-22
Table statistics output

	C	D	E	F	G	H	I	J	K
1					Table Statistics for the Range B3:D5				
2	MathSci	Soc Sci	Grand Total		Test Statistic	Value	DF	Prob.	
3	12	6	22		Pearson Chi-Square	12.384	4	0.015	
4	24	14	51		Continuity Adjusted Chi-Square	10.356	4	0.035	
5	29	28	106		Likelihood Ratio Chi-Square	12.706	4	0.013	
6	65	48	179						
7					Coefficient	Value	Asymptotic Std. Error		
8					Phi	0.263			
9					Contigency	0.254			
10					Cramer's V	0.186			
11					Goodman-Kruskal Gamma	-0.234		0.102	
12					Kendalls tau-b	-0.145		0.065	
13					Stuart's tau-c	-0.132		0.059	
14					Somer's D (C|R)	-0.159		0.071	
15									

To better quantify this relationship, generate a table of expected values:

1 Scroll back to the left, click cell **A8**, then type **Expected Values**.

2 Select the range **B8:D10**.

3 Type =$E3*B$6/$E$6 in cell **B8**, then press [**Enter**] to calculate the expected cell value.

4 Press [**Ctrl**]+**R** to fill the formula values to the right.

5 Press [**Ctrl**]+**D** to fill the formula values down, then click a cell to remove the highlighting.

Figure 7-23
Expected values for the revised table

observed values

expected values

	A	B	C	D	E	F	G	H	
1	Count of Computer	Dept					Table Statistics for the Range		
2	Computer Groups	Bus,Econ	MathSci	Soc Sci	Grand Total		Test Statistic	Value	
3	Mac	4	12	6	22		Pearson Chi-Square	12.384	
4	Mains/Minis	13	24	14	51		Continuity Adjusted Chi-Square	10.356	
5	PC	49	29	28	106		Likelihood Ratio Chi-Square	12.706	
6	Grand Total	66	65	48	179				
7							Coefficient	Value	Asy
8	Expected Values	8.111732	7.988827	5.899441			Phi	0.263	
9		18.80447	18.51955	13.67598			Contigency	0.254	
10		39.0838	38.49162	28.42458			Cramer's V	0.186	
11							Goodman-Kruskal Gamma	-0.234	
12							Kendalls tau-b	-0.145	
13							Stuart's tau-c	-0.132	
14							Somer's D (C|R)	-0.159	
15									

The resulting table (Figure 7-23) gives insight about the differences between departments. The last column is for Soc Sci, and in this column the expected values in D8:D10 do agree with the observed values in D3:D5 (5.899 compared to 6, 13.676 compared to 14, and 28.425 compared to 28). The differences are mainly in the first two columns. In the first column (Bus,Econ), the first two observed values are lower than expected, and the third value is much higher. That is, the Macintosh and mainframe frequencies are low and the PC frequency is high. In the second column, for MathSci, the Macintosh and mainframe frequencies are high and the PC frequency is low. The chi-square summation shows that the relationship mainly involves different patterns of computer usage by Bus,Econ and MathSci. Does this agree with a your intuition? Do you have the impression that business departments use PCs for compatibility with industry?

Suppose that the Pearson chi-square statistic had turned out not to be significant. Would this prove that computer usage and department are independent? In other words, if you do not reject the null hypothesis of independence, should you embrace it and assert that the null hypothesis is true? Consider redoing the survey with a much greater sample size. This would give much greater power, with a greater ability to detect departures from the null hypothesis. Your previous acceptance of the null hypothesis could easily turn into a rejection with a larger sample size. Be careful about asserting that the null hypothesis is true.

Other Test Statistics

Among the other statistics in Figure 7-22, the Likelihood Ratio chi-square statistic is usually close to the Pearson chi-square statistic. Many statisticians prefer using the Likelihood Ratio chi-square because it is used in log-linear modeling—a topic beyond the scope of this book.

Because the χ^2 distribution is a continuous distribution and counts represent discrete values, some statisticians are concerned that Pearson's chi-square statistic is not appropriate. They recommend using the Continuity Adjusted chi-square statistic as an adjustment. Although some statisticians prefer the Continuity Adjusted chi-square, we feel that the Pearson chi-square statistic is more accurate and can be used without adjustment.

Tables with Ordinal Variables

The two-way table you just produced was for two nominal variables. Now let's look at variables that have inherent order, such as education level (freshman, sophomore, junior, senior). For ordinal variables, there are more powerful tests than the Pearson chi-square, which will often fail to give significance for ordered variables.

As an example, consider the Calculus and Enroll B variables. Calculus tells the extent to which calculus is required for a given statistics class (Not req or Prereq). From this point of view, Calculus is an ordinal variable, although it would also be possible to think of it as a nominal variable. When a variable takes on only two values, there really is no distinction between nominal and ordinal, because any two values can be regarded as ordered. The other variable, Enroll B, is a categorical variable that contains measures of the size of the annual course enrollment. In the survey, instructors were asked to check one of eight categories (0–50, 51–100, 101–150, 151–200, 201–300, 301–400, 401–500, 501–) for the number of students in the course.

Testing for a Relationship Between Two Ordinal Variables

To test for a relationship between whether calculus was a required prerequisite for a statistics course and the enrollment count for that course, first form a two-way table for Calculus and Enroll B. You might expect that classes requiring calculus would have smaller enrollments.

To form the table:

1   Click the **SURVEY sheet tab** (you might have to scroll through the worksheets to see it).

2   Click **Data > PivotTable**.

3   Verify that the **Microsoft Excel List or Database option button** is selected, then click **Next**.

4   Verify that the Range box contains **$A$1:$I$393**, then click **Next**.

5   Drag the **Enroll B field button** to the ROW area, the **Calculus field button** to the COLUMN areas, and the **Enroll B field button** to the DATA area.

6   Double-click the **Enroll B field button** in the ROW area, click the **blank item** in the Hide Items list box, then click **OK**.

7   Double-click the **Calculus field button** in the COLUMN area, click the **blank item** in the Hide Items list box, then click **OK**.

8   Click **Finish** in the Step 3 dialog box.

In the table you just created, Excel automatically arranges the Enroll B levels in a combination of numeric and alphabetic order (alphanumeric order), so be careful with category names. For example, if "051–100" were written as "51–100" instead, Excel would place it near the bottom of the table because 5 comes after 1, 2, 3, and 4.

Remember that most of the expected values should exceed 5, so there is cause for concern about the sparsity in the second column between 201 and 500, but because only four of the 16 cells have expected values less than 5, the situation is not terrible. Nevertheless, let's combine enrollment levels from 200 to 500, as specified in rows 7 through 9 of the table you just created.

To combine levels, use the same procedure you did with the computer table:

1   Highlight **A7:A9**, the enrollment row labels for the categories from 200 through 500.

2   Right-click **the selection** to display the shortcut menu, then click **Group and Outline > Group**.

3   Click cell **A2**, type **Enrollment**, and press [**Enter**].

4   Click cell **A7**, type **201-500**, and press [**Enter**].

5   Click cell **B2** and drag it off the table (the pointer appears as ✖).

Now use the CTI Statistical Add-Ins to generate table statistics.

To generate table statistics:

1   Click **CTI > Table Statistics**.

2   Type **B3:C8** in the Input Range box.

3   Click the **Cell option button**, type **F1**, then click **OK**.

Excel produces the table shown in Figure 7-24 (you might have to scroll to see it).

Figure 7-24

Table statistics for Enrollment versus Calculus table

	A	B	C	D	E	F	G	H	I
1	Count of Enroll B	Calculus				Table Statistics for the Range B3:C8			
2	Enrollment	Not req	Prereq	Grand Total		Test Statistic	Value	DF	Prob.
3	001-050	54	20	74		Pearson Chi-Square	10.013	5	0.075
4	051-100	62	15	77		Continuity Adjusted Chi-Square	7.818	5	0.167
5	101-150	41	6	47		Likelihood Ratio Chi-Square	9.762	5	0.082
6	151-200	36	4	40					
7	201-500	53	6	59		Coefficient	Value	Asymptotic Std. Error	
8	501-	38	6	44		Phi	0.171		
9	Grand Total	284	57	341		Contigency	0.169		
10						Cramer's V	0.171		
11						Goodman-Kruskal Gamma	-0.280		0.101
12						Kendalls tau-b	-0.133		0.049
13						Stuart's tau-c	-0.127		0.048
14						Somer's D (C|R)	-0.077		0.029
15									

Ordinal Measures

The table shows three statistics that take account of order: the Goodman-Kruskal Gamma, the Kendall tau-b, and the Stuart tau-c. You can divide each of these statistics by its asymptotic standard error (found in column I) to get an approximate standard normal variable, under the assumption that there is no relationship between rows and columns. When the three statistics are divided by their standard errors, the ratios are −2.77, −2.74, and −2.68, respectively. Comparing these values to the standard normal distribution returns two-sided p-values of 0.0028, 0.0030, and 0.0037. Because each of the three p-values is less than 0.01, you would reject the null hypothesis of no relationship at the 1% level. You conclude that enrollment and a calculus prerequisite are dependent.

Custom Sort Order

With ordinal data you want the values to appear in the proper order when created by the pivot table. If order is alphabetical or the variable itself is numeric, this is not a problem. However, what if the variable being considered has a definite order, but this order is neither alphabetic nor numeric? Consider the Interest variable from the 7SURVEY.XLS file, which measures the degree of interest in a supplementary statistics text. The values of this variable have a definite order (least, low, some, high, most), but this order is not alphabetic or numeric. You could create a numeric variable based on the values of Interest, such as 1 = least, 2 = low, 3 = some, 4 = high, and 5 = most. Another approach is to create a **custom sort order**, which lets you define a sort order for a variable.

You can define any number of custom sort orders. Excel already has some built-in for your use, such as months of the year (Jan, Feb, Mar, ..., Dec), so if your data have a variable with month values, you can sort the data list by months (you can also do this with pivot tables). Try creating a custom sort order for the values of the Interest variable.

To create a custom sort order for Interest:

1   Click **Tools > Options**.

2   Click the **Custom Lists tab**.

3   Click the **List Entries list box**.

4.   Type **least** and press [**Enter**].

5   Type **low** and press [**Enter**].

6   Type **some** and press [**Enter**].

7   Type **high** and press [**Enter**].

8   Type **most** and click **Add**.

Your Custom Lists tab of the Options dialog box should look like Figure 7-25.

Figure 7-25
Custom Lists tab of the
Options dialog box

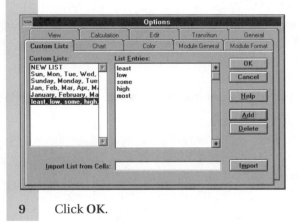

9   Click **OK**.

Now create a pivot table of Interest values to see whether the custom sort order command worked.

To create the pivot table:

1   Click the **SURVEY sheet tab** to return to the SURVEY worksheet.

2   Click **Data > PivotTable**.

3   Verify that the **Microsoft Excel List or Database option button** is selected, then click **Next**.

4   Verify that the Range list box contains **$A$1:$I$393**, then click **Next**.

5   Drag the **Interest field button** to the ROW area, and the **Interest field button** to the DATA area, then click **Finish**.

The resulting pivot table appears, as in Figure 7-26.

Figure 7-26

Interest pivot table with
custom sort order

	A	B	C	D	E	F	G	H	I	J	K	L
1	Count of Interest											
2	Interest	Total										
3	least	22										
4	low	44										
5	some	131										
6	high	110										
7	most	71										
8		14										
9	Grand Total	392										
10												

Notice that Excel automatically sorts the Interest categories in the pivot table in the proper order—least, low, some, high and most—rather than alphabetically. In the exercises that follow, you will analyze the Interest counts using some of the ordinal measures discussed in the previous section.

To save your workbook:

1   Save your workbook on your Student Disk as **7SURVEY.XLS**.

2   Click **File > Exit**.

You can find more information about other pivot table uses in your Excel documentation or in on-line Help. They are an extremely useful tool.

E X E R C I S E S

1. Open the SURVEY.XLS workbook.

   a. Create a pivot table for the departments responding to the survey. Then create a pie chart and bar chart (use the Column chart type) of the total counts reported in the pivot table.

   b. Remove missing values.

   c. Which chart gives a clearer indication of the comparative sizes of the groups?

2. Continue using SURVEY.XLS.

   a. Create a two-way pivot table for Computer vs. Calculus with computer as the ROW field and calculus as the COLUMN field.

   b. Remove missing values.

   c. Use the CTI Statistical Add-Ins to generate table statistics.

   d. Are there problems caused by a sparse row for minicomputers? Exclude the Mini category using the Hide feature, then form the table again.

   e. Is there a significant relationship between the rows and columns? If not, should you conclude that the variables for type of computer used and whether calculus is a prerequisite are independent of each other? Explain.

3. Continue using SURVEY.XLS.

   a. Obtain a two-way table of counts for Interest vs. Calculus with Interest as the ROW field and Calculus as the COLUMN field. (Use either Interest or Calculus as the DATA field.)

   b. Use the ordinal measure Kendall's tau-b to test for significance (make sure that the interest levels are appropriately ordered: least, low, some, high and most; if they are not, see the section "Custom Sort Order" in this chapter). Interest measures the degree of interest in a supplementary text to teach a statistical package. Do you find much of a relationship between the Interest and Calculus variables?

    c. Recreate the pivot tables showing the row percentages and column percentages instead of the counts.

    d. Summarize what you have found in nonstatistical language.

4. Continue using SURVEY.XLS. To see if different departments make software available in different ways, tabulate Available vs. Dept. You might need to get rid of a row and a column to avoid sparse cells, because there are not many students in HealthSc and there are not many departments that ask students to use the software supplied with the text. Do the data show much of a dependence between the department and how students gain access to software?

5. Continue using SURVEY.XLS. Enroll A gives the annual enrollment in the instructor's statistics class (in contrast with Enroll B, which gives annual enrollment for all sections of all instructors for a particular course).

    a. Create a pivot table to see if Enroll A is related to Calculus.

    b. Obtain an ordinal measure such as Kendall's tau-b. What does this measure tell you about the relationship between enrollment and whether or not calculus requires a prerequisite?

    c. If the table is too sparse, you might want to combine some of the categories of Enroll A and obtain a new table.

    d. Save your workbook as E7SURVEY.XLS.

6. Open the JRCOL.XLS workbook.

    a. Create a customized list of teaching ranks sorted in the order: instructor, asst prof, assoc prof, full prof.

    b. Create a pivot table with Rank Hired as the ROW variable and Sex as the COLUMN variable.

    c. Generate the table statistics for this pivot table. Which statistics are appropriate to use with the table? Is there any difficulty with the data in the table? How would you correct these problems?

    d. Group the three professor ranks into one, remove the old grouping, and redo the table statistics. Does this correct the problem from part 6c? What information has been lost? Can you suggest a different grouping of ranks?

    e. Write a report summarizing your results. Would you conclude that the variables for gender and the rank at which the teacher was hired are independent? What are some important factors that you have not looked at in this analysis?

7. Continue using JRCOL.XLS. The PAGE area of the pivot table allows you to view several levels of a table.

    a. Create a pivot table for JRCOL.XLS with Degree as the PAGE field, Rank Hired as the ROW field, and Sex as the COLUMN field. (Because you are obtaining counts, you can use either Sex or Rank Hired as the DATA field.)

    b. Using the drop-down arrows on the PAGE field button, select the table of persons hired with a master's degree.

    c. Generate table statistics for this group. Is the rank when hired independent of gender for candidates with master's degrees?

    d. Group the three ranks of professor into one group and redo part 7b.

    e. Write a report summarizing your conclusions. What are some limitations inherent in these data that make it difficult for you to draw conclusions?

    f. Save your workbook as E7JRCOL.XLS.

8. Open the PCINFO.XLS workbook. In comparing different CPUs you are interested in whether the type of BUS used is independent of the CPU.

   a. Create a pivot table with CPU as the ROW field and BUS as the COLUMN field.

   b. Generate table statistics for the table. Does it appear that BUS and CPU are independent of each other? Is the table sparse? If so, what kind of problems could this cause your analysis?

   c. Eliminate the sparseness of the table by removing the CX/40 and pentium processors. Then recreate the table using the following groups: DX (DX/33, DX/40, DX/50), DX2 (DX2/50, DX2/66), SLC (SLC-2/50, SLC-2/66, SLC-2/80), and SX (SX/25 and SX/33). Generate table statistics for this table. What do you conclude about the relationship between CPU and BUS type?

9. Using the PCINFO.XLS file, create a custom sort order for the CPU type from least powerful to most powerful. The order is: SX/25, SX/33, DX/33, CX/40, DX/40, DX/50, DX2/50, DX2/66 and Pentium/60 (do not include the SLC chips, because they are based on a different architecture).

   a. Create a pivot table with CPU as the ROW field and BUS as the COLUMN field.

   b. Generate table statistics for the pivot table. Because the CPU variable type is ordinal, calculate the p-value for Kendall's tau-b statistic.

   c. Create a SX group (SX/25, SX/33), a DX group (DX/33, DX/40, DX/50) and a DX2 group (DX2/50, DX2/66). Remove the CX/40 and Pentium levels from the row field. Redo the table statistics. How do the values of Kendall's tau-b compare to the value calculated for the sparse table?

   d. Write a report summarizing your conclusions. Are CPU and BUS type independent factors?

   e. Save your workbook as E7PCINFO.XLS.

CHAPTER 8

CORRELATION AND SIMPLE REGRESSION

O BJECTIVES

In this chapter you will learn to:

- Fit a regression line and interpret the coefficients

- Obtain and interpret correlations and their statistical significance

- Understand the relationship between correlation and simple regression

- Understand regression statistics

- Use residuals to check the validity of the assumptions needed for statistical inference

- Obtain a correlation matrix and apply hypothesis tests

- Obtain and interpret a scatterplot matrix

This chapter examines the relationship between two variables using linear regression and correlation. **Linear regression** estimates a linear equation that describes the relationship, whereas **correlation** measures the strength of that linear relationship.

Simple Linear Regression

When you plot two variables against each other, the values on the plot usually don't fall exactly in a perfectly straight line. When you perform a linear regression analysis, you attempt to find the line that best estimates the relationship between two variables (the y, or *dependent variable*, and the x, or *independent variable*). The line you find is called the **fitted regression line**, and the equation that specifies the line is called the **regression equation**.

The Regression Equation

If the data in a scatterplot fall approximately in a straight line, you can use linear regression to find an equation for the regression line drawn over the data. Usually, you will not be able to fit the data perfectly, so some points will lie above and some below the fitted regression line.

The regression line Excel fits will have an equation of the form $y = a + bx$.

Here y is the dependent variable, the one you are trying to predict, and x is the independent, or **predictor**, variable, the one that is doing the predicting. Finally, a and b are called **coefficients**. Figure 8-1 shows a line with a = 10 and b = 2. The short vertical line segments represent the errors, also called **residuals**, which are the gaps between the line and the points. The residuals are the differences between the observed dependent values and the predicted values. Because a is where the line intercepts the vertical axis, a is sometimes called the **intercept** or **constant term** in the model. Because b tells how steep the line is, b is called the **slope**. It gives the ratio between the vertical change and the horizontal change along the line. Here y increases from 10 to 30 when x increases from 0 to 10, so the slope is:

$$b = \frac{\text{vertical change}}{\text{horizontal change}} = \frac{30-10}{10-0} = 2$$

Figure 8-1
The line y = 10 + 2x

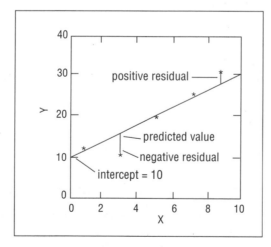

Suppose that x is years on the job and y is salary. Then the y-intercept (x = 0) is the salary for a person with zero years' experience, the starting salary. The slope is the change in salary per year of service. A person with a salary above the line would have a positive residual, and a person with a salary below the line would have a negative residual.

If the line trends downward so that y decreases when x increases, then the slope is negative. For example, if x is age and y is price for used cars, then the slope gives the drop in price per year of age. In this example, the intercept is the price when new, and the residuals represent the difference between the actual price and the predicted price. All other things being equal, if the straight line is the correct model, a positive residual means a car costs more than it should, and a negative residual means that a car costs less than it should (that is, it's a bargain).

Fitting the Regression Line

When fitting a line to data, assume that the data follow a linear model:

$$y = \alpha + \beta x + \varepsilon$$

where α is the "true" intercept, β is the "true" slope, and ε is an error term. When you fit the line, you will be trying to estimate α and β, but you can never know them exactly. The values α and β are called the population values for the constant term and slope, respectively.

Although you can never learn the true slope and intercept, Excel can compute regression estimates if you supply data for x and y. The procedure is called **least squares**, because it chooses the slope and intercept to minimize the sum of squared residuals. Assume n pairs of x and y values: $(x_1, y_1), \ldots, (x_n, y_n)$. If $\bar{x}$ is the sample mean (average) of the x values and $\bar{y}$ is the sample mean of the y values, then the estimated slope is:

$$b = \frac{(x_1 - \bar{x})(y_1 - \bar{y}) + \ldots + (x_n - \bar{x})(y_n - \bar{y})}{(x_1 - \bar{x})^2 + \ldots + (x_n - \bar{x})^2}$$

and the estimated intercept is:

$$a = \bar{y} - b\bar{x}$$

Regression Assumptions

Later in the chapter, you will apply statistical inference to perform hypothesis tests and obtain confidence intervals for the slope and intercept of the fitted regression line. This requires the following assumptions:

- The straight line model is correct.
- The errors are normally distributed with mean 0.
- The errors are independent of each other.
- The errors have constant variance.

Whenever you use regression to fit a line to data, you should consider these assumptions. Fortunately, regression is somewhat robust, so the assumptions do not need to be perfectly satisfied. However, it is necessary to verify that a straight line really does fit the data.

Calculating a Regression Line

Let's try plotting one variable against another and then calculating a regression line that best fits the data. For example, use average ACT (American College Testing) scores to predict graduation rates at Big Ten universities. The data are available in the BIGTEN.XLS file (see Chapter 4 for a description of the data set). The percentage of 1984 freshmen who graduated by 1989 is called Grad Percent, and the ACT average is called ACT (see Table 8-1). The ACT scores are missing for three of the universities.

Table 8-1
Grad percent and ACT for Big Ten schools

College	Graduation Percent	ACT Average
Illinois	76.2	27
Indiana	57.6	24
Iowa	55.4	24
Michigan	76.5	
Mich St	59.7	23
Minnesota	27.0	
Northwestern	86.0	28
Ohio St	46.2	22
Purdue	66.7	23
Wisconsin	59.8	

Computing the Regression Equation by Hand

Before using Excel to do the slope computation, try computing it using the formula described in the section "Fitting the Regression Line." The slope estimate is:

$$b = \frac{(x_1 - \bar{x})(y_1 - \bar{y}) + \dots + (x_n - \bar{x})(y_n - \bar{y})}{(x_1 - \bar{x})^2 + \dots + (x_n - \bar{x})^2}$$

$$= \frac{(27 - 24.43)(76.2 - 63.97) + \dots + (23 - 24.43)(66.7 - 63.97)}{(27 - 24.43)^2 + \dots + (23 - 24.43)^2}$$

$$= 5.448$$

Here n is 7 because there are only seven universities with data on both x and y. Just these seven universities are used in all of the computations, including $\bar{x}$ and $\bar{y}$. The estimate of the intercept is:

$$a = \bar{y} - b\bar{x} = 63.97 - 5.45(24.43) = -69$$

Recall that the linear regression equation is of the form y = a + bx. Because our hand calculations result in b = 5.45 and a = –69, this gives a regression prediction equation of: Graduation Percent = –69 + 5.45 ACT. Now try using Excel to look at a scatterplot of this relationship and then perform the calculations. First copy the data you'll use for the regression into a new worksheet in the workbook.

To copy the data:

1 Open **BIGTEN.XLS** (be sure you select the drive and directory containing your Student files).

2 Click **File > Save As**, select the drive containing your Student Disk, then save your workbook on your Student Disk as **8BIGTEN.XLS**.

3 Highlight the noncontiguous range **A1:A11,C1:C11,E1:E11** (press and hold down the [Ctrl] key as you select the ranges in the C and E columns).

4 Click the **Copy button** on the Standard toolbar.

5 Click **Insert > Worksheet**.

6 Click cell **A1** in the new worksheet (if necessary), then click the **Paste button** on the Standard toolbar.

7 Resize columns A and C so that they look like Figure 8-2.

Figure 8-2
Copied columns, resized

	A	B	C	D	E	F	G	H	I	J	K
1	University	ACT	Grad Percent								
2	Illinois	27	76.2								
3	Indiana	24	57.6								
4	Iowa	24	55.4								
5	Michigan		76.5								
6	Michigan St	23	59.7								
7	Minnesota		27								
8	Northwestern	28	86								
9	Ohio St	22	46.2								
10	Purdue	23	66.7								
11	Wisconsin		59.8								
12											

Before you can perform linear regression on the data, however, you'll have to remove those rows with missing ACT values (rows 5, 7, and 11), because *Excel interprets missing values for ACT as 0's.*

To remove the missing rows:

1 Select **A5:C5**, the cells containing the data for the University of Michigan.

2 Right-click the selection, then click **Delete** in the shortcut menu.

3   Click the **Shift Cells Up option button** if it is not already selected, then click **OK**.

4   Repeat steps 1 through 3 for the data for the Universities of Minnesota and Wisconsin.

Your data list should now look like Figure 8-3.

Figure 8-3
Sheet1 with missing
ACT rows deleted

	A	B	C	D	E	F	G	H	I	J
1	**University**	**ACT**	**Grad Percent**							
2	Illinois	27	76.2							
3	Indiana	24	57.6							
4	Iowa	24	55.4							
5	Michigan St	23	59.7							
6	Northwestern	28	86							
7	Ohio St	22	46.2							
8	Purdue	23	66.7							
9										

You're now ready to create a scatterplot of the relationship between Grad Percent (the y-variable) and ACT (the x-variable).

To create the plot:

1   Select the range **B1:C8** and click the **ChartWizard button** 🔲 on the Standard toolbar.

2   Drag the crosshair pointer over the range **D1:H15**.

3   Verify that the Range text box contains **=$B$1:$C$8** in the Step 1 dialog box, then click **Next**.

4   Double-click the **XY (Scatter)** chart type in the Step 2 dialog box.

5   Double-click **1** in the Step 3 dialog box.

6   Verify that the **Columns data series option button** is selected, that the Use First Column(s) spin box contains **1**, and that the Use First Row(s) spin text box contains **1** in the Step 4 dialog box, then click **Next**.

7   Click **No** for the Add a Legend option button, type **Linear Regression** in the Chart Title text box, **ACT** in the Category (X) text box, and **Grad Percent** in the Value (Y) text box, then click **Finish**.

Excel creates the scatterplot shown in Figure 8-4.

Figure 8-4
Scatterplot of Grad
Percent vs. ACT

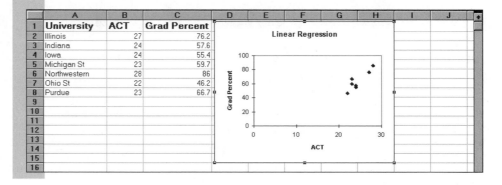

The scatterplot's scale needs some adjustment, because the data points are all crowded beyond 20 on the x-axis and 40 on the y-axis. You also want to add a trend line (that is, the linear regression line).

To format the scatterplot and add a linear regression line:

1   Double-click the **scatterplot** to activate the chart.

2   Double-click the **x-axis** to open the Format Axis dialog box, click the **Scale tab**, change the Minimum scale value from 0 to **20**, then click **OK**.

3   Double-click the **y-axis** to open the Format Axis dialog box, click the **Scale tab**, change the Minimum scale value from 0 to **40**, then click **OK**.

4   Right-click the **data series** (any one of the data points, as in Figure 8-5) and click **Insert Trendline** in the shortcut menu.

Figure 8-5
Right-clicking a data point to open shortcut menu

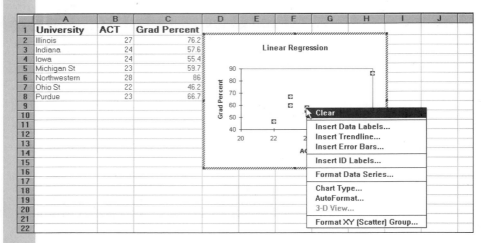

5   Click the **Type tab** and verify that the **Linear Trend/Regression Type** option is selected, then click the **Options tab**.

6   Click the **Display Equation on Chart check box** and the **Display R-squared Value on Chart check box** (Figure 8-6), then click **OK**.

Figure 8-6
Completed Options tab dialog box

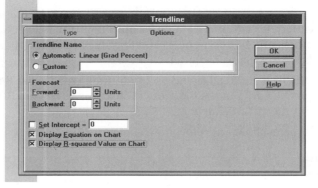

....

Adding a Regression Line to a Scatterplot

- Select the range of data you want to plot and use the Chart Wizard to create a scatterplot.

- Double-click the scatterplot to activate it, and right-click the data series to open the shortcut menu.

- Click Insert Trendline in the shortcut menu.

- Click the type of trend/regression line you want and set the display options for the regression line.

....

Excel displays the scatterplot with the linear regression line, linear regression equation, and R^2 value. See Figure 8-7.

The linear regression equation, given as $y = 5.4481 x - 69.117$, is approximately Grad Percent = $-69 + 5.45$ ACT, as you calculated by hand. The slope of 5.45 indicates that each point in the ACT score is worth about $5\frac{1}{2}$ percentage points in the graduation rate.

Although you can interpret the constant coefficient (that is, the y-intercept) of –69 as the predicted graduation rate when the ACT score is 0, this is absurd, because no university has an ACT average anywhere near 0. The constant coefficient is needed in the prediction equation, but by itself it is meaningless. Note as well that any ACT average of less than 13 will result in the prediction of a negative graduation rate. This does not mean that the linear equation is useless, but that you should be cautious in making any predictions that lie outside the range of the observed data.

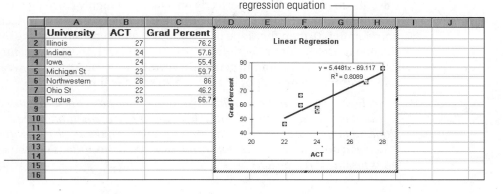

Figure 8-7
Scatterplot of Grad Percent vs. ACT with regression equation and R² value

R² value

The R^2 value is also known as the **coefficient of determination**. It measures the percentage of variation in the values of the dependent variable (in this case, Grad Percent) that can be explained by the change in the independent variable (ACT). R^2 values vary from 0% to 100%. In this example, the regression equation returns an R^2 value of 0.8089, so about 80.9% of the variation in Grad Percent can be explained by the change in ACT score.

Correlation

The slope b (5.45 in our graduation example) in simple linear regression can be any real number, so it is hard to tell from the slope alone whether x and y are strongly related. The **Pearson correlation** is very similar to the slope, except that the correlation is designed to measure the strength of the linear relationship. The correlation r is:

$$r = \frac{(x_1 - \bar{x})(y_1 - \bar{y}) + \dots + (x_n - \bar{x})(y_n - \bar{y})}{\sqrt{(x_1 - \bar{x})^2 + \dots + (x_n - \bar{x})^2}\sqrt{(y_1 - \bar{y})^2 + \dots + (y_n - \bar{y})^2}}$$

Notice that the numerator is exactly the same as the numerator for the slope b in the regression equation. This is important because it means that b = 0 when r = 0 and vice versa.

Correlation and Causality

Correlation indicates the relationship between two variables without assuming that a change in one causes a change in the other. For example, if you learn of a correlation between the number of extracurricular activities and grade-point average for high school students, does this imply that if you raise a student's GPA, he or she will participate in more after-school activities? Or that if you ask the student to get more involved in extracurricular activities, his or her grades will improve as a result? Or is it more likely that if this correlation is true, the type of people who are good students also tend to be the type of people who join after-school groups?

Correlation and Slope

The slope can be any real number, but the correlation must always be between –1 and +1. A correlation of +1 means that all of the data points fall perfectly on a line of positive slope. In such a case, all of the residuals would be 0, and the line would pass right through the points; it would have a perfect fit.

A correlation of –1 means that the data points all fall on a line of negative slope. That is, if r is +1 or –1, then the equation for the line can be used to predict y perfectly from x. Of course, the data do not usually fall perfectly on a line, but highly correlated data will fall close to a straight line, and then you should be able to predict y from x with accuracy. However, if the correlation is 0, then the slope of the line is 0, which means that the line is horizontal. Then the line is useless for prediction because it predicts the same thing for every x, and makes no use of x in the prediction. In general, a correlation near 0 means a weak linear relationship, and a correlation near +1 or –1 means a strong linear relationship.

Note: Remember that the correlation and the slope have the same numerator, so r is 0 when b is 0, and vice versa. Testing the null hypothesis of 0 correlation is the same as testing for 0 slope. In other words, the correlation is significant if the slope is significant, and vice versa. When you do a statistical test for correlation, the assumptions are the same as the assumptions for linear regression.

Computing Correlation by Hand

Can you make a guess about the correlation between Grad Percent and ACT, based on the information in Figure 8-2? The upward trend in the scatterplot indicates a positive correlation, and it should be pretty close to +1, because Grad Percent appears to be highly predictable from ACT. Try computing the correlation by hand before using Excel:

$$r = \frac{(x_1 - \bar{x})(y_1 - \bar{y}) + ... + (x_n - \bar{x})(y_n - \bar{y})}{\sqrt{(x_1 - \bar{x})^2 + ... + (x_n - \bar{x})^2}\sqrt{(y_1 - \bar{y})^2 + ... + (y_n - \bar{y})^2}}$$

$$= \frac{(27 - 24.43)(76.2 - 63.97) + ... + (23 - 24.43)(66.7 - 63.97)}{\sqrt{(27 - 24.43)^2 + ... + (23 - 24.43)^2}\sqrt{(76.2 - 63.97)^2 + ... + (66.7 - 63.97)^2}}$$

$$= 0.899$$

At approximately 0.90, the Pearson correlation is fairly high.

Using Excel to Compute Correlation

Now use Excel to confirm the correlation that you just computed by hand and test the null hypothesis that there is no linear relationship between the variables ACT score and Grad Percent. Excel provides a CORREL function that computes the correlation of the ranges you enter.

To compute the Pearson correlation r with Excel:

1. Select cell **A10** on the Sheet1 worksheet, type **Pearson r** in the cell, then press [**Tab**].
2. Type **=CORREL(B2:B8,C2:C8)** in cell **B10**, then press [**Tab**].

The CORREL function calculates the Pearson correlation coefficient. Excel places the value 0.899 in cell B10. Excel does not have a function to compute the *p*-value associated with this correlation, but one has been provided for you with the CTI Statistical Add-Ins (you will have to load the add-ins following the instructions in Chapter 3 if you have not already done so).

3 Type =CORRELP(B2:B8,C2:C8) in cell C10, then press [**Enter**].

The *p*-value returned in cell C10 equals about 0.006, which is less than 0.05, so the correlation is significant at the 5% level and you can reject the hypothesis of no linear relationship at the 5% level. The CORRELP function returns the two-sided *p*-value, so you accept the alternative hypothesis that the correlation is not zero. Actually, the two-sided *p*-value is twice the one-sided value—in this case about 0.003, so you could accept the alternative hypothesis that the correlation is greater than zero. You could also reject the hypothesis of no relationship at the 1% level because the *p*-value is less than 0.01. This is pretty strong evidence of a relationship between Grad Percent and ACT, but it is not surprising. After all, if a high ACT score means anything, then a university with a higher average ACT should have stronger students, and this should imply better odds for graduation.

Note that the correlation $r = 0.899$ and the coefficient of determination from regression $R^2 = 0.8089 = (.899)^2$. In other words R^2 equals the correlation r, squared. Is this just a coincidence? No; for simple linear regression (linear regression involving one independent variable), the coefficient of determination is always the square of the Pearson correlation coefficient.

Spearman's Rank Correlation Coefficient r_s

A nonparametric equivalent test exists for correlation: **Spearman's *s* statistic**. As for the nonparametric test in Chapter 7, you replace observered values with their ranks and calculate the statistical test (in this case the Pearson correlation r) on the ranks. Spearman's rank correlation, like other nonparametric tests, is less susceptible to the influence of outliers and is better than Pearson's correlation for nonlinear relationships. The downside to the Spearman correlation is that it is not as powerful as the Pearson correlation in detecting significant correlations in situations where the assumptions for linear regression are satisfied.

Excel does not include a function to calculate the Spearman rank correlation, but one has been provided for you in the CTI Statistical Add-Ins. Make sure that the CTI Statistical Add-Ins are installed before performing the following steps.

To calculate the Spearman rank correlation and its corresponding *p*-value:

1 Select cell **A11**, type **Spearman r(s)** in the cell, then press [**Tab**].

2 Type =SPEARMAN(B2:B8,C2:C8) in cell **B11**, then press [**Tab**].

3 Type =SPEARMANP(B2:B8,C2:C8) in cell **C11**, then press [**Enter**].

The worksheet looks like Figure 8-8.

Figure 8-8
Correlations and *p*-values

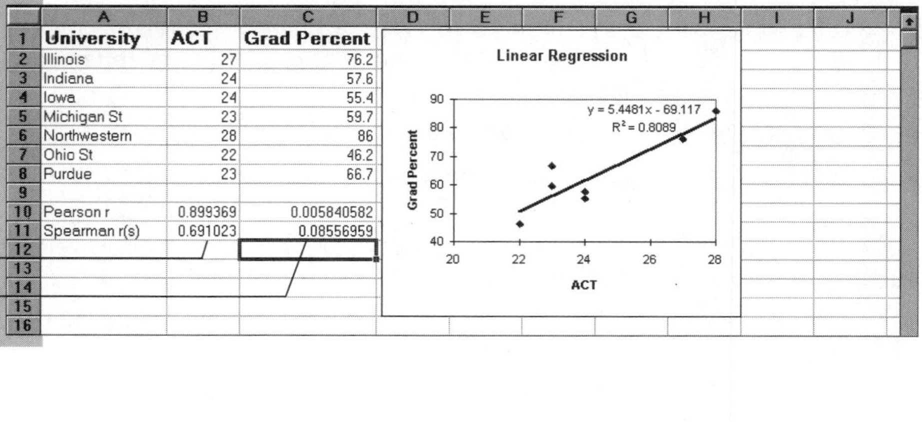

correlations

p-values

The value of the Spearman rank correlation coefficient is 0.691 (cell B11), with a two-sided p-value of about 0.086 (cell C11). Clearly the nonparametric test shows a lesser correlation that is not statistically significant at the 5% level. Nonparametric procedures rarely show statistical significance for sample sizes as small as seven.

To rename the worksheet page and save your work:

1. Right-click the **Sheet1 sheet tab** to open the shortcut menu, then click **Rename**.

2. Type **Regression** in the Name text box, then click **OK**.

3. Click the **Save button** 🖫 to save your file as **8BIGTEN.XLS** (the drive containing your Student Disk should already be selected).

Functions to Calculate Correlation Statistics

- CORREL(range1,range2) calculates the Pearson correlation between the data in cell range1 and cell range2.

- CORRELP(range1,range2) calculates the two-tailed p-value of Pearson correlation between the data in cell range1 and cell range2 (CTI Statistical Add-Ins function).

- SPEARMAN(range1,range2) calculates the Spearman rank correlation between the data in cell range1 and cell range2 (CTI Statistical Add-Ins function).

- SPEARMANP(range1,range2) calculates the two-tailed p-value of the Spearman rank correlation between the data in cell range1 and cell range2 (CTI Statistical Add-Ins function).

Regression Analysis with Excel

You have seen how to plot two variables against each other with a fitted regression line, to calculate the regression equation, and to test how highly correlated one variable is with the other. Now have Excel calculate regression statistics. These let you determine confidence intervals for the slope and constant coefficient, and also check to see whether your regression assumptions are correct.

To calculate these values and tests, use the regression command found in the Analysis ToolPak provided with Excel:

1. If the **Regression worksheet page** is not active, activate it.

2. Click **Tools > Data Analysis** and click **Regression** in the Analysis Tools list box, then click **OK**.

   The Regression dialog box opens.

3. Type **C1:C8** in the Input Y Range text box, then type **B1:B8** in the Input X Range text box.

4. Click the **Labels check box** and the **Confidence Level check box**, then verify that the Confidence Level text box contains **95**.

5. Click the **New Worksheet Ply option button** and type **Regression Stats** in the corresponding text box.

6. Click the **Residuals, Standardized Residuals, Residual Plots**, and **Line Fit Plots check boxes**.

   The Regression dialog box should look like Figure 8-9.

Figure 8-9
Completed Regression
dialog box

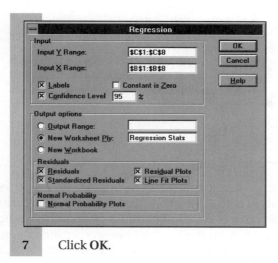

7   Click **OK**.

Statistics from the Analysis ToolPak Regression command appear on the Regression Stats worksheet, as shown in Figure 8-10, which has been formatted with a zoom value of 75% to display the information more clearly (your monitor might require a different zoom value).

Figure 8-10
Worksheet Regression
Stats shown with zoom
set to 75% to show all
features

Regression Statistics ——

ANOVA ——

Parameter Estimates ——

Residual Output ——

plots ——

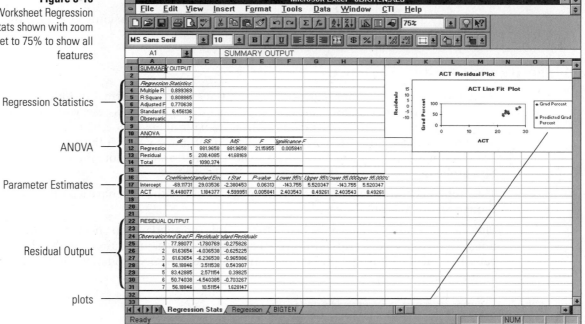

The output is divided into six areas: Regression Statistics, ANOVA, Parameter Estimates, Residual Output, Probability Output (not shown in Figure 8-10), and plots.

....

Creating Regression Statistics

- Activate the worksheet containing the variables you want to use.

- Click Tools > Data Analysis, then click Regression in the Analysis Tools list box.

- Enter the cell range for the dependent variable in the Input Y Range text box and the cell range for the independent variable in the Input X Range text box.

- Choose output options to calculate any combination of residuals, standardized residuals, residual plots, line fit plots, and normal probability plots, then click OK.

....

Regression Statistics

The Regression Statistics section of the output (cells A3:B8) shows summary statistics of the regression. See Figure 8-11, which is formatted for better visibility.

Figure 8-11

Regression Statistics in
cells A3:B8

3	*Regression Statistics*	
4	Multiple R	0.899
5	R Square	0.809
6	Adjusted R Square	0.771
7	Standard Error	6.456
8	Observations	7

You've seen some of these statistics before. The Multiple R value, 0.899, is the absolute value of the Pearson correlation between the observed y (Grad Percent) values and the predicted y values. Except that this value is always positive, for a simple linear regression (involving only one predictor variable) it is equivalent to the Pearson correlation between y and x, in this case 0.899. The Adjusted R Square value is useful for multiple regression and will be covered in Chapter 9.

The Standard Error, 6.456, measures the size of a typical deviation of an observed value (x, y) from the regression line. Think of the standard error as a way of averaging the size of the deviations from the regression line. The typical deviation of an observed point from the regression line in this example is about 6.5.

The Observations value is the size of the sample used in the regression. In this case the regression is based on the values from seven universities.

Verifying Linearity and Constant Variance

Remember to verify the four regression assumptions (linearity, normality, independence, and constant variance) before you accept regression output. In this case, you can get a rough idea of the validity of the assumptions by looking at Figure 8-7. The graph shows that a line fits the data fairly well. You cannot tell from the graph whether the errors are normal, but you can tell that none of the errors is really extreme, and therefore there is not likely to be a problem.

The problem of error dependence generally occurs only if the data have a particular time sequence, which is not the case here.

Plotting the residuals against the predictor variable is a good way to check the assumption of linearity. A U-shaped (or upside-down U) pattern to the residuals, as in Figure 8-12, indicates that the relationship between the variables follows a curve and that the straight line assumption is wrong.

Figure 8-12

Residuals showing a curve

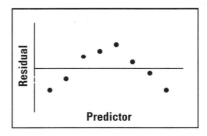

Do the errors have constant variance? If the residuals have a narrow vertical spread for small x and a wide spread for large x, then you might have a problem with nonconstant variance. For example, a plot like Figure 8-13 indicates a problem.

Figure 8-13
Residuals showing
nonconstant variance

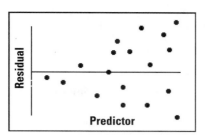

If there are extreme observations (outliers), then they should show up clearly on the plot. Outliers can badly distort the regression results.

The output from the Analysis ToolPak Regression command includes a plot of residuals versus ACT (the predictor variable). The plot, which covers cells J1:O10 on the worksheet, is partially obscured by two other plots generated by Excel.

To work with the ACT Residual Plot and change the horizontal scale:

1 Double-click the **ACT Residual Plot** to activate it.

2 Double-click the **x-axis** to edit the x-axis parameters.

3 Click the **Scale tab** and change the Minimum scale value to **21**, then click **OK**.

Your plot should now look like Figure 8-14.

Figure 8-14
ACT Residual Plot with
x-axis rescaled

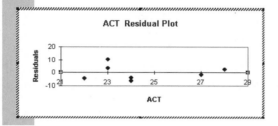

The plot does not show any indication of nonconstant variance or nonlinearity, so there is no problem with the assumptions for constant variance and linearity.

Verifying Normality

You can observe whether the residuals from the regression follow the normal distribution by viewing the normal probability plot (introduced in Chapter 5) of the residuals. If the points of the normal probability plot fall nearly along a straight line you would not reject the hypothesis of normal errors. To create a normal probability plot of the residuals you can use the CTI Statistical Add-Ins.

To create a normal probability plot of the residuals:

1 Click any cell to deactivate the graph, then click **CTI > Normal P-Plot**.

2 Type **C24:C31** in the Input Range text box.

3 Click the **Selection Includes Header Row check box**.

4 Click the **New Sheet option button**, type **Resid Pplot** in the New Sheet text box, then click **OK**.

Excel produces the normal probability plot shown in Figure 8-15.

Figure 8-15
Normal Probability Plot
of the residuals

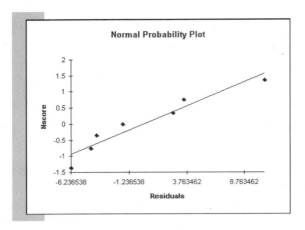

The normal probability plot does not show any great departures from normality, although with only seven observations it would be difficult to detect non-normality. Do not reject the assumption of normality in the distribution of the errors.

Interpreting the Analysis of Variance Table (Optional)

Figure 8-16 show the Analysis of Variance table output from the Analysis ToolPak Regression command.

Figure 8-16
The Analysis of
Variance table in cells
A10:F14

		df	SS	MS	F	Significance F
10	ANOVA					
11		df	SS	MS	F	Significance F
12	Regression	1	881.966	881.97	21.16	0.005840582
13	Residual	5	208.408	41.682		
14	Total	6	1090.37			

Analysis of variance analyzes the variation of graduation percentage by breaking it into two parts: the first due to the regression, and the second due to the residuals, or simple random error.

The values in the *df* column of the table indicate the number of degrees of freedom for each part. The total degrees of freedom are n – 1, or 6 degrees of freedom. One degree of freedom is attributed to the regression. The degrees of freedom for the residual are the difference between the total degrees of freedom and the degrees of freedom attributed to the regression, or 6 – 1 = 5.

The *SS* column gives you the sums of squares. The total sum of squares is the sum of the squared deviations of graduation percentage from its mean. The residual sum of squares is equal to the sum of squared residuals from the regression line. This is the sum that you are trying to minimize in least squares regression. The regression sum of squares is the difference between the total sum of squares and the residual sum of squares. The regression sum of squares (881.97) divided by the total sum of squares (1090.37) equals 0.809—the R^2 value. Note also that the total sum of squares (1090.37) divided by the total degrees of freedom (6) equals 181.79, which is the variance of the y-variable—Grad Percent. The Analysis of Variance table indicates how much of this variance is due to changes in the predictor variable (ACT) and how much is due to random error.

The *MS* (mean square) column shows the ratio of the previous two columns, and the *F* (F-ratio) column shows the ratio of the two mean squares. Note that the mean square error for the residual is equal to the square of the standard error in cell B7 ($6.456^2 = 41.682$). Thus you can use the mean square for the residual to derive the standard error. The significance value returned by the *F*-test equals 0.006, the same probability value given by the Pearson correlation coefficient earlier in this chapter. This is not surprising because the same basic hypothesis is being tested: whether the slope of the regression is zero and whether the correlation is zero. Remember that it means the same for the correlation to be zero as it does for the slope of the regression line to be zero.

Parameter Estimates and Statistics

The file output table created by the Analysis ToolPak Regression command includes the Parameter Estimates Statistics. See Figure 8-17.

Figure 8-17
Parameter Estimates in cells
A16:G18

16		Coefficients	Standard Error	t Stat	P-value	Lower 95%	Upper 95%	Lower 95.000%	Upper 95.000%
17	Intercept	-69.11730769	29.03536302	-2.3805	0.06313	-143.754962	5.520347107	-143.7549625	5.520347107
18	ACT	5.448076923	1.184377161	4.6	0.005841	2.40354348	8.492610366	2.40354348	8.492610366

As you've already seen, the constant coefficient, or Intercept, equals about –69.12, and the slope based on the ACT predictor variables is about 5.448. The standard error for these values is shown in the *Standard Error* column. The *t Stat* column is used for testing the hypothesis that the coefficient is zero, and it is just the ratio of the coefficient to its standard error. If the regression assumptions hold and the population coefficient is 0, then this has the t distribution with $n - 2 = 7 - 2 = 5$ degrees of freedom. The *P-value* column gives the probability of an absolute t exceeding the one observed, assuming that the population value of the coefficient is zero. This p-value for ACT is 0.006, as you would expect, because the ACT parameter measures the slope. The p-value for the intercept is 0.06, so the intercept is not significant at the 5% level; you could not reject a hypothesis stating that the intercept is equal to zero. Finally, the *Lower 95%* and *Upper 95%* columns specify a lower and upper range for the slope parameter. This range is calculated so that you are 95% confident it includes the true slope value.

Note: The confidence intervals appear twice in Figure 8-17. The first pair, a 95% interval, always appears. The second pair always appears, but with the confidence level you specify in the Regression dialog box. In this case, you used the default 95% value, so that interval appears in both pairs.

Conclusions

The assumptions underlying the use of linear regression appear to be satisfied, and you can conclude that there is an increasing relationship between ACT and graduation rate. The graduation rate increases by about 5.5 percentage points when ACT increases by 1 point.

To conclude your work:

1   Click the **Save button** 🖫 to save your file as **8BIGTEN.XLS** (the drive containing your Student Disk should already be selected).

2   Click **File > Close**.

Correlation and Plot Matrices

Correlations and plots are useful when you are trying to assess the relationships between several variables. For each pair of variables it is nice to have the correlation (to measure the strength of the linear relationship) and also to have a plot (to see if the relationship is really linear). Excel can compute the correlations and arrange them in a matrix. You can also use the SPLOM command from the CTI Statistical Add-Ins introduced in Chapter 4 to arrange the corresponding plots in matrix form.

The Correlation Matrix

The Analysis ToolPak includes a command to create a matrix of Pearson correlation coefficients. For this chapter, you'll use the correlation matrix command supplied by the CTI Statistical Add-Ins. The Analysis ToolPak command does not print correlation probabilities, nor does it allow you to calculate Spearman's rank correlation coefficient.

To illustrate the use of a correlation matrix, consider the CALC.XLS workbook. This file contains data collected to see how performance in freshman calculus is related to various predictors (Edge and Friedberg, 1984). The variables include *Calc* (the calculus grade on a scale of 0 to 100 at the end of the first semester), *ACT Math* (the ACT mathematics score), *Alg Place* (an algebra placement test administered in the first week of classes), *Alg2 Grade* (grade in second-year high school algebra), *Calc HS* (1 for those who had calculus in high school and 0 for those who did not), *HS Rank* (high school rank), and *Gender* (0 for women and 1 for men). Try creating a correlation matrix that shows the correlations for all the variables in the file.

To create a matrix of correlation coefficients for the variables in the file:

1   Open **CALC.XLS** (be sure you select the drive and directory containing your Student files).

2   Click **File > Save As**, select the drive containing your Student Disk, then save your workbook on your Student Disk as **8CALC.XLS**.

3   Click **CTI > Correlation Matrix**.

4   Type **A1:E81,G1:H81** in the Input Range text box.

5   Verify that the **Selection Includes Header Row**, **Pearson Correlation**, and **Show P-values check boxes** are selected. Deselect the **Spearman Rank Correlation check box**, as it will not be discussed in this section and takes longer to create.

6   Click the **New Sheet option button**, then type **Corr Matrix** in the New Sheet text box. The Create Matrix of Correlations dialog box should look like Figure 8-18.

Figure 8-18
Completed Create
Matrix of Correlations
dialog box

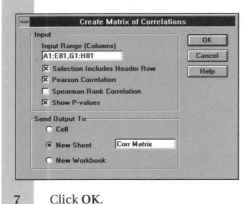

7   Click **OK**.

Creating a Correlation Matrix (Analysis ToolPak)

- Click Tools > Data Analysis and then click Correlation in the Analysis Tools list box.

- Enter the range containing the data in the Input Range text box, and indicate whether the data are grouped by rows or columns.

Creating a Correlation Matrix (CTI Statistical Add-Ins)

The CTI Statistical Add-Ins give you additional output options and automatically update the correlation values if the source data change.

- Click CTI > Correlation Matrix.

- Enter the range containing the data in the Input Range text box.

- Click whether you want to view the Pearson correlations, Spearman rank correlations, or *p*-values.

Correlation of Calc with Others

A matrix of Pearson correlations and associated probabilities appears on the worksheet Corr Matrix. See Figure 8-19.

Figure 8-19
Correlation output worksheet Corr Matrix

correlation matrix —

Pearson Correlations

	Calc HS	ACT Math	Alg Place	Alg2 Grad	HS Rank	Gender Co	Calc
Calc HS	1.000	0.161	0.102	-0.091	0.063	0.014	0.318
ACT Math		1.000	0.427	-0.019	0.443	0.126	0.353
Alg Place			1.000	0.312	0.303	-0.103	0.491
Alg2 Grade				1.000	0.437	-0.446	0.259
HS Rank					1.000	-0.319	0.324
Gender Code						1.000	-0.021
Calc							1.000

matrix of probabilities —

Pearson Probabilities

	Calc HS	ACT Math	Alg Place	Alg2 Grad	HS Rank	Gender Co	Calc
Calc HS	0.000	0.154	0.369	0.421	0.578	0.899	0.004
ACT Math		0.000	0.000	0.865	0.000	0.264	0.001
Alg Place			0.000	0.005	0.006	0.362	0.000
Alg2 Grade				0.000	0.000	0.000	0.020
HS Rank					0.000	0.004	0.003
Gender Code						0.000	0.854
Calc							0.000

Figure 8-19 shows two matrices. The first, in cells A1:H9, is the correlation matrix, which shows the Pearson correlations. The second, the matrix of probabilities in cells A11:H19, gives the corresponding p-values.

The most interesting numbers here are the correlations with the calculus score, because the object of the study was to predict this score. The highest correlation for Calc (column H) appears in cell H5 (0.491), with Alg Place, the placement test score. Next are ACT Math, HS Rank, and Calc HS, which are not impressive predictors, when you consider that the squared correlation gives R^2, the percentage of variance explained by the variable as a regression predictor.

For example, the correlation with HS Rank is 0.324; the square of this is 0.105, so regression on HS Rank would account for only 10.5% of the variation in the calculus score. Another way of saying this is that using HS Rank as a predictor improves by 10.5% the sum of squared errors, as compared with using just the mean calculus score as a predictor. Note that the p-value for this correlation is 0.003 (cell H17), which is less than 0.05, so the correlation is significant at the 5% significance level.

Correlation with a Two-Valued Variable

You might reasonably wonder about using Calc HS here. After all, it assumes only the two values 0 and 1. Does the correlation between Calc and Calc HS make sense? The regression assumptions are also the assumptions for correlations. They require that the errors in the dependent variable be normal. However, there are no such demands on the predictor variable except that the errors be independent normal, with zero population mean and the same population variance, for each of the two values of Calc HS.

You might recognize these as the same assumptions required for the usual two-sample t-test to compare those who did and did not have calculus in high school. The correlation is in fact equivalent, in the sense that the p-value here is the same as from a two-sample t-test. This should not be surprising when you consider that both tests try to measure the relationship of Calc to Calc HS, under the same assumptions.

Bonferroni

The second matrix in Figure 8-19 gives the p-values for the correlations. Except for Gender, all of the correlations with Calc are significant at the 5% level, because all the p-values are less than 0.05.

Some statisticians believe that the *p*-values should be adjusted for the number of tests, because conducting several hypothesis tests raises above 5% the probability of rejecting at least one true null hypothesis. The **Bonferroni** approach to this problem is to multiply the *p*-value in each test by the total number of tests conducted. With this approach, the probability is less than 5% of rejecting one or more of the true hypotheses.

Let's apply this approach to column H of the correlation matrix. Because there are six correlations with Calc, the Bonferroni approach would have us multiply each *p*-value by 6 (equivalent to decreasing the *p*-value required for statistical significance to .05/6 = .0083). Alg2 Grade has a *p*-value of 0.020, and because 6 * (0.020) = 0.120, the correlation is no longer significant from this point of view. Instead of focusing on the individual correlation tests, the Bonferroni approach rolls all of the tests into one big package, with 0.05 referring to the whole package.

Bonferroni makes it much harder to achieve significance, and many researchers are reluctant to use it because it is so conservative. In any case, it should be stressed that this is a controversial area, and professional statisticians argue about it.

Correlations Among Predictors

Some of the correlations among the predictors are interesting. The Alg2 Grade (the grade in second-year high school algebra) has only a weak relationship with ACT Math (the ACT mathematics score), which is surprising. But notice that there is a strong negative relationship with Gender, which means that women had much higher Algebra II grades. The negative correlation favors women, because women were coded 0 and men were coded 1. You can tell that women got higher grades in high school in general because of the negative correlation between Gender and HS Rank. The values of HS Rank run from 1 to 99, with 99 representing the highest grades and 1 representing the lowest. The correlation of Gender with Calc is very small, however, so men and women had nearly the same average in college calculus.

Does Correlation with Calc HS Imply Causation?

The relationships with Calc HS (whether or not calculus was taken in high school) are fairly weak. The correlation between Calc HS and Alg2 Grade is actually negative, but not significant, with a *p*-value of 0.421. The correlations of Calc HS with ACT Math and Alg Place are not significant either. The correlation of 0.318 with Calc is stronger, with a *p*-value of 0.004, but this should not necessarily be interpreted as causal.

Just because the taking of high school calculus and the subsequent college calculus score are related, you cannot conclude that taking calculus in high school causes a better grade in college. The stronger math students tend to take calculus in high school, and these students also do well in college. Only if a fair assignment of students to classes could be guaranteed (so the students in high school calculus are no better or worse than others) could the correlation be interpreted in terms of causation.

Graphing Relationships

The Pearson correlation measures the extent of the linear relationship between two variables. To see if the relationship is really linear between the variables, you should look at the scatterplot. Recall in Chapter 4, you used the CTI Statistical Add-Ins to form a whole matrix of plots called a scatterplot matrix, or SPLOM.

Scatterplot Matrix

To produce a SPLOM using the CALC.XLS data, omit the two-valued predictors Calc HS and Gender, partly because SPLOMs are not as useful for such variables and partly because it becomes increasingly difficult to see the individual plots when there are too many of them.

To create this SPLOM:

1 Click the **Calculus Data sheet tab**.

2 Click **CTI > Scatterplot Matrix**.

3 Type **B1:E81,H1:H81** in the Input Columns text box, then verify that the **Selection Includes Header Row check box** is selected.

4 Click the **New Sheet option button**, then type **SPLOM** in the New Sheet text box. The Create Scatter Plot Matrix dialog box should look like Figure 8-20.

Figure 8-20
Completed Create Scatter
Plot Matrix dialog box

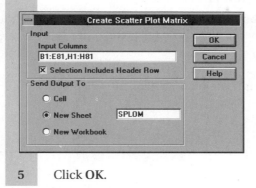

5 Click **OK**.

Looking at SPLOMs on the Screen

Depending on the number of variables you are plotting, SPLOMs can be difficult to view on the screen. If you can't see the entire SPLOM on your screen, consider reducing the value in the Zoom Control box. You can also reduce the SPLOM by selecting it and dragging one of the resizing handles to make it smaller. Figure 8-21 shows it reduced to 90%.

Figure 8-21
SPLOM of Calculus
variables, reduced to
90%

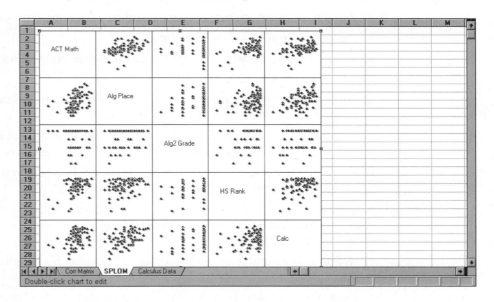

How should you interpret the SPLOM? Each of the five variables is plotted against the other four variables, with the four plots displayed in a row. For example, Calc is plotted as the y-variable against the other four variables in the last row of the SPLOM. ACT Math is the y-variable in the first row. The first plot in the second row is Alg Place vs. ACT Math, the second plot is Alg Place vs. Alg2 Grade, and so on.

Carefully consider the plots in the bottom row, which show Calc against the predictors. Each plot shows a roughly linear upward trend. It would be reasonable to conclude here that correlation and linear regression are appropriate when predicting Calc from ACT Math, Alg2 Grade, Alg Place, and HS Rank.

Recall from Figure 8-19 that Alg Place had the highest correlation with Calc. How is that evident here? A good predictor has good accuracy, which means that the range of y is small for each x. Of the four plots in the bottom row, the plot of Calc against Alg Place has the narrowest range of y values for each x. On the other hand, Alg Place is the best of a weak lot. None of the plots shows that really accurate prediction is possible. None of these plots shows a relationship anywhere near as strong as the relationship between Grad Percent and ACT shown in Figure 8-7.

Now, save your work and exit Excel:

1   Click the **Save button** 💾 to save your file as **8CALC.XLS** (the drive containing your Student Disk should already be selected).

2   Click **File > Exit**.

You will work with multiple predictor variables in Chapter 9.

E X E R C I S E S

1. Open the WHEAT.XLS workbook.

   a. Plot Calories vs. Serving oz (do not include the linear regression line on the plot).

   b. Compute the correlation and *p*-value on the worksheet (you will have to use the CTI Statistical Add-Ins to compute the *p*-value).

   c. Calculate the coefficient of determination without calculating the regression line.

2. Continue using WHEAT.XLS.

   a. Regress Calories on Serving oz.

   b. Plot the residuals against Serving oz, and create a normal probability plot.

   c. Do the assumptions seem to be satisfied?

   d. What are the slope and intercept values for the resulting regression line?

3. Continue using WHEAT.XLS.

   a. Form scatterplot and correlation matrices (Pearson correlation only) for the variables Serving oz, Calories, Protein, Carbo, and Fat.

   b. Why is Fat so weakly related to the other variables? Given that Fat is supposed to be very important in Calories, why is the correlation so weak here?

   c. Would the relationship between Fat and Calories be stronger if we used foods that have a higher fat content?

4. In Exercise 1, the breads in the WHEAT.XLS workbook were low in calories because of high moisture content. One way to avoid this problem is to use a new variable that sums the nutrient weights.

   a. Define a new variable called Total that is the total of the weights of Carbo, Protein, and Fat.

   b. Plot Calories against Total, with the linear regression trend line included, and compute the correlation.

   c. Compute the linear regression of Calories on Total, including the file with residuals, and describe the success in prediction.

   d. Use the CTI Statistical Add-Ins to add labels to the points that indicate fat content. Do the foods with high fat percentage tend to have positive residuals?

   e. Would Total be a better predictor if it accounted for the higher calorie content of fats?

   f. Save your workbook as E8WHEAT.XLS.

5. Open the MUSTANG.XLS workbook, which includes prices and ages of used Mustangs from the *San Francisco Chronicle*, November 25, 1990.

   a. Compute the correlation between Price and Age.

   b. Plot Price against Age, and explain why the correlation is so low.

   c. If you concentrate on just the cars that are less than 10 years old, does the correlation improve?

   d. Excluding the old classic cars, perform a regression of Price against Age and find the drop in Price per year of Age.

   e. Save your workbook as E8MUSTNG.XLS.

6. (Optional) Using the following analysis of variance table for the regression of variable y on variable x, calculate these values: number of observations, variance of y, coefficient of determination, the absolute value of the correlation of y and x, p-value of the correlation of y and x, and standard error (typical deviation of an observed point from the regression line).

ANOVA	df	SS	MS	F	Significance F
Regression	1	129.6429	129.6429	4.918699	0.057
Residual	8	210.8571	26.35714		
Total	9	340.5			

7. Open the CALC.XLS workbook.

   a. Regress Calc on Alg Place and obtain a 95% confidence interval for the slope.

   b. Interpret the slope in terms of the increase in final grade when the placement score increases by one point.

   c. Do the residuals give you any cause for concern about the validity of the model?

   d. Save your workbook as E8CALC.XLS.

8. Open the BOOTH.XLS workbook, which gives total assets and net income for 45 of the largest American banks in 1973.

   a. Plot net income against total assets and notice that the points tend to bunch up toward the lower left, with just a few big banks dominating the upper part of the graph.

   b. Add a linear trend line, and then regress net income against total assets and plot the standard residuals against the predictor values. (The standard residuals appear with the regression output when you select the Standardized Residuals check box in the Regression dialog box.)

   c. Given that the residuals tend to be bigger for the big banks, you should be concerned about the assumption of constant variance. Try taking logs of both variables. Now repeat the plot of one against the other, repeat the regression, and again look at the plot of the residuals against the predicted values. Does the transformation help the relationship? Is there now less reason to be concerned about the assumptions? Notice that there are some banks with strongly positive residuals, indicating good performance, and some banks with strongly negative residuals, indicating below par performance. Indeed, bank #20, Franklin National Bank, has the second most negative residual and failed the following year. Booth (1985) suggests that regression is a good way to locate problem banks before it is too late.

   d. Save your workbook as E8BOOTH.XLS.

9. Open the ALUM.XLS workbook, which contains mass and volume measurements on eight chunks of aluminum from a high school chemistry class.

   a. Plot mass against volume, and notice the outlier.

b. After excluding the outlier, regress mass on volume, without the constant term (select the Constant is Zero check box in the Regression dialog box), because the mass should be 0 when the volume is 0. The slope of the regression line is an estimate of the density (not a statistical word here, but a measure of how dense the metal is) of aluminum.

c. Give a 95% confidence interval for the true density. Does your interval include the accepted true value, which is 2.699?

d. Save your workbook as E8ALUM.XLS.

10. Just to show how one or two values can influence correlations, consider the STATE.XLS data. The data are from the 1986 Metropolitan Area Data Book of the U.S. Census Bureau, and represent 1980 death rates per 100,000 people, for each of the 50 states. Cardio is the cardiovascular death rate, and Pulmon is the pulmonary (lung) death rate.

a. Compute the Pearson and Spearman correlation between them, and also create the corresponding scatterplot using the state variable to label the points (you'll have to use the CTI Statistical Add-Ins to label the points). Which states are outliers on the lower left of the plot? How does the Spearman rank correlation differ from the Pearson correlation?

b. Recompute the correlation without the most extreme point. Recompute again without the next most extreme point. How are the size and significance of the correlations influenced by these deletions? Make a case for the deletions based on the plot and some geography. Why should Alaska be low on both death rates? Does the original correlation give an exaggerated notion of the relation between the two variables? Does the nonparametric correlation coefficient solve the problem? Explain. Would you say that a correlation without a plot can be deceiving?

c. Save your workbook as E8STATE.XLS.

11. Open the FIDELITY.XLS workbook, which contains figures from 1989, 1990, and 1991 for 35 Fidelity sector funds. The source is the *Morningside Mutual Fund Sourcebook 1992, Equity Mutual Funds*. The name of the fund is given in Sector. The variables include the net asset value (NAV) at the end of each of the three years, NAV1989, NAV1990, and NAV1991. Also given is the percentage total return (increase in NAV plus dividends) during the year 1990, TOTL8990, and for the year 1991, TOTL9091.

a. Compute a new variable, NAV8990, as the percentage increase in net asset value during 1990. This is the difference for the year, divided by the value at the beginning of the year, and multiplied by 100.

b. Compute the percentage change NAV9091 for the year 1991, too. Also compute the difference INC8990 = TOTL8990 − NAV8900, the income (dividends) for the year 1990, and similarly INC9091.

c. Find the correlation between the percentage changes in NAV for 1990 and 1991. Find the Pearson correlation of performance in these two years for total return and income. Make the corresponding plots, too, using Sector for labels. You should get a strong correlation for income, but not for NAV. How do you explain this?

12. Continue using the FIDELITY.XLS workbook. The Biotechnology Fund stands out in the plot of NAV9091 vs. NAV8990. It was the only fund that performed well in both years.

a. See what happens to the Pearson correlation if this fund is excluded (copy the data to a second worksheet and delete that row in the new worksheet). The correlation between NAV8990 and NAV9091 should actually be slightly negative with this point excluded. In other words, without this one fund, there is no positive relationship at all in net asset value.

b. Calculate the Pearson p-value for both sets of data.

c. Repeat the analysis with the Spearman rank correlation.

d. If the correlation is this weak, what does it suggest about using performance in one year as a guide to performance in the following year?

e. Save your workbook as E8FIDEL.XLS.

MULTIPLE REGRESSION

O B J E C T I V E S

In this chapter you will learn to:

- Use the F distribution

- Fit a multiple regression equation and interpret the results

- Use plots to aid in the understanding of a regression relationship

- Validate a regression using residual diagnostics

Regression Models

In Chapter 8 you used simple linear regression to predict a dependent variable (y) from a single independent variable (x, a predictor variable). In multiple regression, you predict a dependent variable from several independent variables. For three predictors, x_1, x_2, and x_3, the multiple regression model takes the form:

$$y = \beta_0 + \beta_1 x_1 + \beta_2 x_2 + \beta_3 x_3 + \varepsilon$$

where the coefficients β_0, β_1, β_2, and β_3 are unknown population values that you can estimate and ε is random error, which follows a normal distribution with mean 0 and variance σ^2. Note that the predictors can be functions of variables. The following are also examples of models whose parameters you can estimate with multiple regression:

$$y = \beta_0 + \beta_1 x + \beta_2 x^2 + \beta_3 x^3 + \varepsilon$$
$$y = \beta_0 + \beta_1 x_1 + \beta_2 \sin(x_2) + \beta_3 \cos(x_3) + \varepsilon$$
$$y = \beta_0 + \beta_1 \log(x_1) + \beta_2 \log(x_2) + \varepsilon$$

After computing estimated values for the β coefficients, you can plug them into the equation to get predicted values for y. The estimated regression model is expressed as:

$$y = b_0 + b_1 x_1 + b_2 x_2 + b_3 x_3$$

with b's used instead of β's to distinguish the estimated model from the unknown true model.

The estimated values of the β_i's are chosen so as to reduce the difference between the observed value of y and the predicted value from the regression model. Recall from Chapter 8 that the differences between the actual y values and the predicted y values are called residuals. Linear regression finds least-squares estimates for the coefficients, which means that the coefficients are chosen to minimize the sum of the squared residuals. Residuals are estimates for the true (ε) errors. In determining whether the predictor variables are helpful in predicting the dependent variable, you use the F distribution.

The F Distribution

The F distribution is basic to regression and analysis of variance as studied in this chapter and the next. An example of the F distribution is shown in the instructional template FPDF.XLT.

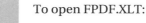

To open FPDF.XLT:

1   Open **FPDF.XLT** (be sure you select the drive and directory containing your Student files).

The template opens to a chart showing the probability density function of the F distribution. See Figure 9-1.

The F distribution has two degrees-of-freedom parameters: the numerator and denominator degrees of freedom. The distribution is usually referred to as $F(m,n)$, that is, the F distribution with m numerator degrees of freedom and n denominator degrees of freedom. The FPDF instructional template opens with an $F(4,9)$ distribution.

Like the chi-square distribution, the F distribution is skewed. To help you better understand the shape of the F distribution, the FPDF template lets you vary the degrees of freedom of the numerator and the denominator by clicking the degrees of freedom scroll arrows.

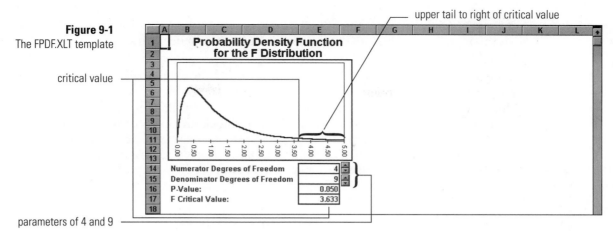

Figure 9-1
The FPDF.XLT template

critical value

upper tail to right of critical value

parameters of 4 and 9

Experiment with the template to view how the distribution of the F changes as you increase the degrees of freedom.

To increase the degrees of freedom in the numerator and denominator:

1   Click the **up spin arrow** in row 14 to increase the numerator degrees of freedom to **5**.

2   Click the **up spin arrow** in row 15 to increase the denominator degrees of freedom to **10**. Then watch how the distribution changes.

In this book, hypothesis tests based on the F distribution always use the area under the upper tail of the distribution to determine the p-value. This template allows you to specify a particular p-value of interest in cell E16. The corresponding critical value appears in cell E17, and a vertical line in the chart indicates the location of the critical value.

To change the p-value:

1   Click cell **E16**, type **0.10**, then press [**Enter**]. This gives you the location of the critical value for the test at the 10% significance level.

Notice that the critical value shifts to the left, telling you that values of the F statistic to the right of this point occur 10% of the time.

Continue working with the F distribution template, trying different parameter values to get a feel for the F distribution.

To close the template:

1   Click **File > Close**.

2   Click **No**, because you don't need to save your work with the template.

Regression Assumptions

When applying the F distribution to results from a multiple regression, you make the same assumptions as in Chapter 8, except that now the form of the model is not necessarily a straight line. Although the other assumptions are the same, they are important enough to repeat here:

- The form of the model is correct (that is, it's appropriate to use a linear combination of the predictor variables to predict the value of the response variable).

- The errors ε are normally distributed with the distribution centered at 0.

- The errors ε are independent.

- The errors ε have a constant variance σ^2.

Any statement in this chapter about hypothesis tests or confidence intervals—or just a statement about the *t* distribution or the *F* distribution—requires that these assumptions be met. The assumptions usually are not satisfied perfectly in practice, but you can get by quite well if they are nearly satisfied.

Simulations have shown that it is sufficient for the model to have approximately the correct form, but there is a problem if plots show a curved relationship between the response variable and the predictors when the model assumes a straight-line relationship. Because three of the assumptions involve the errors, and the residuals are the estimated errors, it is important to look carefully at the residuals. The errors do not have to be perfectly normal, especially if there are a lot of data points, but you should worry if there are extreme values. One stray extreme value (an outlier) can have a major impact on the regression—it can throw everything off.

Using Regression for Prediction

One of the goals of regression is prediction. For example, you could use regression to predict what grade a student will get in a college calculus course (this is the dependent variable, the one being predicted). The predictors (the independent variables) might be ACT or SAT math score, high school rank, and a placement test score from the first week of class. Students with low predictions might be asked to take a lower-level class.

On the other hand, suppose the dependent variable is the price of a four-unit apartment building, and the independent variables are the square footage, the age of the building, the total current rent, and a measure of the condition of the building. Here you might use the predictions to find a building that is undervalued, with a price that is much less than its prediction. This analysis was actually carried out by some students, who found that there was a bargain building available. The owner needed to sell quickly as a result of "cash flow problems."

You can use multiple regression to see how several variables combine to predict the dependent variable. How much of the variability in the dependent variable is accounted for by the predictors? Do the combined independent variables do better or worse than you might expect, based on their individual correlations with the dependent variable? You might be interested in the individual coefficients, and whether or not they seem to matter in the prediction equation. Could you eliminate some of the predictors without losing much prediction ability?

When you use regression in this way, the individual coefficients are important. Rosner and Woods (1988) compiled statistics from baseball box scores, and they regressed runs on singles, doubles, triples, home runs, and walks (walks are combined with hit-by-pitched-ball). Their estimated prediction equation is:

$$\text{runs} = -2.49 + 0.47\ \text{singles} + 0.76\ \text{doubles} + 1.14\ \text{triples} + 1.54\ \text{home runs} + 0.39\ \text{walks}$$

Notice that walks have a coefficient of 0.39, and singles have a coefficient of 0.47, so a walk has more than 80% of the weight of a single. This is in contrast with the popular slugging percentage used to measure the offensive production of players, which gives 0 weight to walks, weight 1 to singles, 2 to doubles, 3 to triples, and 4 to home runs. The Rosner-Woods equation gives relatively more weight to singles, and the weight for doubles is less than twice as much as the weight for singles. Similar comparisons are true for triples and home runs. Do baseball general managers use equations like the Rosner-Woods equation to evaluate ball players? If not, why not?

You can also use regression to see if a particular group is being discriminated against. A company might ask whether women are paid less than men with comparable jobs. You can include a term in a regression to account for the effect of gender. Alternatively, you can fit a regression model for just men, apply the model to women, and see if women have salaries that are less than would be predicted for men with comparable positions. It is now common for such arguments to be offered as evidence in court, and many statisticians have experience in legal proceedings.

Regression Example: Predicting Grades

For a detailed example of a multiple regression, consider the CALC.XLS data from Chapter 8, which were collected to see how performance in freshman calculus is related to various predictors (Edge and Friedberg, 1984).

To perform a multiple regression with the CALC.XLS data, first open the workbook:

1 Open **CALC.XLS** (be sure you select the drive and directory containing your Student files).

2 Click **File > Save As**, select the drive containing your Student Disk, then save your workbook on your Student Disk as **9CALC.XLS**.

Chapter 8 included correlations and a scatterplot matrix for the data. Table 9-1 shows the variables:

Table 9-1
CALC.XLS variables

Range Name	Range	Description
Calc HS	A1:A81	1 for those who had calculus in high school and 0 for those who did not
ACT Math	B1:B81	The ACT mathematics score
Alg Place	C1:C81	An algebra placement test administered in the first week of classes
Alg2 Grade	D1:D81	Grade in second-year high school algebra
HS Rank	E1:E81	High school rank
Gender	F1:F81	F for female and M for male
Gender Code	G1:G81	0 for female and 1 for male
Calc	H1:H81	The calculus grade on a scale of 0 to 100 at the end of the first semester

In Chapter 8, it appeared from the correlation matrix and scatterplot matrix that the algebra placement test is the best individual predictor of the first-semester calculus score (although it is not very successful). Multiple regression gives a measure of how good the predictors are when used together. The model is:

$$\text{Calculus Score} = \beta_0 + \beta_1(\text{Calc HS}) + \beta_2(\text{ACT Math}) + \beta_3(\text{Alg Place})$$

$$+ \beta_4(\text{Alg2 Grade}) + \beta_5(\text{HS Rank}) + \beta_6(\text{Gender Code}) + \varepsilon$$

You can use the Analysis ToolPak Regression command to perform a multiple regression on the data, but the predictor variables must occupy a contiguous range. You will be using columns A, B, C, D, E, and G as your predictor variables, so you need to move column G, Gender Code, next to columns A:E.

To move column G next to columns A:E:

1 Click the **G column header** to select the entire column.

2 Right-click **the selection** to open the shortcut menu, then click **Cut**.

3 Click the *F* **column header**.

4 Right-click **the selection** to open the shortcut menu, then click **Insert Cut Cells**.

You can now identify the contiguous range of columns A:F as your predictor variables.

To perform a multiple regression on the calculus score based on the predictor variables Calc HS, ACT Math, Alg Place, Alg2 Grade, HS Rank, and Gender Code, use the Regression command found in the Analysis ToolPak provided with Excel.

To perform the multiple regression:

1. Click **Tools > Data Analysis**, click **Regression** in the Analysis Tools list box, then click **OK**.

2. Type **H1:H81** in the Input Y Range text box, press [**Tab**], then type **A1:F81** in the Input X Range text box.

3. Click the **Labels check box** and the **Confidence Level check box** to select them, then verify that the Confidence Level box contains **95**.

4. Click the **New Worksheet Ply option button**, click the corresponding text box, then type **Mult Reg**.

5. Click the **Residuals, Standardized Residuals, Residual Plots**, and **Line Fit Plots check boxes** to select them.

The Regression dialog box should look like Figure 9-2.

Figure 9-2
Completed Regression dialog box

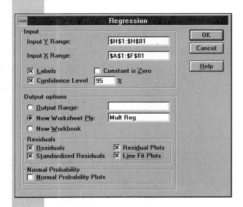

6. Click **OK**.

Excel creates a new sheet, Mult Reg, which contains the summary output and the residual plots.

Interpreting the Regression Output

To interpret the output, look first at the analysis of variance (ANOVA) table found in cells A10:F14. Figure 9-3 shows this range with the columns widened to display the labels. The analysis of variance table shows you whether the fitted regression model is significant.

The analysis of variance table helps you choose between two hypotheses:

H_0: The coefficients of all six predictor variables = 0.
H_a: At least one of the six coefficients $\neq$ 0.

Figure 9-3
ANOVA table from multiple regression, cells A10:F14

10	ANOVA					
11		df	SS	MS	F	Significance F
12	Regression	6	3840.164	640.03	7.19682	4.69578E-06
13	Residual	73	6492.036	88.932		
14	Total	79	10332.2			

There are many different parts to an ANOVA table, some of which you might have considered already in the optional section of the previous chapter. At this point you should just concentrate on the F ratio and its p-value, which tell you whether the regression is significant. This ratio is large when the predictor variables explain much of the variability of the response variable, and hence has a small p-value as measured by the F distribution. A small value for this ratio indicates that much of the variability in y is due to random error (as estimated by the residuals of the model) and is not due to the regression. The next chapter on analysis of variance contains a more detailed description of the ANOVA table.

The F ratio, 7.197, is located in cell E12. Under the null hypothesis you assume that there is no relationship between the six predictors and the calculus score. If the null hypothesis is true, the F ratio in the ANOVA table follows the F distribution, with 6 numerator degrees of freedom and 73 denominator degrees of freedom. You can test the null hypothesis by seeing if this observed F ratio is much larger than you would expect in the F distribution. If you want to get a visual picture of this hypothesis test, use the FPDF instructional template and display the $F(6,73)$ distribution.

The *Significance F* column gives a *p*-value of 4.69×10^{-6} (cell F12), representing the probability that an F ratio with 6 degrees of freedom in the numerator and 73 in the denominator has a value 7.197 or more. This is much less than 0.05, so the regression is significant at the 5% level. You could also say that you reject the null hypothesis at the 5% level and accept the alternative that at least one of the coefficients in the regression is not zero. If the F ratio were not significant, there would not be much interest in looking at the rest of the output.

Multiple Correlation

The regression statistics appear in the range A3:B8, shown in Figure 9-4 (formatted to show column labels).

Figure 9-4
Regression statistics,
cells A3:B8

3	*Regression Statistics*	
4	Multiple R	0.61
5	R Square	0.37
6	Adjusted R Square	0.32
7	Standard Error	9.43
8	Observations	80

The R Square value in cell B5 (0.37) is the coefficient of determination, R^2, discussed in the previous chapter. This value indicates that 37% of the variance in calculus scores can be attributed to the regression. In other words, 37% of the variability in the final calculus score is due to differences among students (as quantified by the values of the predictor variables), and the rest is due to random fluctuation. Although this value might seem low, it is an unfortunate fact that decisions are often made based on weak predictor variables, including decisions about college admissions and scholarships, freshman eligibility in sports, and placement in college classes.

The Multiple R (0.61) in cell B4 is just the square root of the R^2, and it is also known as the **multiple correlation**. It is the correlation between the response variable, calculus score, and the linear combination of the predictor variables as expressed by the regression. If there were only one predictor, this would be the absolute value of the correlation between the predictor and the dependent variable. The Adjusted R Square value in cell B6 (0.32) attempts to adjust the R^2 for the number of predictors. The reason for looking at the adjusted R^2 is that the unadjusted R^2 value either increases or stays the same when you add predictors to the model. If you add enough predictors to the model you can reach some very high R^2 values, but there is not much to be gained by analyzing a data set with 200 observations with a regression model that has 200 predictors, even if the R^2 value is 100%. Adjusting the R^2 compensates for this effect and helps you determine whether adding additional predictors is worthwhile.

An Error Estimate

The Standard Error value, 9.43 (cell B7), is the estimated value of σ, the standard deviation of the error term ε—in other words, the standard deviation of the calculus score once you compensate for differences in the predictor variables. You can also think of the standard error as the typical error for prediction of the 80 calculus scores. Because a span of 10 points corresponds to a difference of one letter grade (A vs. B, B vs. C, etc.), the typical error of prediction is about one letter grade.

Coefficients and the Prediction Equation

At this point you know the model is statistically significant and accounts for about 37% of the variability in calculus scores. What is the regression equation itself and which predictor variables are most important?

You can read the estimated regression model from cells A16:I23, shown in Figure 9-5, where the first column contains labels for the predictor variables.

Figure 9-5
Cells A16:I23 of the
regression output

16		Coefficients	Standard Error	t Stat	P-value	Lower 95%	Upper 95%	Lower 95.000%	Upper 95.000%
17	Intercept	27.9434249	12.43782635	2.2466	0.02769	3.154855925	52.731994	3.154855925	52.73199388
18	Calc HS	7.192302156	2.488207835	2.8906	0.00506	2.23330773	12.151297	2.23330773	12.15129658
19	ACT Math	0.351509377	0.430355761	0.8168	0.41671	-0.506189	1.2092078	-0.506188999	1.209207752
20	Alg Place	0.827044552	0.267521238	3.0915	0.00282	0.293875133	1.360214	0.293875133	1.360213971
21	Alg2 Grade	3.683048871	2.440861323	1.5089	0.13564	-1.18158403	8.5476818	-1.181584029	8.547681771
22	HS Rank	0.11058947	0.116000179	0.9534	0.34356	-0.12059871	0.3417776	-0.12059871	0.34177765
23	Gender Code	2.627113372	2.469150162	1.064	0.29085	-2.29389914	7.5481259	-2.293899142	7.548125885

The *Coefficients* column (B16:B23) gives the estimated coefficients for the model. The corresponding prediction equation is:

Calc = 27.943 + 7.192 (Calc HS) + 0.352 (ACT Math) + 0.827 (Alg Place) + 3.683 (Alg2 Grade) + 0.111 (HS Rank) + 2.627 (Gender Code)

The coefficient for each variable estimates how much the calculus score will change if the variable is increased by 1 and the other variables are held constant. For example, the coefficient 0.352 of ACT Math indicates that the calculus score should increase by 0.352 points if the ACT math score increases by one point and all other variables are held constant.

Some variables, like Calc HS, have a value of either 0 or 1—in this case, to indicate the absence or presence of calculus in high school. The coefficient 7.192 is the estimated effect on the calculus score from taking high school calculus, other things being equal. Because 10 points correspond to one letter grade, the coefficient 7.192 for Calc HS is almost one letter grade.

Using the coefficients of this regression equation you can forecast what a particular student's calculus score might be, given background information on the student. For example, consider a male student who did not take calculus in high school, scored 30 on his ACT Math exam, scored 23 on his algebra placement test, had a 4.0 grade in second-year high school algebra, and was ranked in the 90th percentile in his high school graduation class. You would predict that his calculus score would be:

Calc = 27.943 + 7.192 (0) + 0.352 (30) + 0.827 (23) + 3.683 (4.0) + 0.111 (90) + 2.627 (1)

= 74.87

or about 75 points.

Notice the Gender Code coefficient, 2.627, which shows the effect of gender if the other variables are held constant. Because the males are coded 1 and the females are coded 0, if the regression model is true, a male student will score 2.627 points higher than a female student even when the backgrounds of both students are equivalent (equivalent in terms of the predictor variables in the model).

Whether you can trust that conclusion depends partly upon whether the coefficient for Gender Code is significant. For that you have to determine the precision with which the value of the coefficient has been determined. You can do this by examining the estimated standard deviations of the coefficients, displayed in the *Standard Error* column.

t Tests for the Coefficients

The *t Stat* column shows the ratio between the coefficient and the standard error. If the population coefficient is 0, then this has the *t* distribution with degrees of freedom $n - p - 1 = 80 - 6 - 1 = 73$. Here n is the number of cases (80) and p is the number of predictors (6). The next column, *P-value*, is the corresponding *p*-value, the probability of a *t* this large or larger in absolute value. For example, the *t* for Alg Place is 3.092, so the probability of a *t* this large or larger in absolute value is about 0.003. The coefficient is significant at the 5% level, because this is less than 0.05. In terms of hypothesis testing, you would reject the null hypothesis that the coefficient is 0 at the 5% level and accept an alternative hypothesis. This is a two-tailed test—it rejects the null hypothesis for either large positive or large negative values of *t*—so your alternative hypothesis is that the coefficient is not zero. Notice that only the coefficients for Alg Place and Calc HS are significant. This suggests that you not devote a lot of effort to interpreting the others. In particular, it would not be appropriate to assume from the regression that male students perform better than equally qualified female students.

The range F17:G23 indicates the 95% confidence intervals for each of the coefficients. You are 95% confident that having calculus in high school is associated with an increase in the calculus score of at least 2.23 points and not more than 12.15 points in this particular regression equation.

Is it strange that the ACT math score is nowhere near significant here, even though this test is supposed to be a strong indication of mathematics achievement? Looking back at Figure 8-19, you can see that it has correlation 0.353 with Calc, which is highly significant ($p = 0.001$). Why is it not significant here? The answer involves other variables that contain some of the same information. In using the *t* to test the significance of the ACT Math term, you are testing whether you can get away with deleting this term. If the other predictors can take up the slack and provide most of its information, then the test says that this term is not significant, and therefore not needed in the model. If each of the predictors can be predicted from the others, any single predictor can be eliminated without losing much.

You might think that you could just drop from the model all the terms that are not significant. However, it is important to bear in mind that the individual tests are correlated, so each of them changes when you drop one of the terms. If you drop the least significant term, others might then become significant. A frequently used strategy for reducing the number of predictors involves the following steps:

1. Eliminate the least significant predictor if it is not significant.

2. Refit the model.

3. Repeat steps 1 and 2 until all predictors are significant.

In the exercises, you'll get a chance to rerun this model and eliminate all nonsignificant variables. For now, examine the model and see whether any assumptions have been violated.

Useful Plots

There are a number of useful ways to look at the results produced by multiple linear regression. This section discusses four common plots that can help you assess the success of the regression:

1. Plotting dependent variables against the predicted values shows how well the regression fits the data.

2. Plotting residuals against the predicted values magnifies the vertical spread of the data so that you can assess whether the regression assumptions are justified. A curved pattern to the residuals indicates that the model does not fit the data. If the vertical spread is wider on one side of the plot, then it suggests that the variance is not constant.

3. Plotting residuals against individual predictor variables can sometimes reveal problems that are not clear from a plot of the residuals versus the predicted values.

4. Creating a normal plot of the residuals helps you assess whether the regression assumption of normality is justified.

Dependent vs. Predicted Values

How successful is the regression? To see how well the regression fits the data, plot the actual Calc values against the predicted values stored in B29:B109 (you can scroll down to view the residual output).

To plot the observed calculus scores versus the predicted scores, first place the data on the same worksheet:

1. Select the range **B29:B109** and click the **Copy button** 📋 on the Standard toolbar.

2. Click the **Calculus Data sheet tab**.

3. Select the range **H1:H81**, then click **Insert > Copied Cells** to paste the predicted values into column H.

4. Click the **Shift Cells Right option button** to move the observed calculus scores into column I, then click **OK**.

The predicted calculus scores appear in column H, as shown in Figure 9-6 (formatted to show the column labels).

predicted scores ————————   ———————— observed scores

Figure 9-6
Columns H and I:
predicted calculus scores
and observed scores

	A	B	C	D	E	F	G	H	I	J	K
1	Calc HS	ACT Math	Alg Place	Alg2 Grade	HS Rank	Gender Code	Gender	Predicted Calc	Calc		
2	0	27	21	3.5	68	0	F	75.21286866	62		
3	0	29	16	4	99	0	F	77.05046265	75		
4	1	30	22	4	98	1	M	92.0730654	95		
5	0	34	25	3	90	1	M	84.20016977	78		
6	0	29	22	4	99	0	F	82.01272996	95		
7	1	30	19	4	97	0	F	86.8542289	91		
8	0	29	23	4	79	1	M	83.25509849	72		
9	0	28	15	4	95	0	F	75.42955084	95		
10	0	28	14	4	85	1	M	76.12372496	88		
11	0	31	19	4	82	1	M	80.98170744	97		
12	0	25	12	3	81	1	M	69.28970098	49		
13	0	34	16	3.5	87	1	M	78.26652483	70		
14	0	27	13	4	92	0	F	73.09218395	75		
15	0	28	19	4	89	0	F	78.07419223	78		
16	0	31	25	4	97	0	F	84.97570343	89		
17	0	26	10	3	81	1	M	67.98712125	87		
18	1	24	14	4	91	0	F	79.94641306	79		
19	1	30	18	3	97	1	M	84.97124885	85		
20	0	25	13	2.5	46	1	M	64.40458965	57		
21	0	25	15	3.5	80	1	M	73.5017696	81		
22	0	27	18	3	80	0	F	72.2172842	76		
23	0	27	17	4	89	0	F	76.06859375	88		
24	0	28	21	4	94	0	F	80.28122868	83		
25	1	27	24	3	71	1	M	86.00366182	97		

Mult Reg | Calculus Data

Ready   NUM

Now create a scatterplot of the data in the range H1:I81.

To create the scatterplot of the observed scores versus the predicted scores:

1. Select the range **H1:I81** and click **Insert > Chart > As New Sheet**.

2. Confirm that the range **H1:I81** is entered in the Range text box of the Step 1 dialog box, then click **Next**.

3. Click **XY (Scatter)** as the chart type, then click **Next**.

4. Click **1** (points without lines) as the format, then click **Next**.

5. Confirm that the data series is in **Columns**, the First **1** Column is used for x data, and the First **1** Row is used for legend text, then click **Next**.

6. Click the **No legend option button**, type **Calculus Scores** in the Chart Title text box, **Predicted** in the Category (X) text box, and **Observed** in the Value (Y) text box, then click **Finish**.

The axes of the plot have a lot of white space, so you should rescale the axes to better show the data.

To change the scale of both axes from 0–100 to 40–100:

1　Double-click the **y-axis**.

2　Click the **Scale tab** in the Format Axis dialog box to display scale options.

3　Type **40** in the Minimum text box and **100** in the Maximum text box, then click **OK**.

4　Repeat steps 1 through 3 for the x-axis. The final form of the scatterplot should look like Figure 9-7.

Figure 9-7
Chart sheet showing the scatterplot of observed calculus scores versus predicted calculus scores

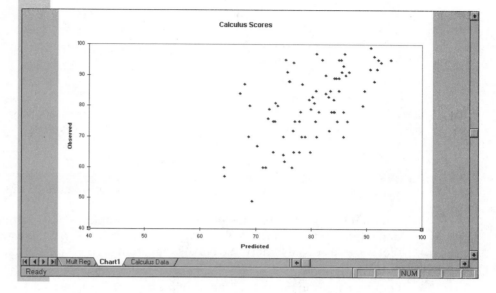

How good is the prediction shown here? Is there a narrow range of observed values for a given predicted value? This plot is a slight improvement on the plot of Calc vs. Alg Place in Figure 8-21. Figure 9-7 should be better, because Alg Place and five other predictors are being used here.

Does it appear that the range of values is narrower for large values of predicted calculus score? If the error variance is lower for students with high predicted values, it would be a violation of the fourth regression assumption, which requires a constant error variance. Consider the students predicted to have a grade of 80 in calculus. These students have actual grades of around 65 to around 95, a wide range. Notice that the variation is lower for students predicted to have a grade of 90. Their actual scores are all in the 80s and 90s. There is a barrier at the top—no score can be above 100—and this limits the possible range. In general, when a barrier limits the range of the dependent variable, it can cause nonconstant error variance. This issue is considered further in the next section.

Plotting Residuals vs. Predicted Values

The plot of the residuals versus the predicted values shows a magnified version of the variation in Figure 9-7, because the residuals are the differences between the actual calculus scores and the predicted values.

To make the plot:

1　Click the **Mult Reg sheet tab** to return to the regression output.

2　Select the range **B29:C109** and click **Insert > Chart > As New Sheet**.

3　Following the Chart Wizard directions as you did for the previous plot, create a scatterplot showing points only with no lines. Use XY (Scatter), format 1, with the same settings as before. Give a chart title of **Residual Plot**, label the x-axis **Predicted** and the y-axis **Residual**.

4   Change the scale of the x-axis from 0–100 to 60–100. Your chart sheet should look like Figure 9-8.

Figure 9-8
Chart sheet showing the
scatterplot of residuals
versus predicted calculus
scores

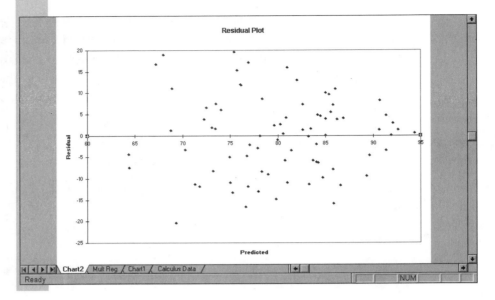

This plot is useful for verifying the regression assumptions. For example, the first assumption requires that the form of the model be correct. Usually, if this assumption is invalid, the plot shows a curved pattern. No curve is apparent here.

If the assumption of constant variance is not satisfied, then it should be apparent in Figure 9-8. Look for a trend in the vertical spread of the data. There appears to be a definite trend toward a narrower spread on the right, and it is cause for concern about the validity of the regression—although regression does have some robustness with respect to the assumption of constant variance.

For data that range from 0 to 100 (such as percentages) the arcsine-square root transformation sometimes helps fix problems with nonconstant variance. The transformation involves creating a new column of transformed calculus scores where:

$$\text{transformed calc score} = \sin^{-1}\left(\sqrt{\text{calculus score}/100}\right)$$

Using Excel, you would enter the formula

    =ASIN(SQRT(x/100))

where x is the value or cell reference of a value that you want to transform.

If you were to apply this transformation here and use the transformed calculus score in the regression in place of the untransformed score, you would find that it helps to make the variance more constant, but the regression results are about the same. Calc HS and Alg Place are still the only significant coefficients, and the R^2 value is almost the same as before. Of course, it is much harder to interpret the coefficients after transformation. Who would understand if you said that each point in the algebra placement score is worth 0.012 points in the arcsine of the square root of the calculus score divided by 100? From this point of view, the transformed regression is useful mainly to validate the original regression. If it is valid and it gives essentially the same results as the original regression, then the original results are valid.

Plotting Residuals vs. Predictor Variables

It is also useful to look at the plot of the residuals against each of the predictor variables, because a curve might show up on only one of those plots or there might be an indication of nonconstant variance. Such plots are created automatically with the Analysis ToolPak Add-Ins.

To view one of these plots:

1 Click the **Mult Reg sheet tab** to return to the regression output.

The plots generated by the add-in start in cell J1 and extend to Z32. Two types of plots are generated: scatterplots of the regression residuals versus each of the regression variables, and the observed and predicted values of the response variable (calculus score) against each of the regression variables. See Figure 9-9 (you might have to scroll up and right to see the charts).

Figure 9-9
Output scatterplots from the Analysis ToolPak

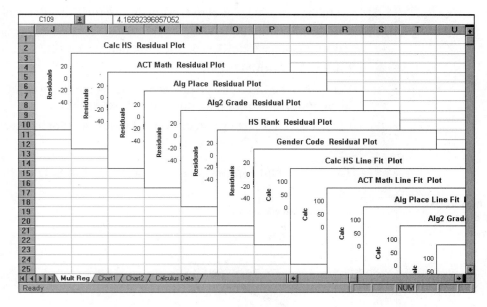

The plots are shown in a cascading format in which the plot title is often the only visible element of a chart. You can view each chart by bringing it to the front of the cascade stack. Try doing this with the plot of the residuals versus Alg Place.

To view the chart:

1 Click the chart **Alg Place Residual Plot** (located in the range L5:Q14).

2 Right-click the selected chart to open the shortcut menu, then click **Bring to Front**.

The scatterplot is now in the front of the plot cascade, as shown in Figure 9-10.

Figure 9-10
Alg Place Residual Plot

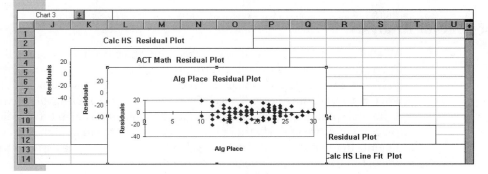

Does the spread of the residual values appear constant for differing values of the algebra placement score? It appears that the spread of the residuals is wider for lower values of Alg Place. This might indicate that you have to transform the data, perhaps using the arcsine transformation just discussed.

Normal Errors and the Normal Plot

What about the assumption of normal errors? Usually, if there is a problem with nonnormal errors, extreme values show up in the plot of residuals versus predicted values. In this example there are no residual values beyond 25 in absolute value, as shown in Figure 9-8. Is this what you would expect with normal data?

How large should the residuals be if the errors are normal? You can decide whether these values are reasonable with a normal probability plot.

To make a normal plot of the residuals, use the CTI Statistical Add-Ins (see Chapter 3 for information on loading the add-ins):

1 Select the range **C29:C109** on the Mult Reg worksheet.

2 Click **CTI > Normal P-Plot**.

3 Verify that the Input Range text box contains the range **C29:C109**, then click the **Selection Includes Header Row check box** to select it.

4 Click the **New Sheet option button**, type **Normal Plot** in the corresponding text box, then click **OK**.

The normal plot appears on a new worksheet, Normal Plot, as shown in Figure 9-11.

Figure 9-11
Normal plot of residuals

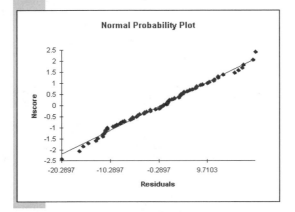

The plot is quite well-behaved. It is fairly straight, and there are not any extreme values (either in the upper-right or lower-left corners) at either end. It appears that there is no problem with the normality assumption.

Summary of Calc Analysis

What main conclusions can you make about the calculus data, now that you have done a regression, examined the regression residual file, and plotted some of the data? With an R^2 of 0.37 and an adjusted R^2 of 0.32, the regression accounts for only about one-third of the variance of the calculus score. This is disappointing, considering all the weight that college scholarships, admissions, placement, and athletics place on the predictors. Only the algebra placement score and whether or not calculus was taken in high school have significant coefficients in the regression. There is a slight problem with the assumption of a constant variance, but that does not affect these conclusions.

To close the workbook and save the changes you've made to 9CALC.XLS:

1 Click **File > Close**.

2 Click **Yes** when prompted to save changes.

Regression Example: Sex Discrimination

Regression can also help you assess whether a particular group is being discriminated against. For example, some of the female faculty at a junior college felt underpaid, and they sought statistical help in proving their case. The college collected data for the variables that influence salary for 37 females and 44 males. The data are stored in the file DISCRIM.XLS. (These data were also discussed in Chapter 4 in the workbook JRCOL.XLS. DISCRIM.XLS orders the data differently so that you can use the Analysis ToolPak Add-Ins more easily).

To open the file:

1 Open **DISCRIM.XLS** (be sure you select the drive and directory containing your Student files).

2 Click **File > Save As**, select the drive containing your Student Disk, then save your workbook on your Student Disk as **9DISCRIM.XLS**.

Table 9-2 shows the variables:

Table 9-2
DISCRIM.XLS variables

Range Name	Range	Description
Gender	A1:A82	F for female and M for male
MS Hired	B1:B82	1 for master's when hired, 0 for no master's when hired
Degree	C1:C82	Current degree: 1 for bachelor's, 2 for master's, 3 for master's plus 30 hours, and 4 for Ph.D
Age Hired	D1:D82	Age when hired
Years	E1:E82	Number of years the faculty member has been at the college
Salary	F1:F82	Current salary when data were collected

In this example, you will use salary as the dependent variable, using four other variables as predictors. One way to see if female faculty have been treated unfairly is to do the regression using just the male data and then apply the regression to the female data. For each female faculty member, this predicts what a male faculty member would make with the same years, age when hired, degree, and master's degree status. The residuals are interesting because they are the difference between what each woman makes and her predicted salary if she were a man. This assumes that all of the relevant predictors are being used, but it would be the college's responsibility to point out all the variables that influence salary in an important way. When there is a union contract, which is the case here, it should be clear which factors influence salary.

Regression on Male Faculty

To do the regression on just the male faculty, and then look at the residuals for the females, use Excel's AutoFilter capability and copy the male columns to a new worksheet.

To create a worksheet of salary information for male faculty only:

1 Click **Data > Filter > AutoFilter**.

2 Click the **Gender drop-down arrow**, then click **M**.

3 Select the range **A1:F82**, then click the **Copy button** on the Standard toolbar.

4 Right-click the **Salary Data sheet tab** to open the shortcut menu, then click **Insert**.

5 Click **Worksheet** in the New list box, then click **OK**.

6 Click the **Paste button** on the Standard toolbar.

7 Double-click the **Sheet1 sheet tab**, type **Male Data** in the Name text box, then click **OK**.

The salary data for male faculty now occupy the range A1:F45 on the Male Data worksheet. Now you are ready to analyze this subset of data.

Using a SPLOM to See Relationships

To get an idea of the relationships among the variables, it is a good idea to compute a correlation matrix and plot the corresponding scatterplot matrix.

To create the SPLOM:

1. Select the range **B1:F45**.

2. Click **CTI > Scatterplot Matrix**.

3. Verify that the range **B1:F45** appears in the Input Columns text box and that the **Selection Includes Header Row check box** is selected.

4. Click the **New Sheet option button**, type **Male SPLOM** in the New Sheet text box, then click **OK**.

   In order to see the entire SPLOM you might have to reduce the zooming factor.

5. Click the **Zoom control box**, type **85** to see the whole scatterplot matrix (you might want to enter a different zoom factor depending upon your monitor's resolution), then press [**Enter**].

The SPLOM appears as shown in Figure 9-12.

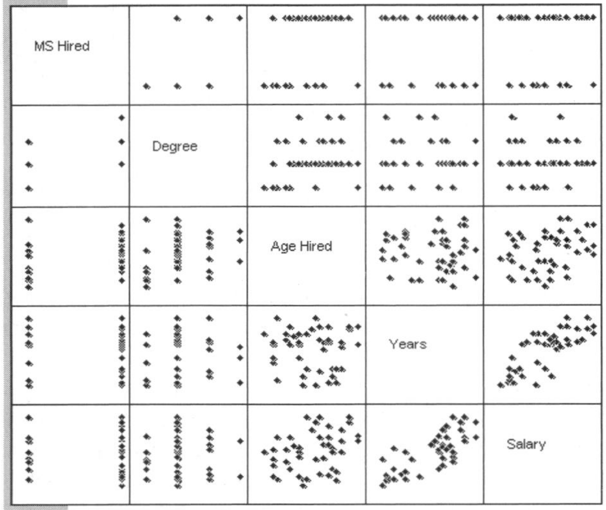

Figure 9-12
SPLOM of variables for male faculty salary data

Recall that you examined a scatterplot matrix of Salary, Years, and Age Hired for all 81 cases in Chapter 4 (Figure 4-26); the plot here shows the same patterns.

Years employed is a good predictor, because the range of salary is fairly narrow for each value of years employed (although the relationship is not perfectly linear). Age at which the employee was hired is not a very good predictor, because there is a wide range of salary values for each value of age hired. There is not a significant relationship between the two predictors, years employed and age hired. What about the other two predictors? Looking at the plots of salary against degree and MS hired, it is clear that neither of them is closely related to salary. The people with higher degrees do not seem to be making higher salaries. Those with a master's degree do not seem to be making much more, either. Therefore the correlations of degree and MS hired with salary should be low.

You might have some misgivings about using degree as a predictor: After all, it is only an ordinal variable. There is a natural order to the four levels, but it is arbitrary to assign the values 1, 2, 3, and 4. This says that the spacing from bachelor's to master's (1 to 2) is the same as the spacing from master's plus 30 hours to Ph.D (3 to 4). You could instead assign the values 1, 2, 3, and 5, which would mean greater space from master's plus 30 hours to the Ph.D. In spite of this arbitrary assignment, ordinal variables are frequently used as regression predictors. Usually, it does not make a significant difference whether the numbers are

1, 2, 3, and 4 or 1, 2, 3, and 5. In the present situation, you can see from Figure 9-12 that salaries are about the same in all four degree categories, which implies that the correlation of salary and degree is close to 0. This is true no matter what spacing is used.

Correlation Matrix

The SPLOM shows the relationships between salary and the other variables. To quantify this relationship create a correlation matrix of the variables.

To form the correlation matrix:

1   Click the **Male Data sheet tab**; the range B1:F45 should still be selected.

2   Click **CTI > Correlation Matrix**.

3   Verify that the range **B1:F45** appears in the Input Range (Columns) text box.

4   Verify that the **Selection Includes Header Row**, **Pearson Correlation**, and **Show P-values check boxes** are selected. Deselect the **Spearman Rank Correlation check box**.

5   Click the **New Sheet option button**, type **Male Corr Matrix** in the New Sheet text box, then click **OK**. The resulting correlation matrix appears on its own sheet, as shown in Figure 9-13.

Figure 9-13
Correlation matrix

	A	B	C	D	E	F	G	H	I	J	K	L
1	Pearson Correlations											
2		MS Hired	Degree	Age Hired	Years	Salary						
3	MS Hired	1.000	0.520	0.219	-0.099	0.009						
4	Degree		1.000	0.215	-0.103	-0.072						
5	Age Hired			1.000	-0.064	0.325						
6	Years				1.000	0.765						
7	Salary					1.000						
8												
9	Pearson Probabilities											
10		MS Hired	Degree	Age Hired	Years	Salary						
11	MS Hired	0.000	0.000	0.153	0.525	0.952						
12	Degree		0.000	0.161	0.505	0.643						
13	Age Hired			0.000	0.681	0.032						
14	Years				0.000	0.000						
15	Salary					0.000						

You might wonder why the variable Age Hired is used instead of employee age. The problem with using the employee age is one of collinearity. **Collinearity** means that one or more of the predictor variables are highly correlated with each other. In this case, the age of the employee is highly correlated with the number of years employed because there is some overlap between the two (people who have been employed more years are likely to be older). This means that the information those two variables provide is somewhat redundant. On the other hand, you can tell from Figure 9-13 that the relationship between years employed and age when hired is negligible because the p-value is 0.681 (cell E13). Using the variable Age Hired instead of Age gives the advantage of having two nearly uncorrelated predictors in the model. When predictors are only weakly correlated, it is much easier to interpret the results of a multiple regression.

The correlations for salary show a strong relationship to the number of years employed, and some relationship to age when hired, but there is little relationship to a person's degree. This is in agreement with the SPLOM in Figure 9-12.

Multiple Regression

What happens when you throw all four predictors into the regression pot?

To specify the model for the regression:

1   Click the **Male Data sheet tab**.

| | 2 | Click **Tools > Data Analysis**, click **Regression** in the Analysis Tools list box, then click **OK**. The Regression dialog box might contain the options you selected for the previous regression. |

2 Click **Tools > Data Analysis**, click **Regression** in the Analysis Tools list box, then click **OK**. The Regression dialog box might contain the options you selected for the previous regression.

3 Type **F1:F45** in the Input Y Range text box, press [**Tab**], then type **B1:E45** in the Input X Range text box.

4 Verify that the **Labels check box** is selected, and that the **Confidence Level check box** is selected and contains a value of **95**.

5 Click the **New Worksheet Ply option button** and type **Male Reg** in the corresponding text box (replace the current contents if necessary).

6 Verify that the **Residuals, Standardized Residuals, Residual Plots**, and **Line Fit Plots check boxes** are selected.

7 Click **OK**.

The first portion of the summary output is shown in Figure 9-14, formatted to show column labels.

Figure 9-14
Regression output

	A	B	C	D	E	F	G	H	I
1	SUMMARY OUTPUT								
2									
3	*Regression Statistics*								
4	Multiple R	0.8557441							
5	R Square	0.732298							
6	Adjusted R Square	0.7048413							
7	Standard Error	3168.4337							
8	Observations	44							
9									
10	ANOVA								
11		*df*	*SS*	*MS*	*F*	*Significance F*			
12	Regression	4	1071001259	2.7E+08	26.6711	1.05518E-10			
13	Residual	39	391519916.9	1E+07					
14	Total	43	1462521176						
15									
16		*Coefficients*	*Standard Error*	*t Stat*	*P-value*	*Lower 95%*	*Upper 95%*	*Lower 95.000%*	*Upper 95.000%*
17	Intercept	12900.669	3178.167572	4.05915	0.00023	6472.224168	19329.11	6472.224168	19329.11291
18	MS Hired	744.48209	1304.008741	0.57092	0.57133	-1893.121953	3382.086	-1893.121953	3382.08614
19	Degree	-783.52875	746.0876662	-1.0502	0.3001	-2292.632017	725.5745	-2292.632017	725.5745125
20	Age Hired	373.73542	83.19837919	4.4921	6.1E-05	205.4509745	542.0199	205.4509745	542.019856
21	Years	606.17594	64.53147029	9.39349	1.5E-11	475.648849	736.703	475.648849	736.7030304

Interpreting the Regression Output

The R^2 of 0.732 shows that the regression takes care of 73.2% of the variance of salary. However, when this is adjusted for the number of predictors (four), the adjusted R^2 is about 0.705 = 70.5%. The standard error is 3168.43, so salaries vary roughly plus or minus $3,000 from their predictions. The overall F ratio is about 26.67, with a p-value in cell F12 of 1.06×10^{-10}, which rules out the hypothesis that all four population coefficients are 0. Looking at the coefficient values and their standard errors, you see that the Degree coefficient and MS Hired coefficient have values that are not much more than 1 times their standard errors. In other words, their t statistics are much less than 2, and their p-values are much more than 0.05. Therefore, they are not significant at the 5% level. On the other hand, years employed and age when hired do have coefficients that are much larger than their standard errors, with t's of 9.39 and 4.49, respectively. The corresponding p-values are significant at the 0.1% level.

The coefficient estimate of 606 for years employed indicates that each year on the job is worth $606 in annual salary, if the other predictors are held fixed. Correspondingly, because the coefficient for Age Hired is about $374, all other factors being equal, an employee who was hired a year older than another employee will be paid an additional $374.

Residual Analysis of Discrimination Data

Now check the assumptions under which you performed the regression.

To create a plot of residuals versus predicted salary values:

1 Select the range **B27:C71** and click **Insert > Chart > As New Sheet**.

2 Following the Chart Wizard directions, create a scatterplot showing points only with no lines. Give a chart title of **Male Residual Plot**, label the x-axis **Predicted** and the y-axis **Residual**.

3 Change the scale of the x-axis from 0–45000 to 20000–45000. Your chart sheet should look like Figure 9-15.

Figure 9-15
Chart sheet showing the scatterplot of residuals versus predicted salaries for males

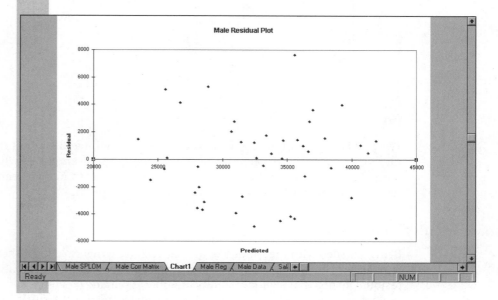

There does not appear to be a problem with nonconstant variance. At least, there is not a big change in the vertical spread of the residuals as you move from left to right. However, there are two points that look questionable. The one at the top has a residual value near 8000 (indicating that this individual is paid $8,000 more than predicted from the regression equation), and at the bottom of the plot an individual is paid about $6,000 less than predicted from the regression.

Except for these two, the points have a somewhat curved pattern—high on the ends and low in the middle—of the kind that is sometimes helped by a log transformation. As it turns out, the log transformation would straighten out the plot, but the regression results would not change much. For example, if log(salary) is used in place of salary, the R^2 value changes only from 0.732 to 0.733. When the results are unaffected by a transformation, it is best not to bother, because it is much easier to interpret the untransformed regression.

Normal Plot of Residuals

What about the normality assumption? Are the residuals reasonably in accord with what is expected for normal data?

To create a normal probability plot of the residuals:

1 Click the **Male Reg sheet tab**.

2 Select the range **C27:C71**.

3 Click **CTI > Normal P-Plot**.

| 4 | Verify that the Input Range text box contains the range **C27:C71**, then click the **Selection Includes Header Row check box** to select it. |
| 5 | Click the **New Sheet option button**, type **Male Nplot** in the corresponding text box, then click **OK**. |

The normal plot appears on a new worksheet, Male Nplot, as shown in Figure 9-16.

Figure 9-16
Normal plot of salary
residuals for male faculty

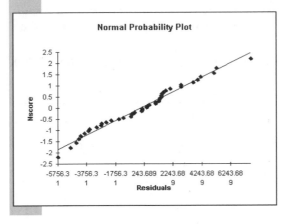

The plot is reasonably straight, although there is a point at the upper right that is a little farther to the right than expected. This point belongs to the employee whose salary is $8,000 more than predicted, but it does not appear to be too extreme. You can conclude that the residuals seem consistent with the normality assumption.

Are Female Faculty Underpaid?

Being satisfied with the validity of the regression on males, let's go ahead and apply it to the females to see if they are underpaid. The idea is to look at the residuals, because a negative residual means that a female faculty member is making less than the salary that would be predicted if she were a male. Your ultimate goal is to choose between two hypotheses:

H_0: The population mean salaries of females are equal to the salaries predicted from the population model for males.

H_a: The population mean salaries are less than the salaries predicted from the population model for males.

To obtain statistics on the salary for females relative to males, you must create new columns of predicted values and residuals.

To create new columns of predicted values and residuals:

1	Click the **Salary Data sheet tab** (you might have to scroll through the workbook to find it).
2	Click **Data > Filter > AutoFilter** to turn off the AutoFilter.
3	Click cell **G1**, type **Pred Sal**, press [**Tab**], type **Resid**, then press [**Enter**].
4	Select the range **G2:H82**.
5	In cell **G2**, type =**12900.67+744.4821*MS_Hired–783.529*Degree+373.7354*Age_Hired +606.1759*Years** (the regression equation for males) and press [**Tab**] (recall that you can use range names in this fashion because the data are arranged in a list and the range names have been predefined for you).
6	Type =**F2–G2** in cell **H2** and press [**Enter**].
7	Click **Edit > Fill > Down** or press [**Ctrl**]+**D**. The data should appear as in Figure 9-17.

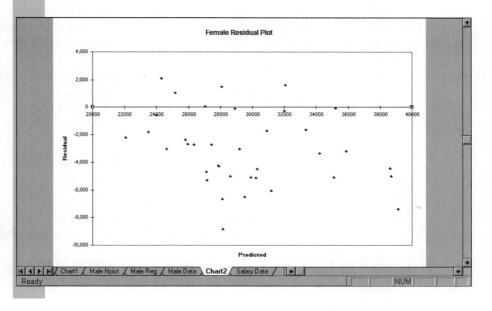

Figure 9-17
Predicted salaries and residuals

	Gender	MS Hired	Degree	Age Hired	Years	Salary	Pred Sal	Resid
2	F	1	2	35	0	26,209	25158.83	1,050
3	F	0	1	37	0	23,253	25945.35	-2,692
4	F	0	1	31	1	26,399	24309.11	2,090
5	F	0	1	25	1	19,876	22066.7	-2,191
6	F	1	2	32	1	21,619	24643.8	-3,025
7	F	1	2	41	2	23,602	28613.6	-5,012
8	F	0	1	39	2	23,602	27905.17	-4,303
9	F	1	2	37	2	22,447	27118.66	-4,672
10	F	0	1	37	2	21,864	27157.7	-5,294
11	F	1	2	39	2	23,602	27866.13	-4,264
12	F	1	2	27	3	23,413	23987.48	-574
13	F	0	1	38	3	19,313	28137.61	-8,825
14	F	1	2	38	3	21,455	28098.57	-6,644
15	F	1	2	42	4	25,072	30199.68	-5,128
16	F	0	1	40	4	22,981	29491.26	-6,510
17	F	1	2	24	4	21,669	23472.45	-1,803
18	F	1	2	33	5	24,740	27442.24	-2,702
19	F	1	2	30	5	23,602	26321.04	-2,719
20	F	1	3	40	6	24,772	29881.04	-5,109
21	F	1	2	39	6	25,784	30290.83	-4,507
22	F	1	2	36	6	26,120	29169.62	-3,050
23	F	1	2	27	6	23,449	25806.01	-2,357
24	F	0	1	38	8	25,110	31168.49	-6,058
25	F	1	2	25	11	29,598	28089.41	1,509

To see whether females are paid about the same salary that would be predicted if they were males, create a scatterplot of residuals versus predicted salary.

To create the scatterplot:

1 Select the range **G1:H38** (the female faculty) and click **Insert > Chart > As New Sheet**.

2 Follow the Chart Wizard directions and create a scatterplot showing points only with no lines. Give a chart title of **Female Residual Plot**, label the x-axis **Predicted** and the y-axis **Residual**.

3 Change the scale of the x-axis from 0–40000 to 20000–40000. Your chart sheet should look like Figure 9-18.

Figure 9-18
Chart sheet showing the scatterplot of residuals versus predicted salary for females

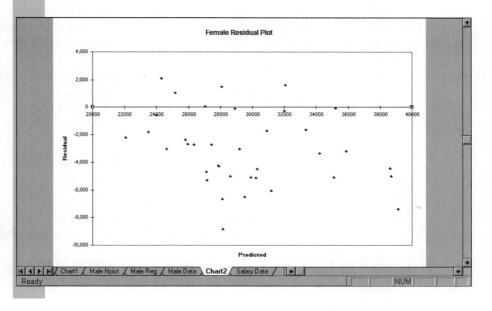

Out of 37 female faculty, only five have salaries greater than what would be predicted if they were males, while 32 have salaries less than predicted. Calculate the descriptive statistics for the females to determine the average discrepancy in salary.

To calculate descriptive statistics for female faculty salaries:

1. Click the **Salary Data sheet tab** to return to the data worksheet.

2. Click **Tools > Data Analysis**, click **Descriptive Statistics**, then click **OK**.

3. Type **H1:H38** in the Input Range text box and verify that the **Grouped by Columns option button** is selected.

4. Click the **Labels in First Row check box** and the **Summary Statistics check box** to select them.

5. Click the **New Worksheet Ply option button**, type **Female Stats** in the corresponding text box, then click **OK**.

The output (after increasing the width of column A) is shown in Figure 9-19.

	A	B
1	Resid	
2		
3	Mean	-3063.64
4	Standard Error	437.6355
5	Median	-3049.62
6	Mode	#N/A
7	Standard Deviation	2662.033
8	Sample Variance	7086418
9	Kurtosis	-0.50251
10	Skewness	0.129762
11	Range	10914.5
12	Minimum	-8824.61
13	Maximum	2089.886
14	Sum	-113355
15	Count	37
16	Confidence Level(95.000%)	857.7485

Based on the descriptive statistics you can conclude that the female faculty are paid on average $3,063.64 less than equally qualified male faculty (as quantified by the predictor variables). The largest discrepancy is a female faculty member who is paid $8,824.61 less than expected (cell B12). On the other hand, another female is paid $2,090 more than expected (cell B13).

To understand the salary deficit better, you can plot residuals against the relevant predictor variables. Start by plotting the female salary residuals versus age when hired (you could plot residuals versus years but you would see no particular trend).

To plot the residuals against Age Hired:

1. Click the **Salary Data sheet tab** to return to the data worksheet.

2. Select the range **D1:D38** and **H1:H38** (use the [Ctrl] key to select noncontiguous ranges), then click **Insert > Chart > As New Sheet**.

3. Follow the Chart Wizard directions and create a scatterplot showing points only with no lines. Give a chart title of **Residuals vs Age Hired**, label the x-axis **Age Hired** and the y-axis **Residuals**.

4. Change the scale of the x-axis from 0–50 to 20–50. Your chart should look like Figure 9-20.

Figure 9-20
Chart sheet showing the
scatterplot of residuals
versus Age Hired

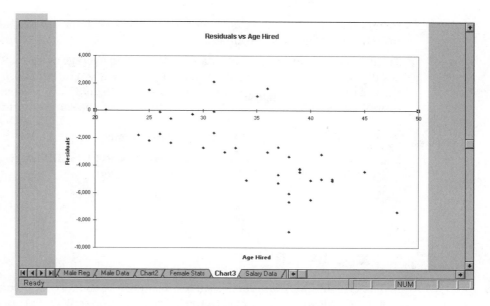

There seems to be a downward trend to the scatterplot. Add a linear regression line to the plot, regressing residuals versus age when hired.

To add a linear regression line to the plot:

1   Right-click **the data series** (any one of the data points in Figure 9-20) and click **Insert Trendline** in the shortcut menu.

2   Click the **Type tab**, verify that the **Linear Trend/Regression Type option** is selected, then click **OK**. Your plot should now look like Figure 9-21.

Figure 9-21
Scatterplot of residuals
versus Age Hired with
linear trend line

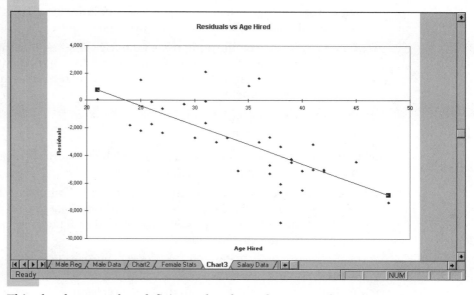

This plot shows a salary deficiency that depends very much on the age at which a female was hired. Those who were hired under the age of 25 have residuals that average around 0, or a little below. On the other hand, those who were hired over the age of 40 are underpaid by more than $5,000, on the average. The most underpaid female has a deficit of nearly $9,000.

Drawing Conclusions

Why should it make a difference when females are hired? Perhaps older males are given credit for experience when they are hired, but older females are not. The college might be able to justify this for women who were not working in a relevant occupation, but it is hard to believe that it would be true of all women. All of the females who were hired over the age of 36 have very negative residuals.

To summarize, the female faculty are underpaid an average of about $3,000. However, there is a big difference depending on how old they were when hired. Those who were hired after the age of 40 have an average deficit of more than $5,000. Nevertheless, when the case was eventually settled out of court, each woman received the same amount.

To save and close the 9DISCRIM workbook:

1 Click the **Save button** 🖫 to save your file as **9DISCRIM.XLS** (the drive containing your Student Disk should already be selected).

2 Click **File > Exit**.

E X E R C I S E S

1. The PCSURV.XLS data set has survey results on 35 models of personal computers, taken from *PC Magazine*, February 9, 1993. The magazine sent 17,000 questionnaires to randomly chosen subscribers, and the data are based on 8,176 responses. There are five columns: Company, the name of the company; Reliability, the overall reliability; Repairs, satisfaction with repair experience; Support, satisfaction with technical support; and Buy Again, future likelihood of buying from the vendor.

   a. To get an idea of the relationships, obtain the correlation matrix and the scatterplot matrix for the four numeric variables.

   b. Using the CTI Statistical Add-Ins, identify any outliers that stand out on the plots (you will have to either enlarge the scatterplot matrix and ungroup the plots, or recreate the plots of interest in a larger scale). Which vendors have the highest Buy Again scores? The lowest Buy Again scores? Do the big companies like IBM, AT&T, Apple, Compaq, and Digital Equipment Corporation have high ratings?

   c. Regress Buy Again on the other three numeric variables. Plot the residuals to check the assumptions. How successful are the other variables at predicting the willingness of customers to buy again? Does it appear that the regression assumptions are satisfied? Summarize your conclusions.

   d. Save your workbook as E9PCSURV.XLS.

2. Open the workbook WHEAT.XLS.

   a. Generate the correlation matrix for the variables Calories, Carbo, Protein, and Fat. Also create the corresponding scatterplot matrix.

   b. Regress Calories on the other three variables and obtain the residual output. How successful is the regression? It is known that carbohydrates have 4 calories per gram, protein has 4 calories per gram, and fats have 9 calories per gram. How do the coefficients compare with the known values?

   c. Explain why the coefficient for Fat is the least accurate, in terms of its standard error and in comparison with the known value of 9. (*Hint:* Examine the data and notice that the Fat content is specified with the least precision.)

   d. Plot the residuals against the predicted values. Is there an outlier? Use the CTI Statistical Add-Ins to label the points of the scatterplot to see which case is most extreme. Do the calories add up correctly for this case? That is, when you multiply the carbohydrate content by 4, the protein content by 4, and the fat content by 9, does it add up to more calories than are stated on the package? Notice also that another case has the same values of Carbo, Protein, and Fat, but the Calories value is 10 higher. How do you explain this? Would a company understate the calorie content?

   e. Save your workbook as E9WHEAT.XLS.

3. The WHEATDAN.XLS data set is a slight modification of WHEAT.XLS, with an additional case, the Apple Danish from McDonald's. The reason for its inclusion is that it has a substantial fat content, in contrast to the foods in WHEAT.XLS. Because none of the foods in WHEAT.XLS have much fat, it is impossible to see from WHEAT.XLS how much fat contributes to the calories in the foods.

   a. Repeat the regression of Exercise 2 for WHEATDAN.XLS and see if the coefficient for Fat is now estimated more accurately. Use both the known value of 9 for comparison and the standard error of the regression that is printed in the output.

   b. Save your workbook as E9WHEATD.XLS.

4. BASE26.XLS includes data from major league baseball games played Tuesday, July 29, 1992, as reported the next day in the Bloomington, Illinois, *Pantagraph*. For each of the 26 major league teams, the data include Runs, Singles, Doubles, Triples, Home Runs, and Walk HBP. This last variable combines walks and hit-by-pitched-ball.

   a. Regress Runs on the other variables and compare with the results obtained by Rosner and Woods (1988), as quoted in the beginning of this chapter. Are the differences explainable in terms of the standard errors of the coefficients?

   b. Do the Rosner-Woods coefficients make more sense in terms of which should be largest and which should be smallest?

   c. How would you expect your answers to change if you obtained several more days of data?

   d. Save your workbook as E9BASE26.XLS.

5. The HONDACIV.XLS data include Price, Age, and Miles for used Honda Civics advertised in the *San Francisco Chronicle*, November 25, 1990. Notice that there is a problem with missing data, because Miles was not included in many of the advertisements.

   a. Because of problems that can occur with missing data, copy only the rows with non-missing values for mileage to a new worksheet. In this new worksheet, find correlations and plots to relate Price to the predictors Age and Miles (you will have to copy Age to the column that is adjacent to the Miles column).

   b. Regress Price on Age and Miles, and examine the residuals.

   c. Notice that there is one car that is much older than the others, and its high residual suggests that it might be an outlier. Do a new regression without this observation to see how much difference it makes. Which regression is better? Interpret your regression results in terms of the effect of one more year of Age on the price.

   d. What about Miles? Why does Miles not seem to matter? Notice that when you look at the data, the Miles values tend to be low relative to the age of the car. Would you advertise the miles on your car if it were high relative to the car's age? If people advertise only low mileage, then how would this affect the regression?

6. Repeat Exercise 5 using Log Price instead of Price.

   a. Does this improve the multiple correlation? Does the old car no longer have a high residual, so there is no longer any need to do a new regression without this car? Is it still true that Miles is not significant in the regression?

   b. When Log Price is used as the dependent variable, the regression can be interpreted in terms of percentage drop in Price per year of age, instead of a fixed drop per year of Age when Price is used as the dependent variable. Does it make more sense to have the price drop by 16.5% each year or to have the price drop by $721 per year? In particular, would an old car lose as much value per year as it did when it was young?

   c. Save your workbook as E9HONDAC.XLS.

7. Open the workbook CARS.XLS. The data file is based on material collected by Donoho and Ramos (1982) and has been used by statisticians to compare the abilities of different statistics packages. The workbook contains observations from 392 car models on the following eight variables: MPG (miles per gallon), Cylinders (number of cylinders), Engine

Disp (engine displacement in cu. in.), Horsepower, Weight (vehicle weight in lbs.), Accelerate (time to accelerate from 0 to 60 mph in seconds), Year (model year), and Origin (origin of car: American, European, or Japanese).

    a. Create a correlation matrix (excluding the Spearman rank correlation) and a scatterplot matrix of the seven quantitative variables.

    b. Regress MPG on Cylinders, Engine Disp, Horsepower, Weight, Accelerate, and Year.

    c. Note that the regression coefficients for Engine Disp and Horsepower are nonsignificant. Compare this to the p-values for these variables in the correlation matrix. What accounts for the lack of significance? (*Hint*: Look at the correlations between Engine Disp, Horsepower, and Weight.)

    d. Create a scatterplot of the regression residuals versus the predicted values. Based on the scatterplot, do the assumptions of the regression appear to be violated? What would you recommend?

8. In the workbook CARS.XLS create a new column, Log MPG, of the logarithm of the miles per gallon.

    a. Redo the regression in part 7b. How does the significance level of the F ratio compare with the untransformed regression?

    b. Plot the regression residuals against the predicted values. Has the transformation improved this scatterplot compared to the previous exercise?

    c. Plot the residuals versus each of the predictor variables. Do problems still remain with the residuals? Identify any problems and describe what steps you would take to solve them.

9. Using the CARS.XLS regression of the previous exercise, try to reduce the number of predictor variables in the model. Use this algorithm:

    a. Perform the regression.

    b. If any coefficients in the regression are nonsignificant, redo the regression with the least-significant variable removed.

    c. Continue until all coefficients remaining are significant.

Using this method, find a model with all coefficients significant that will predict the Log MPG.

10. In the workbook CARS.XLS, regress the variable Log MPG that you created in Exercise 8 on Cylinders, Engine Disp, Horsepower, Weight, Accelerate, and Year for only the American cars (you will have to copy the data to a new worksheet using the AutoFilter function).

    a. Analyze the residuals of the model. Do they follow the assumptions reasonably well?

    b. Apply the results of this regression to the Japanese models to calculate the predicted MPG for the Japanese cars. Create a plot of residuals versus predicted MPGs for the Japanese cars.

    c. Summarize your conclusions, answering the question whether Japanese cars have a different MPG after correcting for the other factors.

    d. Save your workbook as E9CARS.XLS.

11. Repeat the previous exercise to compare the European models with their predictions.

12. Open the 9CALC.XLS workbook that you created in this chapter.

    a. Refer to the regression of calculus scores upon the predictor variables. Remove the least significant coefficient and refit the model. Continue this process until you have a model in which all the parameters are significant.

    b. How does the R^2 value for this reduced model compare to the full model that you started the process with?

    c. What have you gained by reducing the model? What have you lost? Explain.

CHAPTER 10

ANALYSIS OF VARIANCE

O BJECTIVES

In this chapter you will learn to:

- Compare several groups graphically

- Compare the means of several groups using analysis of variance

- Correct for multiple comparisons using the Bonferroni test

- Find which pairs of means differ significantly

- Compare analysis of variance to regression analysis

- Perform a two-way analysis of variance

- Create and interpret an interaction plot

- Check the validity of assumptions

One-way Analysis of Variance

In Chapter 6 you used the *t*-test to compare two treatment groups, such as two groups taught by two different methods. What if there are four treatment groups? You might have 40 subjects split into four groups, with each group receiving a different treatment. The treatments might be four different drugs, four different diets, or four different advertising videos. Analysis of variance provides a test to determine whether to accept or reject the hypothesis that the means are all equal. If you reject this hypothesis, you can then use the CTI Statistical Add-Ins to compare each pair of means to see which are different. The technique of comparing all the pairs of treatments, checking each for significance, is called **multiple comparisons**.

Analysis of Variance Assumptions

What assumptions must you make for these tests? As in regression, there is a model for the data. For the *i*th treatment group, the model is:

$$y = \mu_i + \varepsilon$$

where ε is a normally distributed error with population mean 0 and variance σ^2. This model says that all of the observations in a treatment group have the same treatment mean μ_i because the error term has mean 0. All of the errors are assumed to be independent as well. This type of model is called a **means model**.

A summary of these assumptions shows them to be similar to those used for regression analysis:

- All of the observations in a treatment group have the same mean.

- The errors are normally distributed.

- The errors are independent.

- The errors have constant variance σ^2.

The similarity to regression is no accident. As you will see later in this chapter, analysis of variance is a special case of regression.

To verify these assumptions, it is helpful to make a plot that shows the distribution of observations in each of the treatment groups. If the plot shows large differences in the spread among the treatment groups, there might be a problem of nonconstant variance. If the plot shows outliers, there might be a problem with the normality assumption. Independence could also be a problem if time is important in the data collection, in which case consecutive observations might be correlated. However, there are usually no problems with the independence assumption in the analysis of variance.

Example of One-way Analysis: Comparing Hotel Prices

Some professional associations are reluctant to hold meetings in New York City because of high hotel prices and taxes. The American Statistical Association, for example, has not met in New York since 1973. Are hotels in New York City more expensive than hotels in other major cities?

To answer this question, let's look at hotel prices in four major cities: Los Angeles, New York City, San Francisco, and Washington, D.C. For each city, a random sample of eight hotels was taken from the *1992 Mobil Travel Guide to Major Cities* and stored in the workbook HOTEL.XLS. The workbook contains the following variables, shown in Table 10-1:

Table 10-1	Range Name	Range	Description
HOTEL.XLS variables	City	A1:A33	Location of the hotel
	Hotel	B1:B33	Name of the hotel
	Stars	C1:C33	*Mobil Travel Guide* rating, on a scale from 1 to 5
	Price	D1:D33	Cost for a single room

To open HOTEL.XLS, be sure that Excel is started and the windows are maximized, and then:

1   Open **HOTEL.XLS** (be sure you select the drive and directory containing your Student files).

2   Click **File > Save As**, select the drive containing your Student Disk, then save your workbook on your Student Disk as **10HOTEL.XLS**.

The hotel workbook looks like Figure 10-1.

Figure 10-1
10HOTEL.XLS workbook

Los Angeles data

San Francisco data

Washington D.C. data

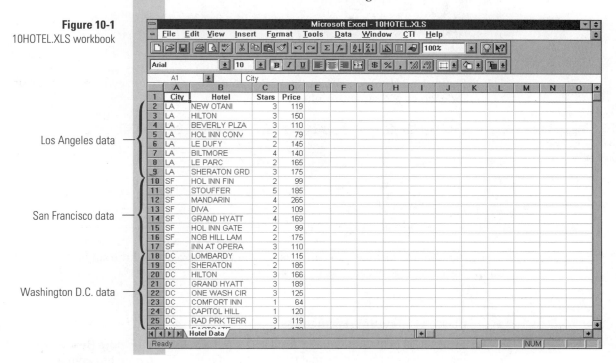

Graphing the Data to Verify Assumptions

It is best to begin with a graph that shows the distribution of price in each of the four cities. In Chapter 6 you used the Multiple Histogram command from the CTI Statistical Add-Ins to create such a plot for data grouped by gender. Unfortunately, the different categories of the data must be in different columns to use this command. The hotel price data for the four cities are all in a single column (column D), so you need to unstack the data to place the data for each city in a separate column. The CTI Statistical Add-Ins provide the capability to unstack a column containing data from different categories into different columns.

To unstack the Price column (D) using the City column (A) as the category:

1   Click **CTI > Unstack Column**.

2   Type **Price** in the Data Values Range text box (this is the range name), then press [**Tab**].

3   Type **City** in the Categories Range text box.

4   Click the **Selection Includes Header Row check box** to select it.

5   Click the **New Sheet option button** and type **Price Data** in the corresponding text box. Your dialog box should look like Figure 10-2.

Figure 10-2
Completed Unstack Column
dialog box

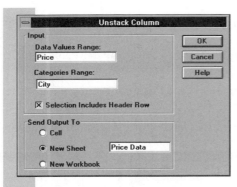

6   Click **OK**.

Figure 10-3 shows the unstacked Price column, now organized with the hotel prices for each city in separate columns.

Figure 10-3
Unstacked price information
on worksheet Price Data

	A	B	C	D	E	F	G	H	I	J	K	L
1	LA	SF	DC	NY								
2	119	99	115	170								
3	150	185	185	135								
4	110	265	166	185								
5	79	109	189	250								
6	145	169	125	250								
7	140	99	64	170								
8	165	175	120	210								
9	175	110	119	215								
10												

With the data in this format, you can create a multiple histogram of the single-room price with City as the category variable.

To create the graph:

1   Click **CTI > Multiple Histograms**.

2   Type **A1:D9** in the Input Range text box, click the **Selection Includes Header Row check box** to select it, then verify that the Number of Bins in Histogram box contains **15**.

3   Click the **New Sheet option button**, type **Price Histograms** in the corresponding text box, then click **OK**.

Figure 10-4 shows the distribution of a price for a single room in these four cities (the Zoom Control setting has been changed to 95% to display all four charts; you might need a different zoom factor, depending on your monitor).

Figure 10-4
Multiple histogram of Price
with City as the categorical
variable

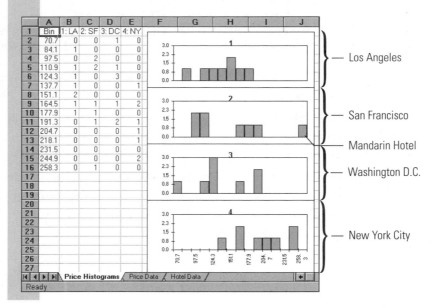

What do these plots tell you about the analysis of variance assumptions? One of the assumptions states that the variance is the same in each group. If one city has prices that are all bunched together while another has a very wide spread of prices, unequal variances could be a problem, but that is not the case here. The plot shows similar spreads in the four groups.

What about the assumption of normal data? The analysis of variance is robust to the normality assumption, so only major departures from normality would cause trouble. In particular, an extreme outlier would have a major effect on the mean and variance for one of the cities. Figure 10-4 shows one hotel in San Francisco that is much more expensive than the others in that city (Chart 2 has a single bar on the far right). If you examine that value, you'll discover that the Mandarin costs more than $250 per night (the highest in the whole sample), but the other hotels in San Francisco all cost less than $200. Of all the 32 hotels, this one is the most influential in the sense that its removal would have the greatest effect on its group mean and variance. Excluding this hotel would change its city mean and variance more than excluding any of the other hotels. It would be a good idea to do the analysis both with and without this hotel. If exclusion makes little difference in the results, there is not much to worry about. On the other hand, if exclusion makes a big difference, your results rest on shaky ground. After all, the hotels were chosen randomly; the Mandarin was included only by chance.

In Figure 10-4, it appears that New York (Chart 4) has the highest mean. Still, there is a lot of overlap between the New York prices and the others. The Mandarin in San Francisco is more expensive than any of the New York City hotels. Do you think that New York City is significantly more expensive than the other cities? Stay tuned.

Computing the Analysis of Variance

Now that you have a visual image of how the price data are distributed in the four cities, you can examine the data using ANOVA.

Note: The acronym ANOVA is often used for analysis of variance.

To obtain the analysis of variance:

1   Click the **Price Data sheet tab**.

2   Click **Tools > Data Analysis**, click **Anova: Single Factor** in the Analysis Tools list box, then click **OK**.

3   Type **A1:D9** in the Input Range text box, and verify that the **Grouped By Columns option button** is selected.

4   Click the **Labels in First Row check box** to select it, then verify that the Alpha box contains **0.05**.

5   Click the **New Worksheet Ply option button** and type **Price ANOVA** in the corresponding text box. Your dialog box should look like Figure 10-5.

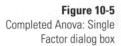
Figure 10-5
Completed Anova: Single Factor dialog box

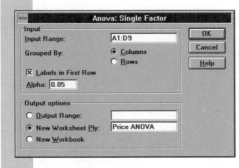

6   Click **OK**.

Figure 10-6 shows the resulting analysis of variance output, with some of the columns slightly widened.

Figure 10-6

Analysis of Variance output

	A	B	C	D	E	F	G	H	I	J	K
1	Anova: Single Factor										
2											
3	SUMMARY										
4	Groups	Count	Sum	Average	Variance						
5	LA	8	1083	135.375	980.8393						
6	SF	8	1211	151.375	3414.839						
7	DC	8	1083	135.375	1771.125						
8	NY	8	1585	198.125	1649.554						
9											
10											
11	ANOVA										
12	Source of Variation	SS	df	MS	F	P-value	F crit				
13	Between Groups	21145.38	3	7048.458	3.60703	0.02549	2.946685				
14	Within Groups	54714.5	28	1954.089							
15											
16	Total	75859.88	31								
17											

ANOVA table

....

Analyzing Data Using One-way Analysis of Variance

- If the data consist of a column of values and a column of categories, unstack the column using the CTI Statistical Add-Ins Unstack Column command.

- Click Tools > Data Analysis, then click ANOVA: Single Factor in the Analysis Tools list box.

- Enter the input range and specify whether the different levels of the group are organized by row or column.

....

Interpreting the Analysis of Variance Table

In performing an analysis of variance you determine what part of the variance you should attribute to randomness and what part you can attribute to other factors (in this case, differences between cities). Analysis of variance does this by splitting the total sum of squares into two parts: a part attributed to differences between groups and a part due to random error or random chance.

The ANOVA table displays the total sum of squares (the sum of squared deviations from the mean), which also can be described as the error sum of squares when each observation is predicted by the overall mean. It is proportional to the variance, because if $\bar{y}$ is the mean of all the observations, then:

$$\text{variance} = \frac{1}{n-1}\sum(y-\bar{y})^2 = \frac{1}{n-1}(\text{total sum of squares})$$

Here n is the number of observations (32 in this example) and $\bar{y}$ is the average hotel price based on hotels from all four cities. Because the variance of the response variable y is proportional to the total sum of squares, analyzing the total sum of squares can be described as analyzing the variance.

For each hotel, designate the mean hotel price for the city as $\hat{y}$. Cells D5:D8 of Figure 10-6 contain the four city means:

LA	135.375
SF	151.375
DC	135.375
NY	198.125

Are city means significantly different or can you effectively estimate the mean price for hotels in any of the four cities by using the overall mean?

This is where splitting the total sum of squares is helpful. To split the total sum of squares into two parts, the ANOVA table takes advantage of the formula:

....

$$\sum(y-\bar{y})^2 = \sum(y-\hat{y})^2 + \sum(\hat{y}-\bar{y})^2$$

On the left of the equals sign is the total sum of squares, that is, the sum of errors using the overall mean to predict an individual hotel's price. In Figure 10-6, the total sum of squares is 75,859.88 (cell B16).

The first term on the right of the equals sign in the equation above is the **error sum of squares**, the sum of squared errors when the city mean is used to predict the price of the hotel rooms. It is also called the **within-groups sum of squares** or, in this case, the within-cities sum of squares because the price of each hotel room is compared with the average price for that city. Cell B14 in Figure 10-6 contains the within-groups sum of squares, 54,714.50.

The second term on the right of the equals sign is the city sum of squares, the sum of the squared differences between the city means and the mean of all observations. This is also known as the **between-groups sum of squares**, or, for this example, the between-cities sum of squares.

The sum has $n = 32$ terms, but all of the terms for a given city are the same. You could also write the second term in the analysis of variance as:

$$8(\hat{y}_{LA} - \bar{y})^2 + 8(\hat{y}_{SF} - \bar{y})^2 + 8(\hat{y}_{DC} - \bar{y})^2 + 8(\hat{y}_{NY} - \bar{y})^2$$

There are eight observations in a city, and the first term in each set of parentheses is the city mean. For these data, the between-groups sum of squares is 21,145.38 (cell B13).

Let's try to relate these sums of squares back to the question of whether you should use the overall mean rather than each city mean to estimate the price for a single room. If the city means are very different from each other, there would be a large value for the between-groups sum of squares. On the other hand, if the city means are all the same, the between-groups sum of squares would be zero. The argument goes the other way, too; a large value for the between-groups sum of squares could indicate that the city means are very different whereas a small value might show that they are not so different. Of course, you would still have to test those conclusions.

A large value for the between-groups sum of squares could be due to a large number of groups, so you need to know how many groups there are. The degrees of freedom (df) column in the ANOVA table (cells C13:C16) tells you that. The df for the city factor (in this case the between-groups term) is the number of groups minus 1, or $4 - 1 = 3$ (cell C13). The degrees of freedom for the total sum of squares is the total number of observations minus 1, or $32 - 1 = 31$ (cell C16). The remaining degrees of freedom are assigned to the error term (the within-groups term) and are equal to $31 - 3 = 28$ (cell C14).

The Mean Square (*MS*) column (cells D13:D14) shows the sum of squares divided by the degrees of freedom; you can think of the entries in this column as variances. The first value, 7048.458 (cell D13), measures the variance between cities; the second value, 1954.089 (cell D14), measures the variance within cities. The within-groups mean square also estimates σ^2—the variance of the error term ε, discussed earlier in this chapter. If the variability in the hotel prices between cities is large relative to the variability of hotel prices within each city, this indicates that average hotel price is not the same for each city.

The *F* value, 3.607 (cell E13), is the ratio of the two variances; it has the $F(3,28)$ distribution if the assumptions are satisfied and the city population means are the same. Here 3 is the numerator degrees of freedom and 28 is the denominator degrees of freedom. The *F* distribution is used to choose between two hypotheses:

H_0 : The true hotel price means (μ_1, μ_2, μ_3, and μ_4) are the same for all four cities.
H_a : The four means (μ_1, μ_2, μ_3, and μ_4) are not all the same.

Reject the hypothesis of equal city means if the *F* ratio is in the upper 5% of the $F(3,28)$ distribution.

The *p*-value is 0.025 (cell F13), which is less than the alpha value 0.05 that you specified in the dialog box, so reject the hypothesis of equal city means at the 5% level. Also notice that the calculated *F* statistic, 3.607, is greater than the *F* critical value 2.95 (cell G13), indicating that the calculated *F* statistic is in the upper 5% of the $F(3,28)$ distribution. At the 5% significance level, it appears that the expected price of a hotel room is not the same in the four cities.

Figure 10-7 shows the *F* distribution with 3 numerator degrees of freedom, 28 denominator degrees of freedom, and an *F* ratio of 3.607 (you don't have to generate this figure, although you could by using the FPDF.XLT instructional template discussed in Chapter 9). As you can see, 3.607 is an extreme value for this distribution.

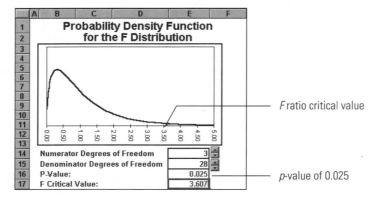

Basically, the analysis of variance test compares two variances: the between-city variance vs. the within-city variance. If the between-city variation is much higher than the within-city variation, the cities are significantly different. This should make sense intuitively. For the *F* ratio to be high enough to reject the equal-means hypothesis, the city means should differ enough so that the variance between cities is high compared with the variance within cities.

Although the output does not show it, you can use the values in the ANOVA table to derive some of the same statistics you used in regression analysis. For example, the ratio of the between-groups sum of squares to the total sum of squares equals R^2, the coefficient of determination discussed in some depth in Chapters 8 and 9. In this case $R^2 = 21,145.38 / 75,859.88 = 0.2787$. So, about 27.9% of the variability in hotel prices is explained by the city of origin.

Means and Standard Deviations

On the top of the output (Figure 10-6), before the analysis of variance table, are the counts, sums, averages, and variances for the four cities (cells A4:E8). Notice that by coincidence DC and LA have exactly the same mean.

What about the spread of the data? You can use the variance values in column E to calculate the standard deviation, which would represent the "typical" deviation from the city's mean-price value.

To calculate the standard deviation for each city:

1. Click cell **F4**, type **Std Dev**, then press **[Enter]**.

2. Select the range **F5:F8**, type **=SQRT(E5)** in cell F4, then press **[Enter]**.

3. Click **Edit > Fill > Down** or press **[Ctrl]+[D]**. Excel enters the rest of the standard deviations into column F.

In accord with Figure 10-4, there is not a lot of variation among the standard deviations. The four cities each have about the same spread with the typical deviations from the mean hotel price ranging from $30 to $60. On the other hand, the mean price for a single room in a New York City hotel is a lot higher than the other three means. Is the difference significant? To find out you can create a matrix of pairwise mean differences.

Multiple Comparisons: Bonferroni Test

The ANOVA table has led you to reject the hypothesis that the mean single-room price is the same in all four cities and to accept the alternative that the four means (μ_1, μ_2, μ_3, and μ_4) are not all the same. Looking at the mean values, you might be tempted to conclude that the price for New York City hotel rooms is significantly higher and leave it at that. This assumption would be wrong because you haven't tested for this specific hypothesis. Moreover, the price for a single room in San Francisco is higher than the price for a room in either Los Angeles or Washington, D.C. Is this difference also significant?

Excel does not provide a function to test pairwise mean differences, but one has been provided for you with the CTI Statistical Add-Ins.

To create a matrix of paired differences:

1 Click the **Price Data sheet tab** to return to the worksheet containing the price data.

2 Click **CTI > Means Matrix (1-way ANOVA)**.

3 Type **A1:D9** in the Input Range text box, then verify that the **Selection Includes Header Row check box** is selected.

4 Click the **Show P-values check box** to select it, then click the **Use Bonferroni Correction check box** that appears.

5 Click the **New Sheet option button**, then type **Price Means** in the corresponding text box.

Your dialog box should look like Figure 10-8.

Figure 10-8
Completed Create Matrix of Mean Values dialog box

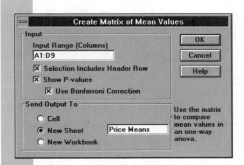

6 Click **OK**.

Among the four cities, there are six pairwise differences, which are listed in the Pairwise Mean Diff table as shown in Figure 10-9.

Figure 10-9
Pairwise mean differences matrix and *p*-values

	A	B	C	D	E	F
1	Pairwise Mean Diff.(row - column)					
2		LA	SF	DC	NY	
3	LA	0	-16	0	-62.75	
4	SF		0	16	-46.75	
5	DC			0	-62.75	
6	NY				0	
7	MSE = 1954.089					
8	Pairwise Comparison Probabilities (Bonferroni Correction)					
9		LA	SF	DC	NY	
10	LA	0.000	1.000	1.000	0.050	
11	SF		0.000	1.000	0.261	
12	DC			0.000	0.050	
13	NY				0.000	
14						

Pairwise Mean Diff table

Bonferroni Correction table

- If the data consist of a column of values and a column of categories, unstack the values column using the CTI Statistical Add-Ins Unstack Column command.

- Click CTI > Means Matrix (1-way ANOVA).

- Enter the input range and specify whether you want *p*-values printed and whether to use the Bonferroni correction on the *p*-values.

You can tell from the Pairwise Mean Diff table that the mean cost for a single hotel room in Los Angeles is $16 less than the mean cost in San Francisco. The largest difference is between Los Angeles or Washington, D.C. and New York City, with a single room in these cities costing about $63 less than a single room in New York City hotels. Note that the output includes the mean squared error value from the ANOVA table, 1954.089 (cell A7), which is the estimate of the variance of hotel prices.

You also requested in the dialog box to create a table of *p*-values for these mean differences using the Bonferroni correction factor. Recall from Chapter 8 that the Bonferroni procedure is a conservative method for calculating the probabilities by multiplying the *p*-value by the total number of comparisons. Because the *p*-values are much higher than you would see if you compared the cities with *t*-tests, it is harder to get significant comparisons with the Bonferroni procedure. However, the Bonferroni procedure has the advantage of giving fewer false positives than *t*-tests would give.

With the Bonferroni procedure, the chances of finding at least one significant difference among the means is less than 5% if each of the four population means is the same. On the other hand, if you do six *t*-tests to compare the four cities at the 5% level, there is much more than a 5% chance of getting significance in at least one of the six tests if all four population means are the same. There are other methods available to help you adjust the *p*-value for multiple comparisons, including Tukey's and Scheffé's, but the Bonferroni method is the easiest to implement in Excel, which does not provide a correction procedure.

Note: Essentially, the difference between the Bonferroni procedure and a t-test is that for the Bonferroni procedure the 5% applies to all six comparisons together, but for t-tests the 5% applies to each of the six comparisons separately. In statistical language, the Bonferroni procedure is testing at the 5% level experimentwise, whereas the t-test is testing at the 5% level pairwise.

The pairwise comparison probabilities show that only the two biggest differences are significant (highlighted in red for those with a color monitor). Only the differences between NY and LA and between NY and DC have *p*-values at 0.05, so only these two differences are significant at the 5% level.

Although the NY vs. SF difference is not significant, it is interesting that the other two differences, NY vs. DC and NY vs. LA, are significant. Recall that the data collection was partly motivated by high prices in New York City.

When to Use the Bonferroni Correction

As the size of the means matrix increases, the number of comparisons increases as well. Consequently, the *p*-values for the pairwise differences are greatly inflated. As you can imagine, there might be a point where there are so many comparisons in the matrix that it is nearly impossible for any *one* of the comparisons to be statistically significant using the Bonferroni correction factor. Many statisticians are concerned about this problem and feel that although the Bonferroni correction factor does guard well against incorrectly finding significant differences, it is also too conservative and misses true differences in pairs of mean values.

In such situations, statisticians make a distinction between paired comparisons that are planned before the data are analyzed and those that occur only after looking at the data. For example, the planned comparisons here are the differences in hotel room price between

New York City and the others. You should be careful with new comparisons that you come up with after you have collected the data. You should hold these comparisons to a much higher standard than the comparisons you've planned to make all along. This distinction is important in order to ward off the effects of data "snooping" (unplanned comparisons). Based on some simulations by Carmer and Swanson (1973), some statisticians make the following recommendations when you are analyzing the paired means differences in your analysis of variance:

1. Conduct an F-test for equal means.

2. If the F statistic is significant at the 5% level, make any planned comparisons you want without correcting the p-value. For data snooping, use a correction factor such as Bonferroni's on the p-value.

3. If the F statistic for equal means is not significant, you can still consider any planned comparisons, but only with a correction factor to the p-value. Do not analyze any unplanned comparisons.

It should be emphasized that the validity of this approach is questioned by some statisticians at the same time it is embraced by others.

The Boxplot

Earlier you used multiple histograms to compare the distribution of hotel prices among the different cities. The boxplot is also very useful for this task, because it shows the broad outline of the distributions and displays the median for the four cities. Recall that if the data are very badly skewed, the mean might be strongly affected by outlying values. The median would not have this problem.

To create a boxplot of price versus city:

1   Click the **Price Data sheet tab** to return to the worksheet containing the price data.

2   Click **CTI > Boxplots**.

3   Type **A1:D9** in the Input Columns text box and verify that the **Selection Includes Header Row check box** is selected.

4   Click the **New Sheet option button**, type **Price Boxplot** in the corresponding text box, then click **OK**.

The resulting boxplot shown in Figure 10-10 gives a slightly different perspective.

Figure 10-10
Boxplot of prices versus city

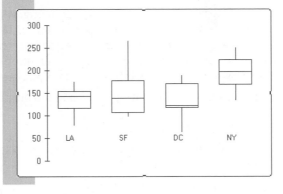

Compare the medians, indicated by the middle horizontal line in each box. The median for San Francisco is below the median for LA, even though you discovered from the pairwise mean difference matrix that the mean price in San Francisco is $16 above the mean in Los Angeles. The reason for the difference is the Mandarin Hotel, which is much more expensive than the other San Francisco hotels. This hotel has a big effect on the San Francisco mean price, but not on the median. The median is much more robust to the effect of outliers. On the other hand, although the cost of the Mandarin Hotel is much higher than the others, it is not shown as an outlier in the plot. To qualify as an outlier, its distance above the box would have to exceed the box's height by 50%.

This discussion might make you curious about what would happen if the Mandarin Hotel were excluded; you will have the opportunity to find out in the exercises.

One-way Analysis of Variance and Regression

You can think of analysis of variance as a special form of regression. In the case of analysis of variance, the predictor variables are discrete. Still, you can express an analysis of variance in terms of regression and in doing so can get additional insights into the data. To do this you have to reformulate the model.

Earlier in this chapter you were introduced to the means model:

$$y = \mu_i + \varepsilon$$

for the ith treatment group. An equivalent way to express this relationship is with the **effects model**:

$$y = \mu + \alpha_i + \varepsilon$$

Here μ is the overall grand mean, α_i the effect of the ith treatment group on the grand mean, and ε is a normally distributed error with population mean 0 and variance σ^2.

Let's apply this equation to the hotel data. In this data set there are four groups representing the four cities, so you would expect this model to have a grand mean term μ and four effect terms, α_1, α_2, α_3, and α_4, representing the four cities. There is a problem, however: you have five parameters in your model but you are estimating only four mean values. This is an example of an **overparametrized model**, where you have more parameters than response values. As a result, there is an infinite number of possible values for the parameters that will "solve" the equation. To correct this problem, you have to reduce the number of parameters. Statistical packages generally do this in one of two ways: they either constrain the values of the effect terms so that the sum of the terms is zero, or they define one of the effect terms to be zero. Let's apply this second approach to the hotel data and perform the analysis of variance using regression modeling.

Indicator Variables

To complete the analysis of variance using regression modeling, you have to create indicator variables for the data. Indicator variables take on values of either 1 or 0, depending upon whether the data belong to a certain treatment group or not. For example, you can create an indicator variable where the variable values are 1 if the observation comes from a hotel in San Francisco or 0 if the observation comes from a hotel not in San Francisco.

You'll use the indicator variables to represent the terms in the effects model.

	To create indicator variables for the hotel data:
1	Click the **Hotel Data sheet tab** (you might have to scroll to see it) to return to the worksheet containing the hotel data.
2	Click **CTI > Make Indicator Variables**.
3	Type **City** in the Categories Range text box, then verify that the **Selection Includes Header Row check box** is selected.
4	Click the **Cell option button**, then type **F1** in the corresponding text box.
	Your dialog box should look like Figure 10-11.

Figure 10-11
Completed Make Indicator
Variables dialog box

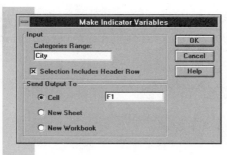

5   Click **OK**.

As shown in Figure 10-12, the procedure creates four columns of indicator variables for the four possible values of City (column A).

Figure 10-12
Indicator variables in
columns F:I

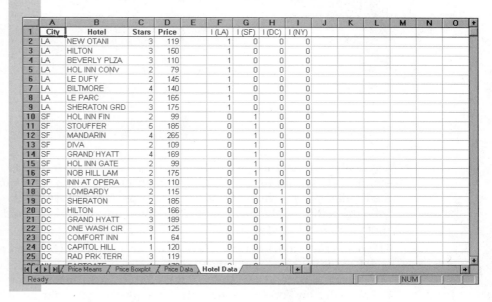

The values in column F, labeled I(LA), are equal to 1 if the values in the row come from a hotel in Los Angeles, and 0 if they do not. Similarly, the values for the next three columns are 1 if the observations come from San Francisco, Washington, D.C., and New York City, respectively, and 0 otherwise.

Fitting the Effects Model

With these columns of indicator variables you can now fit the effects model to the hotel pricing data.

To fit the effects model using regression analysis:

1   Click **Tools > Data Analysis**, click **Regression** in the Analysis Tools list box, then click **OK**.

2   Type **Price** in the Input Y Range text box, press [**Tab**], then type **F1:H33** in the Input X Range text box.

Recall that you have to remove one of the effect terms (by setting it to 0) to keep from overparametrizing the model. For this example, remove the New York effect term (you could have removed any one of the four city effect terms).

3   Click the **Labels check box** to select it, because the range includes a header row.

4   Click the **New Worksheet Ply option button**, then type **Price Effects** in the corresponding text box.

5 Verify that all four **Residuals check boxes** are deselected, then click **OK**.

The regression output appears as in Figure 10-13 (the columns are resized to show the labels).

Figure 10-13
Effects model regression
output

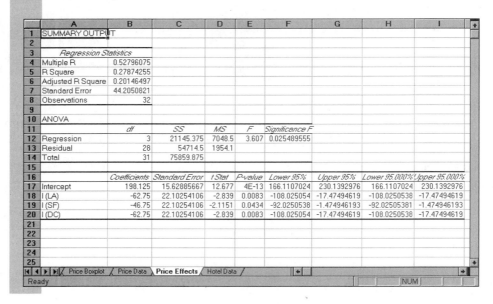

The analysis of variance table produced by the regression (cells A10:F14) and shown in Figure 10-13 should appear familiar to you because it is equivalent to the ANOVA table created earlier and shown in Figure 10-6. There are two differences: the Between Groups row from the earlier ANOVA table is the Regression row in this table, and the Within Groups row is now termed the Residual row.

The parameter values of the regression are also familiar. The intercept coefficient, 198.125 (cell B17), is the same as the mean price in New York. The values of the LA, SF, and DC effect terms now represent the difference between the mean hotel price in these cities and the price in New York. Note that this is exactly what you calculated in the matrix of paired mean differences shown in Figure 10-9. The p-values for these coefficients are the uncorrected p-values for comparing the paired mean differences between these cities and New York. If you multiplied these p-values by 6 (the number of paired comparisons in the paired-mean differences matrix) you would have the same p-values shown in Figure 10-9.

Can you see how the use of indicator variables allowed you to create the effects model? Consider the values for I(LA). For any non-Los Angeles hotel, the value of the indicator variable is 0, so the effect term is multiplied by 0, thereby having no impact on the estimate of the hotel price. It is only for Los Angeles hotels that the effect term is present.

As you can see, using regression analysis to fit the effects model gives you much of the same information as the one-way analysis of variance.

The model you've considered suggests that the average price for a single room at a hotel in New York City is significantly higher than the average price for a single room in either Los Angeles or Washington, D.C., but not necessarily higher than one in San Francisco. You can expect to pay about an average of $198 for a single room in New York City, and $63 less than this in LA or DC. The San Francisco single-room average cost is about $47 less than the cost in New York City. This might explain why the American Statistical Association has not met in New York City since 1973.

To save and close the 10HOTEL.XLS workbook:

1 Click the **Save button** 🖫 to save your file as 10HOTEL.XLS (the drive containing your Student Disk should already be selected).

2 Click **File > Close**.

Two-way Analysis of Variance

One-way analysis of variance compares several groups corresponding to a single categorical variable (for example, city). This one categorical variable is called a **factor**. One-way analysis of variance involves a single factor.

You can use two-way analysis of variance to learn about the effects of two factors. In agriculture, for example, you might be interested in the effects of both potassium and nitrogen on the growth of potatoes. In medicine you might want to study the effects of two factors—medication and dose—on the duration of headaches. In education you might want to study the effects of grade level and gender on the time required to learn a skill. A marketing experiment might consider the effects of advertising dollars and advertising medium (television, magazines, etc.) on sales.

Recall earlier in the chapter you looked at the means model for a one-way analysis of variance. Two-way analysis of variance models can also be expressed as a means model:

$$y_{ijk} = \mu_{ij} + \varepsilon_{ijk}$$

where y is the response variable, μ_{ij} is the mean for the ith level of one factor and the jth level of the second factor. Within each combination of the two factors, you might have multiple observations called **replicates**. Here ε_{ijk} is the error for the ith level of the first factor, jth level of the second factor, and kth replicate following a normal distribution with mean 0 and variance σ^2.

The model is more commonly presented as an effects model where:

$$y_{ijk} = \mu + \alpha_i + \beta_j + \alpha\beta_{ij} + \varepsilon_{ijk}$$

Here y is the response variable, μ is the overall mean, α_i the effect of the ith treatment for the first factor and β_j is the effect of the jth treatment for the second factor. The term $\alpha\beta_{ij}$ represents the *interaction* between the two factors, that is, the effect that the two factors have upon each other. For example, in an experiment where the two factors are advertising dollars and advertising medium, the effect of an increase in sales might be the same regardless of what advertising medium is used, or it might vary depending upon the medium. When the increase is the same regardless of the medium, the interaction is 0, whereas in the second situation there is an interaction between advertising dollars and medium.

A Two-factor Example

To see how different factors affect the value of a response variable, consider an example of the effects of four different assembly lines (A, B, C, and D) and two shifts (a.m. or p.m.) on the production of microwave ovens for an appliance manufacturer. Assembly line and shift are the two factors; the assembly line factor has four levels, and the shift factor has two levels. Each combination of the factors, line, and shift, is called a **cell**, so there are $4 * 2 = 8$ cells. The response variable is the total number of microwaves assembled in a week for one assembly line operating on one particular shift. For each of the eight combinations of assembly line and shift, six separate weeks worth of data are collected.

You can describe the mean number of microwaves created per week with the effects model where:

> Mean number of microwaves = overall mean + assembly line effect +
> shift effect + interaction + error

Now let's formulate a possible model of how the mean number of microwaves produced could vary between shifts and assembly lines. Let the overall mean number of microwaves produced for all shifts and assembly lines be 240 per week. Now let the four assembly line effects be A, +66 (that is, assembly line A produces on average 66 more microwaves than the overall mean); B, –2; C, –100; D, +36. Let the two shift effects be p.m., –6; and a.m., +6. Notice that the four assembly line effects add up to zero, as do the two shift effects. This follows from the need to constrain the values of the effect terms to avoid overparametrization, as was discussed with the one-way effects model earlier in this chapter.

If you exclude the interaction term from the model, the population cell means (the mean number of microwaves produced) look like this:

	A	B	C	D
p.m.	300	232	134	270
a.m.	312	244	146	282

These values are obtained by adding the overall mean + the assembly line effect + the shift effect for each of the eight cells. For example, the mean for the p.m. shift on assembly line A is:

$$\text{overall mean} + \text{assembly line effect} + \text{shift effect} = 240 + 66 - 6 = 300$$

Without interaction, the difference between the a.m. and the p.m. shifts is the same (12) for each assembly line. You can say that the difference between a.m. and p.m. is 12, no matter which assembly line you are talking about. This works the other way, too. For example, the difference between line A and line C is the same (166) for both the p.m. shift (300 – 134) and the a.m. shift (312 – 146). You might understand these relationships better from a graph. Figure 10-14 shows a plot of the eight means with no interaction (you don't have to produce this plot).

Figure 10-14
Plot of means without interaction

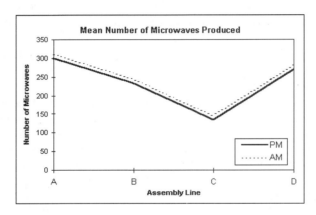

The cell means are plotted against the assembly line factor using separate lines for the shift factor. This is called an **interaction plot**, which you'll create later in this chapter.

Because there is a constant spacing of 12 between the two shifts, the lines are parallel. The pattern of ups and downs for the p.m. shift is the same as the pattern of ups and downs for the a.m. shift.

What if interaction is allowed? Suppose that the eight-cell population means are as follows:

	A	B	C	D
p.m.	295	235	175	200
a.m.	317	241	142	220

In this situation the difference between the shifts varies from assembly line to assembly line, as shown in Figure 10-15. This means that any inference on the shift effect must take into account the assembly line. You might claim that the a.m. shift generally produces more microwaves, although this is not true for assembly line C.

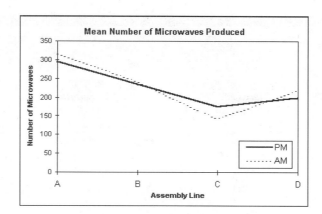

Figure 10-15

Plot of means with interaction

The assumptions for two-way ANOVA are essentially the same as those for one-way ANOVA. For one-way ANOVA all observations on a treatment were assumed to have the same mean, but here all observations in a cell are assumed to have the same mean. The two-way ANOVA assumes independence, constant variance, and normality, just as in one-way ANOVA (and regression).

Example of Two-way Analysis: Comparing Soft Drinks

The file COLA.XLS contains data describing the effects of cola (Coke, Pepsi, Shasta, and generic) and type (diet or regular) on the foam volume of cola soft drinks. Cola and type are the factors; cola has four levels, and type has two levels. There are, therefore, eight combinations, or cells, of cola brand and soft drink type. For each of the eight combinations, the experimenter purchased and cooled a six-pack, so there are 48 different cans of soda. Then the experimenter chose a can at random, poured it in a standard way into a standard glass, then measured the volume of foam.

Why would it be wrong to test all of the regular Coke first, then the diet Coke, etc.? Although the experimenter might make every effort to keep everything standardized, trends that influence the outcome could appear. For example, the temperature in the room or the conditions in the refrigerator might change during the experiment. There could be subtle trends in the way the experimenter poured and measured the cola. If there are trends, it would make a difference which brand was poured first, so it is best to pour the 48 cans in random order.

The COLA.XLS workbook contains the following variables, shown in Table 10-2:

Table 10-2

COLA.XLS variables

Range Name	Range	Description
Can_No	A1:A49	The number of the can (1–6) in the six-pack
Cola	B1:B49	The cola brand
Type	C1:C49	Type of cola: regular or diet
Foam	D1:D49	The foam content of the cola
Cola_Type	E1:E49	The brand and type of the cola

To open the COLA.XLS data:

1 Open **COLA.XLS** (be sure you select the drive and directory containing your Student files).

2 Click **File > Save As**, select the drive containing your Student Disk, then save your workbook on your Student Disk as **10COLA.XLS**.

The workbook looks like Figure 10-16.

Figure 10-16

10COLA.XLS workbook

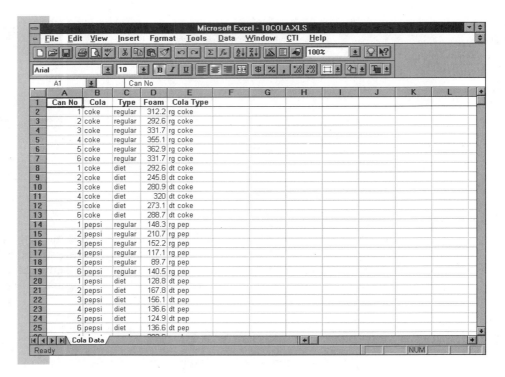

	A	B	C	D	E	F	G	H	I	J	K	L
1	Can No	Cola	Type	Foam	Cola Type							
2	1	coke	regular	312.2	rg coke							
3	2	coke	regular	292.6	rg coke							
4	3	coke	regular	331.7	rg coke							
5	4	coke	regular	355.1	rg coke							
6	5	coke	regular	362.9	rg coke							
7	6	coke	regular	331.7	rg coke							
8	1	coke	diet	292.6	dt coke							
9	2	coke	diet	245.8	dt coke							
10	3	coke	diet	280.9	dt coke							
11	4	coke	diet	320	dt coke							
12	5	coke	diet	273.1	dt coke							
13	6	coke	diet	288.7	dt coke							
14	1	pepsi	regular	148.3	rg pep							
15	2	pepsi	regular	210.7	rg pep							
16	3	pepsi	regular	152.2	rg pep							
17	4	pepsi	regular	117.1	rg pep							
18	5	pepsi	regular	89.7	rg pep							
19	6	pepsi	regular	140.5	rg pep							
20	1	pepsi	diet	128.8	dt pep							
21	2	pepsi	diet	167.8	dt pep							
22	3	pepsi	diet	156.1	dt pep							
23	4	pepsi	diet	136.6	dt pep							
24	5	pepsi	diet	124.9	dt pep							
25	6	pepsi	diet	136.6	dt pep							

Graphing the Data to Verify Assumptions

Before performing a two-way analysis of variance on the data, you should plot the data values to see if there are any major violations of the assumptions. As a first step, create a multiple histogram of the eight cells of cola and type. First you must unstack the data using the Unstack command on the CTI menu.

To unstack the foam data:

1. Click **CTI > Unstack Column**.

2. Type **Foam** in the Data Values Range text box, press [**Tab**], then type **Cola_Type** in the Categories Range text box (these are the range names).

3. Click the **Selection Includes Header Row check box** to select it.

4. Click the **New Sheet option button**, type **Foam Data** in the corresponding text box, then click **OK**.

The unstacked foam data appear in a new worksheet called Foam Data, as shown in Figure 10-17.

Figure 10-17

Unstacked foam data

	A	B	C	D	E	F	G	H	I	J	K	L
1	rg coke	dt coke	rg pep	dt pep	rg sha	dt sha	rg gen	dt gen				
2	312.2	292.6	148.3	128.8	292.6	292.6	156.1	167.8				
3	292.6	245.8	210.7	167.8	253.6	253.6	253.6	249.7				
4	331.7	280.9	152.2	156.1	362.9	214.6	273.1	187.3				
5	355.1	320	117.1	136.6	280.9	269.2	175.6	210.7				
6	362.9	273.1	89.7	124.9	249.7	312.2	284.8	292.6				
7	331.7	288.7	140.5	136.6	214.6	312.2	300.4	327.8				
8												

To create multiple histograms of the foam data:

1. Click **CTI > Multiple Histograms**.

2. Type **A1:H7** in the Input Range text box and click the **Selection Includes Header Row check box** to select it, then verify that the Number of Bins in Histogram is **15**.

3. Click the **New Sheet option button**, type **Foam Histograms** in the corresponding text box, then click **OK**.

Eight histograms are shown in the Foam Histograms worksheet, one for each cell. Figure 10-18 shows the first few, using a 90% zoom factor.

Figure 10-18
Multiple histograms of the cola data

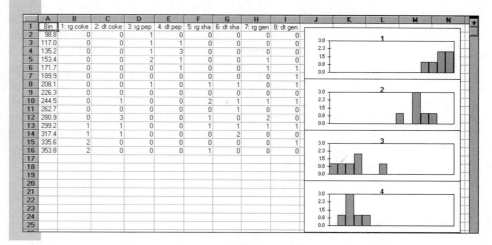

Due to the number of charts, you will either have to reduce the zoom factor on your worksheet or scroll vertically through the worksheet to see all the plots.

Do you see major differences in spread among the eight groups? If so, it would suggest a violation of the equal variances assumption, because all of the groups are supposed to have the same population variance. Although there are differences in the sample variances, they are not significant enough to cause concern. The natural variation in spreads is quite large, so the difference in spreads would have to be very large to be significant. You should also look for outliers, because extreme observations can make a big difference in the results. An outlier could be the result of a strange can of cola, a wrong observation, a recording error, or an error in entering the data. However, no outliers are apparent in the plots.

To gain further insight into the distribution of the data, create boxplots of each of the eight combinations of brand and type:

1   Click the **Foam Data sheet tab** to return to the foam data.

2   Click **CTI > Boxplots**.

3   Type **A1:H7** in the Input Range text box and verify that the **Selection Includes Header Row check box** is selected.

4   Click the **New Sheet option button**, type **Foam Bplots** in the corresponding text box, then click **OK**.

As shown in Figure 10-19, the boxplots also indicate that while individual cells of brand and type might have outlying values, the range of the foam values is generally the same for each of the eight cells.

Figure 10-19
Boxplots of Foam versus Cola Type

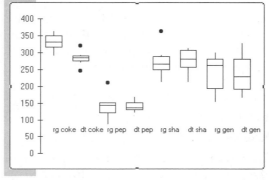

An advantage of the boxplot over the multiple histograms is that it is easier to view the relative change in foam volume from diet to regular for each brand of cola. The first two boxplots represent the range of foam values for regular and diet Coke respectively, after which come the Pepsi values, Shasta values, and, finally, the generic values. Notice in the plot that the same pattern occurs for both the diet and regular colas. Coke is the highest, Pepsi is the lowest, while Shasta and generic are in the middle. The difference in the foam between the diet and regular soda does not depend much on the cola brand. This suggests that there is no interaction between the cola effect and the type effect.

Can you draw preliminary conclusions regarding the effect of type (diet or regular) on foam volume based on this plot? Does it appear that there is much difference between the diet and regular colas? Because the foam levels do not appear to differ much between diet and regular, you can expect that the test for the type effect in a two-way analysis of variance will not be significant. On the other hand, look at the differences among the four brands of colas. The foamiest can of Pepsi is below the least foamy can of Coke, so you might expect that there will be a significant cola effect.

With a good visual picture of the foam data, let's perform a two-way analysis of variance.

Using Excel to Perform a Two-way Analysis of Variance

The Analysis ToolPak provides two versions of the two-way analysis of variance. One is for situations in which there is no replication of combination of factor levels. That would be the case in this example if the experimenter had tested only one can of soda for each of the eight cells rather than purchasing a six-pack with six cans (that is, with replication) for each cell. Therefore, you should use the Analysis ToolPak function to perform a two-way analysis of variance with replication.

A second important consideration is that the number of cans for each cell of brand and type must be the same. Specifically you cannot use data that have five cans of diet Coke and six cans of regular Coke. Data with the same number of replications per cell are called **balanced data**. If the number of replicates is different for different combinations of brand and type, you cannot use the Analysis ToolPak's two-way analysis of variance command.

Finally, to perform a two-way analysis of variance with replication on these data, they must be in a specific format before the Analysis ToolPak can use it; Figure 10-20 shows the format. You can format the foam data this way using the Make Two Way Table command included with the CTI Statistical Add-Ins.

Figure 10-20

Foam data in format for two-way analysis of variance

	A	B	C	D	E
1	Foam		Cola		
2	Type	coke	pepsi	shasta	generic
3	regular	312.2	148.3	292.6	156.1
4		292.6	210.7	253.6	253.6
5		331.7	152.2	362.9	273.1
6		355.1	117.1	280.9	175.6
7		362.9	89.7	249.7	284.8
8		331.7	140.5	214.6	300.4
9	diet	292.6	128.8	292.6	167.8
10		245.8	167.8	253.6	249.7
11		280.9	156.1	214.6	187.3
12		320	136.6	269.2	210.7
13		273.1	124.9	312.2	292.6
14		288.7	136.6	312.2	327.8

The data are formatted so that the first factor (the four cola brands) is displayed in the columns, and the second factor (diet or regular) is shown in the rows of the table. Replications (the six cans in each pack) occupy six successive rows. Each cell in the two-way table is the value of the foam volume for a particular can.

To restructure the data using the CTI Statistical Add-Ins:

1 Click the **Cola Data sheet tab**.

2 Click **CTI > Make Two Way Table**.

3 Type **Cola** in the Factor 1 Range text box (the columns in the two-way table), then type **Type** in the Factor 2 Range text box (the rows), and then type **Foam** in the Response Values Range text box (the values).

4   Click the **Selection Includes Header Row check box** to select it.

5   Click the **New Sheet option button**, then type **Two Way Table** in the corresponding text box. The complete dialog box should look like Figure 10-21.

Figure 10-21
Completed Create Two Way
Table dialog box

6   Click **OK**.

The structure of the data on the Two Way Table worksheet now resembles Figure 10-20, and you can now perform a two-way analysis of variance on the foam data.

To calculate the two-way analysis of variance:

1   Click **Tools > Data Analysis**, click **Anova: Two-Factor With Replication** in the Analysis Tools list box, then click **OK**.

2   Type **A2:E14** in the Input Range text box, press [**Tab**], type **6** in the Rows per sample text box, then verify that the Alpha box contains **0.05**.

3   Click the **New Worksheet Ply option button** and type **Foam Two Way** in the corresponding text box.

Your dialog box should look like Figure 10-22.

Figure 10-22
Completed Anova:
Two-Factor With Replication
dialog box

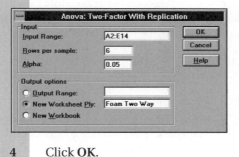

4   Click **OK**.

Analyzing Data with Two-way Analysis of Variance

- If the data consist of a column of values and two columns of categories, restructure the data into a two-way table using the CTI Statistical Add-Ins command Make Two Way Table.

- If there is only one value per combination of row and column, use Anova: Two-Factor Without Replication. If there is more than one value per combination of row and column use Anova: Two-Factor With Replication. If there are blanks for one or more of the cells, you cannot use the Analysis ToolPak to perform two-way ANOVA.

- Click Tools > Data Analysis, then click Anova: Two-Factor Without Replication or Anova: Two-Factor With Replication.

Interpreting the Analysis of Variance Table

The analysis of variance table appears as in Figure 10-23, with the columns resized to show the labels (you might have to scroll to see this part of the output).

Figure 10-23

Cells A23:G30 of the Two Way Analysis of Variance output

23	ANOVA						
24	*Source of Variation*	*SS*	*df*	*MS*	*F*	*P-value*	*F crit*
25	Sample	1880.003	1	1880.003	1.022122	0.318093	4.08474
26	Columns	183750.5	3	61250.17	33.30054	5.84E-11	2.838746
27	Interaction	4903.38	3	1634.46	0.888625	0.455293	2.838746
28	Within	73572.58	40	1839.315			
29							
30	Total	264106.5	47				
31							

There are three effects now, whereas the one-way analysis had just one. The three effects are Sample for the type effect (row 25), Columns for the cola effect (row 26), and Interaction for the interaction between type and cola (row 27). The Within row (row 28) displays the Within sum of squares, also known as the error sum of squares.

The value for the sample sum of squares, 1880.003 (cell B25), is proportional to the sum of the squared differences between the two type means (diet or regular) and the overall mean. A large value for the sample sum of squares relative to the total sum of squares, 264,106.5 (cell B30), indicates that much of the variability in foam volume depends upon whether the cola is diet or regular. Similarly, the columns sum of squares, 183,750.5 (cell B26), is proportional to the sum of the squared differences between the cola brand means and the overall mean. A large value indicates that there are some significant differences in foam volume between cola brands.

If the brand and diet account for most of the variance between colas, the interaction sum of squares is small. However, if the sum of squares for the interaction, 4903.38 (cell B27), is large relative to the total sum of squares, this indicates that an interaction exists.

Finally, the Within, or error sum of squares, 73,572.58 (cell B28), is the sum of the squared differences between the observations (each particular can) and the eight cell means (brand and type). This is the sum of squares due to error or random fluctuation in foam volume.

The type effect has two levels (diet or regular), so it has only one degree of freedom (cell C25), whereas the cola effect has four levels (Coke, Pepsi, Shasta or generic) and therefore has three degrees of freedom (cell C26). The product of these two numbers—1 and 3—gives the degrees of freedom for the interaction term, 3 (cell C27).

The mean square values in column D (cells D25:D28) are the ratios of the sums of squares to the degrees of freedom (column C to column D); they express the influence of the different effects in terms of variance. For example, the mean square for the column or cola effect, 61,250.17 (cell D26), is the variance among the four brands of cola, whereas the mean square for the sample effect, 1880.03 (cell D25), is the variance between the two types of cola. The Within or error term, 1839.315 (cell D28), is an estimate of σ^2—the variance in foam volume after accounting for the factors of cola brand, type, and the interaction between the two. In other words, after accounting for these effects in your model, the typical deviation—or standard deviation—in foam volume is $\sqrt{1840} = 42.9$.

You want to determine whether the variability between cola brands or between cola types is large relative to the variability due to random error. To find out, look towards the *F* ratio in column E, which is the ratio of the mean square of the effects (brand, type, or interaction) to the mean square for the Within or error term. A very large *F* ratio indicates that the variance between factor levels of the effect is much higher than the variance that can be attributed to random error, and this indicates that the difference between the factor levels is more than can be explained by the random fluctuations in foam volume.

You can determine whether the effects are significant by referring to the *p*-values (cells F25:F27) shown for each of the three effects in the model. Examine the interaction *p*-value first, which is 0.455 (cell F27)—much bigger than 0.05 and not even close to being significant at the 5% level. Once you've confirmed that no interaction exists, you can look at the main effects.

The column or cola effect is highly significant, with a p-value given as 5.84×10^{-11} (cell F26). This is less than 0.05, so there is a significant difference among colas at the 5% level (because the p-value is less than 0.001, there is significance at the 0.1% level, too). On the other hand, the p-value is 0.318 for the sample or type effect (cell F25), so there is no significant difference between diet and regular colas.

These quantitative conclusions from the analysis of variance are in agreement with the qualitative conclusions drawn from the boxplot (Figure 10-19), namely that there is a significant difference in foam volume between colas but not necessarily between diet and regular colas of the same brand. Nor does there appear to be an interaction between cola brand and type in how they influence foam volume.

Finally, how much of the total variation in foam volume has been explained by the two-way ANOVA model? Recall that the coefficient of determination (R^2 value) is equal to the fraction of the total sum of squares that is explained by the treatment sum of squares. In this case that value is:

$$(1880.003 + 183{,}750.5 + 4903.38) / 264{,}106.5 = 0.721$$

So about 72% of the total variation in foam volume can be attributed to the cola brand, the type (diet or regular), and the interaction between cola and type. Only about 28% is attributed to random error.

The Interaction Plot

The interaction plot is especially useful for seeing interaction in analysis of variance. You can create an interaction plot for these data using a pivot table.

To set up the pivot table for creating an interaction plot for foam volume versus cola and type:

1　Click the **Cola Data sheet tab** to return to the data.

2　Click **Data > PivotTable** to start the PivotTable Wizard.

3　Click **Next** for the first step because you are creating a pivot table from an Excel data list.

4　Verify in Step 2 that the Range text box contains **A1:E49**, then click **Next**.

In Step 3 of the PivotTable Wizard you will set up the basic structure of the two-way table.

5　Drag the **Type field button** to the ROW area of the table.

6　Drag the **Cola field button** to the COLUMN area of the table.

7　Drag the **Foam field button** to the DATA area of the table so that **Sum of Foam** appears in the DATA area.

8　Double-click the **Sum of Foam button** to open the PivotTable Field dialog box.

9　Click **Average** in the Summarize by list box, then click **OK**.

10　Click **Next** to move to the Step 4 of 4 dialog box.

11　Deselect the **Grand Totals for Columns** and **Grand Totals for Rows check boxes**, then click **Finish**.

The pivot table shows the eight cell means for the eight combinations of cola and type, as indicated in Figure 10-24.

Figure 10-24
Pivot table of eight cell means

	A	B	C	D	E
1	Average of Foam	Cola			
2	Type	coke	generic	pepsi	shasta
3	diet	283.5166667	239.3166667	141.8	275.7333333
4	regular	331.0333333	240.6	143.0833333	275.7166667

Next create a line plot of the cell means.

To create a line plot of the cell means:

1 Select the range **A1:E4** and click **Insert > Chart > As New Sheet**.

2 In Step 1 of the Chart Wizard, verify that the range **A1:E4** is entered in the Range text box, then click **Next**.

3 Double-click the **Line** chart type.

4 Double-click **2** (the format for two lines without symbols).

5 Verify that the data series is in **Rows**, and that **1** appears in both spin boxes, then click **Finish**.

Your interaction chart should look like Figure 10-25.

Figure 10-25
Interaction plot of foam versus cola with type (diet or regular) appearing as separate lines in the plot

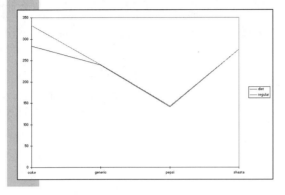

Creating an Interaction Plot

- Create a pivot table of the means for each combination of the two factors.
- Create a line plot using the means calculated in the pivot table.
- If the lines appear parallel, it indicates an absence of interaction. If the lines are not parallel, an interaction might exist.

The plot shows that the diet and regular colas are very close, except for Coke. For the absence of interaction, you are looking for the difference between diet and regular to be the same for each cola brand. This means that the lines would move in parallel, always with the same vertical distance. Of course, there is a certain amount of random variation, so the lines will not be perfectly parallel. The lines in Figure 10-24 are not perfectly parallel, but the analysis of variance found no evidence of significant interaction.

Summary

To summarize the results from the plots and the analysis of variance, there is no significant interaction and no significant difference between regular and diet colas, but there is a significant difference among cola brands. Coke has the highest foam volume, Pepsi has the lowest, and the other two are in between.

To save and close the 10COLA.XLS workbook:

1 Click the **Save button** 🖫 to save your file as 10COLA.XLS (the drive containing your Student Disk should already be selected).

2 Click **File > Exit**.

EXERCISES

1. The FOURYR.XLS workbook has information on 24 colleges from the 1992 edition of *U.S. News and World Report*'s "America's Best Colleges," which lists 140 national liberal arts colleges, 35 in each of four quartiles. A random sample of six colleges was taken from each quartile. The data set includes College, the name of the college; Group, the quartile ranging from 1 to 4; Top10, the percentage of freshmen who were in the top 10% of their high school classes; Spending, the dollars spent by the college per pupil; and Tuition, a year's tuition at the college.

   a. Create a multiple histogram of the Tuition for different group levels (you'll have to unstack the Tuition column first).

   b. Perform a one-way ANOVA to compare Tuition in the four quartiles. Does Group appear to have a significant effect?

   c. Create a matrix of paired mean differences. Is the first quartile significantly more expensive than the others? Is the bottom quartile significantly less expensive than the others?

   d. Summarize your results, asking the question: "On the average, is it more expensive to attend a more highly rated college?"

2. The FOURYR.XLS workbook of Exercise 1 also includes the variable Computer, taken from the *Computer Industry Almanac 1991*. This represents the number of students per computer station, which includes microcomputers, mainframe terminals, and workstations. A low value indicates that the college has plentiful access to computers. Unfortunately, data are not available for all colleges in the sample. Copy the set of colleges for which you have computer data to a new worksheet.

   a. Create a boxplot for Computer versus Group. There is an outlier present. Which college is it?

   b. Perform a one-way analysis of variance to compare quartiles. Are there significant differences among the four groups?

   c. Remove the outlier observation that you noted in part 2a and redo the analysis of variance. Does removal of this observation make much difference in the results? If you find no significant difference, does this mean that access to computers is independent of college quartile? Would you make such an assertion based on data from only 14 colleges?

3. Try using regression analysis instead of ANOVA for the data in Exercise 1.

   a. Regress Tuition on Group (do not use indicator variables). Is the R^2 almost as high as for the ANOVA?

   b. Interpret the Group regression coefficient in terms of the drop in tuition when you move to a higher group number. Conversely, how much extra does it cost to attend a college in a more prestigious group (with a lower group number)?

   c. To test the adequacy of the regression model as compared to the ANOVA model, you can do an *F* test. For testing the null hypothesis that the regression model fits, the *F* ratio is:

$$\frac{(\text{SSE Regression} - \text{SSE ANOVA})/2}{\text{MSE ANOVA}}$$

   In the formula, SSE is the error sum of squares and MSE is the mean square value for error.

   Note: In the Analysis ToolPak regression output, the error sum of squares is the Residual sum of squares in cell C13. In the Analysis of Variance table, it is the Within Groups sum of squares in cell B14; the mean square value for error is the Within Groups Mean Squares in cell D14.

The numerator is divided by 2 because the SSE for regression has two more degrees of freedom than the SSE for ANOVA. You should wind up with an F ratio that is less than 1, which says that you can accept the smaller model, because you would never reject the null hypothesis if F is less than 1. In other words, you can accept the regression model as a good fit. This is convenient, because the regression model has such a simple interpretation—that it costs about $2,000 to move up one group in prestige.

d. Save your workbook as E10FOUR.XLS.

4. The BASEINFD.XLS workbook data set has statistics on 106 major league baseball infielders at the start of the 1988 season. The data include Salary, LN Salary (the logarithm of salary), and Position.

a. Create multiple histograms and boxplots to see the distribution of Salary for each position. Notice how skewed the distributions are.

b. Also make the plots in part 4a for LN Salary to see if the distributions are better.

c. Perform a one-way ANOVA of LN Salary on Position to see if there is any significant difference of salary among positions. How do you explain your results?

5. The BASEINFD.XLS workbook of Exercise 4 also contains RBI Aver, the average runs batted in per time at bat.

a. Create multiple histograms and boxplots of this variable against Position to verify that one-way ANOVA would be appropriate.

b. Perform a one-way ANOVA of RBI Aver against Position.

c. Create a matrix of paired mean differences to compare infield positions (use the Bonferroni correction factor). Which positions differ significantly? Can you explain why?

d. Save your workbook as E10BASE.XLS.

6. The HOTEL2.XLS data set is taken from the same source as the HOTEL.XLS data, except that an effort was made to keep the data balanced for a two-way ANOVA. This means that the random sample was forced to have the same number of hotels in each of 12 cells of city and rating, (four levels of city and three levels of stars). Hotels were excluded unless they had two, three, or four stars, and only two hotels are included per cell. Therefore, the sample has 24 hotels. Included in the file is a variable, City Stars, that indicates the 12 cells.

a. Start by making boxplots and multiple histograms of Price vs. City Stars. Do you see any differences compared with the results for the previous hotel data?

b. Do a two-way ANOVA for Price vs. Stars and City. (You will have to create a two-way table that has Stars as the row variable and City as the column variable). Is there significant interaction? Are the main effects significant?

c. Based on the means for the three levels of Stars, give an approximate figure for the additional cost per star. Give the means for the four cities, and compare these with the means for the previous hotel data.

7. Consider again the data of Exercise 6. In terms of the means of Exercise 6, the gap between two-star hotels and three-star hotels was nearly the same as the gap between three-star and four-star hotels. This suggests a linear relationship.

a. Graph Price vs. Stars, and label each point with the value of City. Does it appear that in each city the relationship between Price and Stars is linear?

b. Do a linear regression of Price vs. Stars for each of the four cities. Do the slopes appear to be the same for the different cities?

c. Save your workbook as E10HOTL2.XLS.

8. In the One-way Analysis of Variance section of this chapter, the Mandarin Hotel was an outlier. Repeat the one-way ANOVA for the HOTEL.XLS data without the Mandarin Hotel. Does it make much difference in the results? Save your workbook as E10HOTEL.XLS.

9. The HONDA25.XLS workbook contains the prices of used Hondas and indicates the age (years) and whether the transmission is 5-speed or automatic.

   a. Perform a two-sample t-test for the Price data based on the Transmission type.

   b. Perform a one-way ANOVA with Price as the dependent variable and Transmission as the grouping variable.

   c. Compare the value of the t statistic in the t-test to the value of the F ratio in the F-test. Do you find that the F ratios for ANOVA are the same as the squares of the t values from the t-test, and that the p-values are the same?

   d. Use one-way ANOVA to compare the ages of the Hondas for the two types of transmissions. Does this explain why the difference in Price is so large?

10. Perform two regressions of Price vs. Age for the HONDA25.XLS workbook discussed in Exercise 9: the first for automatic transmissions and the second for 5-speed transmissions. Compare the two linear regression lines. Do they appear to be the same? What problems do you see with this approach? Save your workbook as E10HONDA.XLS.

11. The workbook HONDA12.XLS contains a subset of the HONDA25.XLS workbook in which the age variable is made categorical and has the values 1-3, 4-5, and 6 or more. Some observations of HONDA25.XLS have been removed to balance the data. The variable Trans indicates the transmission, and the variable Trans Age indicates the combination of transmission and age class.

   a. Create a multiple histogram and boxplot of Price vs. Trans Age. Does the constant variance assumption for a two-way analysis of variance appear justified?

   b. Perform a two-way analysis of variance of Price on Age and Trans (you will have to create a two-way table using Age as the row variable and Trans as the column variable).

   c. Create an interaction plot of Price vs. Trans and Age (you will need to create a pivot table of means for this).

   d. Save your workbook as E10HON12.XLS.

TIME SERIES

O B J E C T I V E S

In this chapter you will learn to:

- Plot a time series

- Compare a time series to lagged values of the series

- Use the autocorrelation function to determine the relationship between past and current values

- Use moving averages to smooth out variability

- Use simple exponential smoothing and two-parameter exponential smoothing

- Recognize seasonality and adjust data for seasonal effects

- Use three-parameter exponential smoothing to forecast future values of a time series

- Optimize the simple exponential smoothing constant

Time Series Concepts

A **time series** is a sequence of observations taken at evenly spaced time intervals. The sequence could be daily temperature measurements, weekly sales figures, monthly stock market prices, quarterly profits, or yearly power-consumption data. Time series analysis involves looking for patterns that help us understand what is happening with the data and help to predict future observations. For some time series data (for example, monthly sales figures), you can identify patterns that change with the seasons. This seasonal behavior is important in forecasting.

Usually the best way to start analyzing a time series is by plotting the data against time to show trends, seasonal patterns, and outliers. If the variability of the series changes with time, the series might benefit from a transformation that stabilizes the variance. Constant variance is assumed in much of time series analysis, just as in regression and analysis of variance, so it pays to see first if a transformation is needed. The logarithmic transformation is one such example that is especially useful for economic data. For example, if there is growth in power consumption over the years, then the month-to-month variation might also increase proportionally. In this case, it might be useful to analyze either the log or the percentage change, which should have a variance that changes little over time.

The Dow in the '80s

To illustrate these ideas, consider the DOW.XLS workbook, which contains the monthly closing Dow Jones averages for 1981 through 1990. Time series analysis can shed light on how the Dow Jones has changed, for better or for worse, over this ten-year period. The DOW.XLS workbook contains the following variables and reference names, as shown in Table 11-1:

Table 11-1

DOW.XLS range names

Range Name	Range	Description
Year	A1:A121	The year
Month	B1:B121	The month
Dow_Jones	C1:C121	The monthly average of the Dow Jones

To open DOW.XLS, be sure that Excel is started and the windows are maximized, and then:

1 Open **DOW.XLS** (be sure you select the drive and directory containing your Student files).

2 Click **File > Save As**, select the drive containing your Student Disk, then save your workbook on your Student Disk as **11DOW.XLS**.

Plotting the Dow Jones Time Series

Before doing any computations it is best to explore the time series graphically.

To plot the Dow Jones average:

1 Select the range **A1:C121**.

2 Click **Insert > Chart > As New Sheet** to open the Chart Wizard.

3 Verify that the Range text box contains the range **A1:C121** in Step 1, then click **Next**.

4 Double-click the **Line** chart type in Step 2.

5 Double-click format **2** (lines without symbols) in Step 3.

6 Verify that the **Data Series in Columns option button** is selected, that the first **2** columns are used for category (X) labels, and that the first **1** row is used for legend text, then click **Next**.

7 Click the **No Legend option button**, type **Dow Jones Averages 1981-1990** in the Chart Title text box, press [**Tab**], type **Year** in the Category (X) text box, press [**Tab**], type **Dow Jones** in the Value (Y) text box, then click **Finish**.

Because the labels crowd each other, the x-axis of the chart is difficult to read. You can format the chart so it shows tick marks and labels at 12-month intervals (which makes sense because each category level on the x-axis represents a month).

To reduce the number of categories on the x-axis:

1 Double click the **x-axis** to open the Format Axis dialog box, then click the **Scale tab**.

2 Type **12** in the Number of Categories between Tick-Mark Labels text box and **12** in the Number of Categories between Tick Marks text box, then click **OK**.

The plot of the Dow Jones average in the '80s is shown in Figure 11-1.

Figure 11-1
Time plot of the Dow Jones average

October 1987 drop

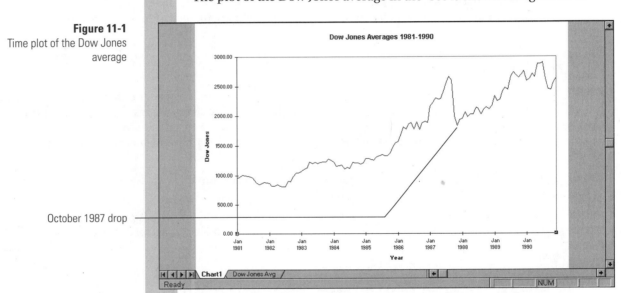

Note: If your axis labels don't look the same, your monitor might not be able to accommodate labels in this form.

The plot in Figure 11-1 shows the amazing growth from near 1,000 to near 3,000 in just ten years. Notice also the big drop that occurred in October 1987, when the Dow dropped 600 points. Throughout the '80s there were prophets of doom saying to get out of the market because the sky was about to fall. Yet in spite of the crash of October 1987, average stockholders profited by just retaining their holdings for these ten years.

As you examine the data, notice that the Dow ranges from around 800 to slightly over 1,000 in 1981, but that the range is about 2,400 to 2,900 in 1990. In ten years the Dow increased by about a factor of three, and the range of the data also increased by a factor of three. Specifically, the range of the data in 1990 is 2905 – 2442 = 463, which is about three times the 1981 range of 1004 – 850 = 154. So as the Dow increased in value, it also increased in variability.

Analyzing the Change in the Dow

For people who own stocks, the changes in the Dow Jones average from month to month are important. Your next step in examining these data is to calculate the month-to-month change.

To calculate the change in the Dow:

1 Click the **Dow Jones Avg sheet tab** to return to the data.

2 Click cell **D1**, type **Diff**, then press [**Enter**].

3 Select the range **D3:D121** (note, *not* D2:D121).

4 Type **=C3-C2** in cell D3, then press [**Enter**].

5 Click **Edit > Fill > Down** or press [**Ctrl**]+**D** to enter the rest of the differences in the range D3:D121.

Now that you have calculated the differences in the Dow from one month to the next, you can plot those differences.

To plot the change in the Dow versus time:

1 Select the range **A1:B121**, press and hold [**Ctrl**], then select the range **D1:D121**.

2 Click **Insert > Chart > As New Sheet** to open the Chart Wizard.

3 Verify that the Range text box contains **A1:B121, D1:D121**, then click **Next**.

4 Double-click the **Line** chart type.

5 Double-click format **3** (symbols without lines).

6 If Excel warns you that the Number must be between 0 and 3, click **OK**. This has no effect on your chart.

7 Verify that the **Data Series in Columns option button** is selected, click the **Columns up spin arrow** so that the first **2** columns are used for Category (X) labels, verify that the first **1** rows are used for legend text, then click **Next**.

8 Click the **No Legend option button**, type **Dow Changes 1981-1990** in the Chart Title text box, press [**Tab**], type **Year** in the Category (X) text box, press [**Tab**], type **Difference** in the Value (Y) text box, then click **Finish**.

9 Once again reformat the **x-axis scale** as you did for the plot of the Dow Jones average so that tick marks and labels appear at 12-month intervals. Your final chart should look like Figure 11-2.

Figure 11-2
Chart of differences in the Dow Jones average in the 1980s

October 1987 outlier

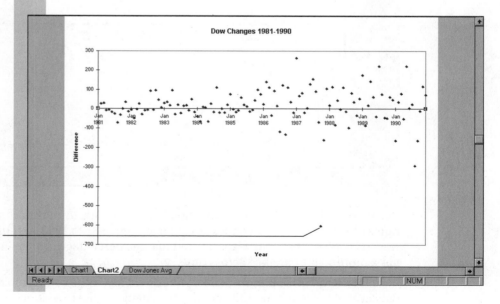

The plot shows steadily increasing variance in the monthly change of the Dow over the ten years. Notice the outlier, the drop of 600 points in October 1987.

Analyzing the Percentage Change in the Dow

Because the variance in the monthly change in the Dow increased through the '80s, it might be useful to transform the values to stabilize the variance, because many statistical procedures require constant variance. As you saw in Chapter 3, a logarithmic transformation is often used in situations where the variance of a data set changes in proportion to the size of the data values. Another transformation you might consider is the percentage change in the Dow by month. Let's calculate this value and see whether it stabilizes the variance.

To calculate the percentage change in the Dow and then chart the data:

1 Click the **Dow Jones Avg sheet tab** to return to the data.

2 Click cell **E1**, type **Perc Diff**, then press [**Enter**].

3 Select the range **E3:E121** (*not* E2:E121).

4 Type **=100*D3/C2** in cell E3 (this is the percentage change from the previous month), then press [**Enter**].

5 Click **Edit > Fill > Down** or press [**Ctrl**]+**D** to enter the rest of the differences in the range E3:E121.

6 Select the range **A1:B121**, press and hold [**Ctrl**], then select the range **E1:E121**.

7 Click **Insert > Chart > As New Sheet**.

8 Use the Chart Wizard to plot the values in the cell range A1:B121, E1:E121. Select the **Line** chart type, format **3**, make sure that the first **2** columns are used for Category Axis labels, give a Chart Title of **Dow % Changes 1981-1990**, a Category (X) label of **Year**, and a Value (Y) label of **% Change**. Reduce the axis scale to a yearly scale as before. The plot on your chart sheet should look like Figure 11-3.

Figure 11-3
Plot of percentage change in the Dow per month

± 10% range

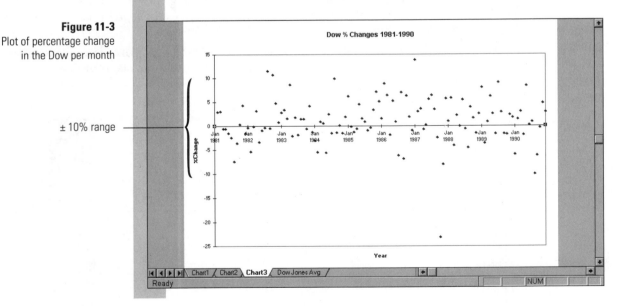

The variance is nearly stable across the plot. The month-to-month changes are mostly in the range of ±10%, and this fact does not change from 1981 to 1990, except for the outlier occurring in October 1987. You've discovered that the percentage change in the Dow per month has a more stable variance across time.

Because the variance is more stable, you should feel more comfortable about applying statistical analysis to the monthly percentage change in the Dow Jones average. As a first step, calculate the average monthly percentage change in the Dow during the '80s.

To calculate descriptive statistics for the percentage change in the Dow:

1   Click the **Dow Jones Avg sheet tab** to return to the data.

2   Click **Tools > Data Analysis**, click **Descriptive Statistics** in the Analysis Tools list box, then click **OK**.

3   Type **E1:E121** in the Input Range text box, verify that the **Group by Columns option button** is selected, then click the **Labels in First Row check box**.

4   Click the **New Worksheet Ply option button**, then type **Per Chng Stats** in the corresponding text box.

5   Click the **Summary statistics check box** to select it, then click **OK**. Figure 11-4 shows the resulting statistics, formatted so that the labels are readable.

Figure 11-4
Descriptive statistics for the percentage change in the Dow

	A	B	C	D	E	F	G	H	I	J	
1	*Perc Diff*										
2											
3	Mean	0.979886									
4	Standard Error	0.439387									
5	Median	0.853089									
6	Mode	#N/A									
7	Standard Deviation	4.793147									
8	Sample Variance	22.97426									
9	Kurtosis	4.870956									
10	Skewness	-0.86781									
11	Range	37.03959									
12	Minimum	-23.2159									
13	Maximum	13.82368									
14	Sum	116.6064									
15	Count	119									
16	Confidence Level(95.000%)	0.861181									
17											

Descriptive statistics for the monthly percentage change in the Dow appear in Figure 11-4. The average percentage change in the Dow during the 1980s is 0.98% (cell B3)—or slightly less than a 1% increase per month. This indicates that on average you would expect the Dow to increase about 1% each month over the previous month's level. However, the standard deviation of the percentage change is 4.79% (cell B7). Because the mean is much less than the standard deviation, the percentage change in any given month could easily be negative. However, because of the positive mean change, the long-run changes are upward. Would you be willing to invest long-term in the market if you could expect an upward trend of about 1% per month? Does it make sense that stock market investors tended to be happy with the Reagan and Bush administrations? On the other hand, would you be careful about investing in the Dow over the short term if the average percentage change from one month to another were negative?

Lagged Values

Often in time series you will want to compare the value observed at one time point to a value observed one or more time points earlier. In the Dow Jones data, for example, you might be interested in comparing the monthly value of the Dow with the value it had the previous month. Such prior values are known as **lagged values**. Lagged values are an important concept in time series analysis. You can lag observations for one or more time points. In the example of the Dow Jones average, the lag 1 value is the value of the Dow one month prior, the lag 2 value is the Dow Jones average two months prior, and so forth.

You can calculate lagged values by letting the values in rows of the lagged column be equal to values one or more rows above in the unlagged column. Let's add a new column to the DOW.XLS workbook consisting of Dow Jones averages lagged one month.

To create a column of lag 1 values for the Dow Jones average:

1   Click the **Dow Jones Avg sheet tab** to return to the data.

2 Right-click the C column header so that the entire column is selected and the short-cut menu opens. Click **Insert** in the shortcut menu.

3 Click cell **C1**, type **Lag1 Dow** and press [**Enter**].

4 Select the range **C3:C121** (*not* C2:C121).

5 Type **=D2** in cell C3 (this is the value from the previous month), then press [**Enter**].

6 Click **Edit > Fill > Down** or press [**Ctrl**]+**D** to enter the rest of the values in the lagged column. See Figure 11-5.

Figure 11-5
The lagged column of Dow
Jones averages

C	D
Lag1 Dow	**Dow Jones**
	947.27
947.27	974.58
974.58	1003.87
1003.87	997.75
997.75	991.75
991.75	976.88
976.88	952.34
952.34	881.47

lagged value ——⌐ 952.34   881.47 ⌐—— observed value

As shown in Figure 11-5, each row of the lagged Dow Jones average is equal to the value of the Dow one row, or one month, prior. You could have created a column of lag 2 values by selecting the range C4:C121 and letting C4 be equal to D2 and so on. Note that for the lag 2 values you have to start two rows down as compared to one row down for the lag 1 values. The lag 3 values would have been put into the range C5:C121.

How do the values of the Dow compare to its values one month prior? To see the relationship between the Dow and the lag 1 value of the Dow, create a scatterplot.

To create a scatterplot of the Dow versus the lagged Dow:

1 Select the range **C1:D121**.

2 Click **Insert > Chart > As New Sheet**.

3 Use the Chart Wizard to plot the values in the cell range C1:D121. Select the **XY(Scatter)** chart type, format **1**, do not add a legend to the plot, enter a Chart Title of **Lagged Dow Values**, a Category (X) label of **Prior Month Value**, and a Value (Y) label of **Dow Jones Average**. The plot on your chart sheet should look like Figure 11-6.

Figure 11-6
Scatterplot of Dow Jones
average versus Dow Jones
average for prior month

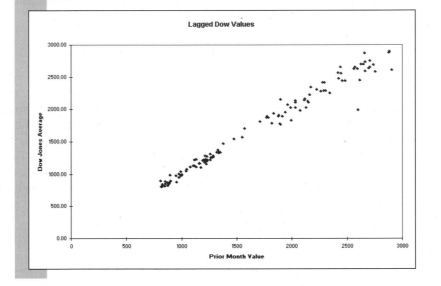

As shown in Figure 11-6, there is a strong positive relationship between the value of the Dow and the value of the Dow in the previous month. This means that a high value for the Dow in one month implies a high (or above average) value in the following month; a low value in one month indicates a low (or below average) value in the next month. In time series analysis we study the correlations among observations, and these relationships are sometimes helpful in predicting future observations. In this example, the Dow is strongly correlated with the value of the Dow in the previous month. The Dow Jones might also be correlated with observations two, three, or more months earlier. To discover the relationship between a time series and other lagged values of the series, statisticians calculate the autocorrelation function.

The Autocorrelation Function

If there is some pattern in how the values of your time series change from observation to observation, you could use it to your advantage. Perhaps a below-average value in one month means that it is more likely the series will be high in the next month. Or maybe the opposite is true—a month in which the series is low makes it more likely that the series will continue to stay low for awhile.

The **autocorrelation function (ACF)** is useful in finding such patterns. It is similar to a correlation of a data series with its lagged values. The ACF value for lag 1 (denoted by r_1) calculates the relationship between the data series and its lagged values. The formula for r_1 is

$$r_1 = \frac{(y_2-\bar{y})(y_1-\bar{y}) + (y_3-\bar{y})(y_2-\bar{y}) + \ldots + (y_n-\bar{y})(y_{n-1}-\bar{y})}{(y_1-\bar{y})^2 + (y_2-\bar{y})^2 + \ldots + (y_n-\bar{y})^2}$$

Similarly, the formula for r_2, the ACF value for lag 2 is:

$$r_2 = \frac{(y_3-\bar{y})(y_1-\bar{y}) + (y_4-\bar{y})(y_2-\bar{y}) + \ldots + (y_n-\bar{y})(y_{n-2}-\bar{y})}{(y_1-\bar{y})^2 + (y_2-\bar{y})^2 + \ldots + (y_n-\bar{y})^2}$$

The general formula for calculating the autocorrelation for lag k is:

$$r_k = \frac{(y_{k+1}-\bar{y})(y_1-\bar{y}) + (y_{k+2}-\bar{y})(y_2-\bar{y}) + \ldots + (y_n-\bar{y})(y_{n-k}-\bar{y})}{(y_1-\bar{y})^2 + (y_2-\bar{y})^2 + \ldots + (y_n-\bar{y})^2}$$

Before considering the autocorrelation of the Dow Jones data, apply these formulae to a smaller data set, as shown in Table 11-2:

Table 11-2
Data set

Observation	Y_n	Y_{n-1}	Y_{n-2}
1	6		
2	4	6	
3	8	4	6
4	5	8	4
5	0	5	8
6	7	0	5

The mean value $\bar{y}$ is 5, y_1 is 6, y_2 is 4, y_3 is 8 and so on through y_n, which is equal to 7. To find the lag 1 autocorrelation, use the formula for r_1 so that

$$r_1 = \frac{(4-5)(6-5) + (8-5)(4-5) + \ldots + (7-5)(0-5)}{(6-5)^2 + (4-5)^2 + \ldots + (7-5)^2}$$

$$= \frac{-14}{40} = -0.35$$

In the same way, the value for r_2, the lag 2 ACF value, is

$$r_2 = \frac{(8-5)(6-5) + (5-5)(4-5) + \ldots + (7-5)(5-5)}{(6-5)^2 + (4-5)^2 + \ldots + (7-5)^2}$$

$$= \frac{-12}{40} = -0.30$$

The values for r_1 and r_2 imply a negative correlation between the current observation and its lag 1 and lag 2 values (that is, the previous two values). So a low value at one time point indicates high values for the next two time points, whereas a low value indicates high values for the next two observations. Knowing how to compute r_1 and r_2, you should be able to compute r_3, the lag 3 autocorrelation. Your answer should be 0.275, a positive correlation, indicating that values of this series are positively correlated with observations three time points earlier.

Recall from earlier chapters that a constant variance is needed for statistical inference in simple regression and also for correlation. The same holds true for the autocorrelation function. The ACF can be misleading for a series with unstable variance, so it might first be necessary to transform for a constant variance before using the ACF.

Applying the ACF to the Dow Jones Average

Now apply the ACF to the Dow Jones average in the 1980s. Neither Excel nor the Analysis ToolPak provides an ACF function, but one has been provided with the CTI Statistical Add-Ins.

To compute the autocorrelation function for the Dow Jones average:

1   Click the **Dow Jones Avg sheet tab** to return to the data.

2   Click **CTI > ACF Plot**.

3   Type **Dow_Jones** in the Input Range text box (or **D1:D121**), then verify that the **Selection Includes Header Row check box** is selected. The ACF command allows you to specify the largest lag for which you want the autocorrelation.

4   Enter **18** in the Calculate ACF up through lag spin box to calculate the autocorrelations between the Dow and values of the Dow up to 18 months earlier.

5   Click the **New Sheet option button**, then type **Dow Jones ACF** in the corresponding text box. Your dialog box should look like Figure 11-7.

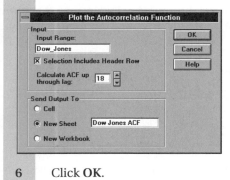

Figure 11-7
Completed Plot the Autocorrelation Function dialog box

6   Click **OK**.

The output shown in Figure 11-8 lists the lags from 1 to 18 in column A and gives the corresponding autocorrelations in the next column.

The lower and upper ranges of the autocorrelations are shown in columns C and D and indicate how low or high the correlation needs to be for significance at the 5% level. Autocorrelations that lie outside this range are shown in red (if you have a color monitor). The plot of the ACF values and confidence width gives a visual picture of the patterns in the data. The two curves indicate the width of the 95% confidence interval of the autocorrelations.

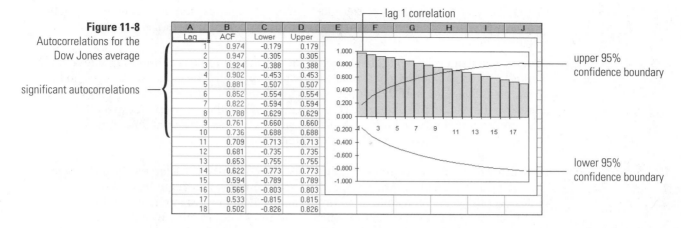

Figure 11-8
Autocorrelations for the
Dow Jones average

significant autocorrelations

Lag	ACF	Lower	Upper
1	0.974	-0.179	0.179
2	0.947	-0.305	0.305
3	0.924	-0.388	0.388
4	0.902	-0.453	0.453
5	0.881	-0.507	0.507
6	0.852	-0.554	0.554
7	0.822	-0.594	0.594
8	0.788	-0.629	0.629
9	0.761	-0.660	0.660
10	0.736	-0.688	0.688
11	0.709	-0.713	0.713
12	0.681	-0.735	0.735
13	0.653	-0.755	0.755
14	0.622	-0.773	0.773
15	0.594	-0.789	0.789
16	0.565	-0.803	0.803
17	0.533	-0.815	0.815
18	0.502	-0.826	0.826

upper 95%
confidence boundary

lower 95%
confidence boundary

Calculating the ACF for a Time Series

- Choose ACF Plot from the CTI menu.

- Enter the input range, indicate whether the range includes a header row, and enter the largest lag for which you want the autocorrelation.

- Specify where you want to send the output.

The autocorrelations are very high for the lower lag numbers, and they remain significant (that is, they lie outside the 95% confidence width boundaries) through lag 10. Specifically, the correlation between the Dow Jones average and the value for the Dow in the previous month is 0.974 (cell B2). The correlation between the Dow and the Dow's value 12 months prior is the correlation for lag 12, 0.681 (cell B13). This is typical for a series that has a strong trend upward or downward. Given the increase in the Dow during the '80s, it shouldn't be surprising that high values of the Dow are correlated with previous high values. In such a series, if an observation is above the mean, then its neighboring observations are also likely to be above the mean, and the autocorrelations with nearby observations are high. In fact, when there is a trend, the autocorrelations tend to remain high even for high lag numbers.

Other ACF Patterns

Other time series show different types of autocorrelation patterns. Figure 11-9 shows four examples of time series: trend, cyclical, oscillating, and random, along with their associated autocorrelation functions.

Figure 11-9
Four sample time series with
their ACF patterns

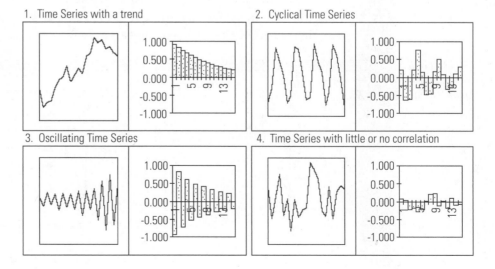

1. Time Series with a trend

2. Cyclical Time Series

3. Oscillating Time Series

4. Time Series with little or no correlation

You have already seen the first example with the Dow Jones data. The trend need not be increasing; a decreasing trend also produces the type of ACF pattern shown in the first example in Figure 11-9.

The seasonal or cyclical pattern shown in the second example is common in sales data that follow a seasonal pattern (such as sales of winter apparel). The length of the cycle in this example is five, indicated in the ACF by the large positive autocorrelation for lags 5 and 10. Because the data in the time series follow a cycle of length 5, you would expect that values five units apart would be highly correlated with each other. Seasonal time series models are covered more thoroughly later in this chapter.

The third example shows an oscillating time series. In this case, a large value is followed by a low value and then by another large value. An example of this might be winter and summer sales over the years for a large retail toy company. Winter sales might always be above average because of the holiday season, whereas summer sales might always be below average. This pattern of oscillating sales might continue and could follow the pattern shown in example 3. The ACF for this time series has an alternating pattern of positive and negative autocorrelations.

Finally, if the observations in the time series are independent or nearly independent, there is no discernible pattern in the ACF and all the autocorrelations should be small, as shown in the fourth example.

There are many other possible patterns of behavior for time series data besides the four examples shown here.

Applying the ACF to the Percentage Change in the Dow

Having looked at the autocorrelation function for the Dow, let's look at the ACF for the monthly percentage change. You could argue that the percentage change is what really matters in the study of the stock market. The ACF tells you whether the percentage change in the Dow can be related to percentage changes observed in previous months.

To calculate the autocorrelation for the percentage change in the Dow:

1   Click the **Dow Jones Avg sheet tab** to return to the data.

2   Click **CTI > ACF Plot**.

3   Type **F3:F121** in the Input Range text box, then click the **Selection Includes Header Row check box** to deselect it because this selection does not include a header row (if you include the blank cell, F2, you will get an erroneous value for the autocorrelation).

4   Enter **18** in the Calculate ACF up through lag spin box.

5   Click the **New Sheet option button**, type **Per Chng ACF** in the corresponding text box, then click **OK**. The ACF plot appears as in Figure 11-10.

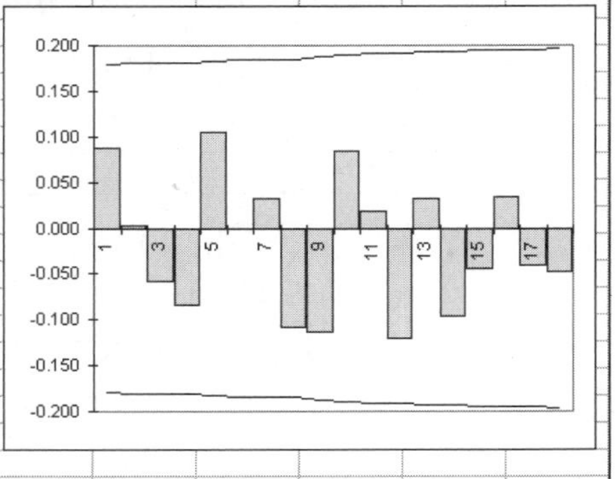

Figure 11-10
Output of the ACF macro for the percentage change

	A	B	C	D
1	Lag	ACF	Lower	Upper
2	1	0.087	-0.180	0.180
3	2	0.003	-0.181	0.181
4	3	-0.060	-0.181	0.181
5	4	-0.085	-0.182	0.182
6	5	0.105	-0.183	0.183
7	6	-0.002	-0.185	0.185
8	7	0.032	-0.185	0.185
9	8	-0.110	-0.185	0.185
10	9	-0.115	-0.187	0.187
11	10	0.085	-0.189	0.189
12	11	0.017	-0.191	0.191
13	12	-0.122	-0.191	0.191
14	13	0.032	-0.193	0.193
15	14	-0.098	-0.193	0.193
16	15	-0.045	-0.195	0.195
17	16	0.034	-0.195	0.195
18	17	-0.042	-0.195	0.195
19	18	-0.049	-0.196	0.196

The autocorrelations of Figure 11-10 show nothing significant; in fact, none of the autocorrelations is even close to being significant. This suggests that there is not much correlation between past and future changes in the market. This is characteristic of the **random walk model** for price movements, which says that market changes in different time periods are independent random variables with mean = 0. Past changes are useless in predicting the future.

On the other hand, the ten years of untransformed Dow Jones data showed a definite upward trend, with changes in different time periods not averaging to 0. So the random walk model is not valid for the monthly Dow Jones averages. These ten years of past changes suggest that the future is predictable—that changes will be generally positive and that it pays to invest in the market. Although the recent past and the long-term trend over the history of the Dow Jones average have been kind to investors, the upward trend is not necessarily permanent. If the market loses its upward trend over an extended period, then it might require a change of terminology. Stock market "investors" might need to be called stock market "gamblers"!

Moving Averages

As you saw with the percentage change in the Dow Jones average, time series data can fluctuate unpredictably from one observation to the next. To smooth the unpredictable ups and downs of a time series, you can form averages of consecutive observations. For example, if you wanted to see if the present observation is an improvement over the recent past, you can compute the average of the last ten observations and compare that value to the present observation.

If you calculate the average of the last ten months, and you do this every month, you are forming a **moving average**. Specifically, to calculate the ten-month moving average for values prior to the observation, y_n, you define the moving average, $y_{ma(10)}$, such that

$$y_{ma(10)} = \frac{\left(y_{n-1} + y_{n-2} + y_{n-3} + y_{n-4} + y_{n-5} + y_{n-6} + y_{n-7} + y_{n-8} + y_{n-9} + y_{n-10}\right)}{10}$$

Excel provides the ability to add a moving average to a scatterplot using the Insert Trendline command. Let's add a ten-month moving average to the percentage change in the Dow in the 1980s.

To add a moving average to a scatterplot:

1   Click the **Chart3 sheet tab** to go to the chart sheet containing the scatterplot of % Change in the Dow versus Year (this is the chart shown in Figure 11-3).

2   Right-click the **data series** (any data value) on the chart to select it and open the shortcut menu.

3   Click **Insert Trendline** in the shortcut menu.

4   Click the **Type tab** if it is not already in the forefront, click the **Moving Average** trend type, then click the **Period up spin arrow** until **10** appears as the period. Your dialog box should look like Figure 11-11.

Figure 11-11
Completed Trendline
dialog box

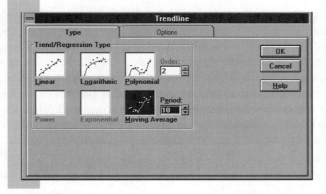

5   Click **OK**, then click outside the chart to deselect the data series. The moving average curve appears as in Figure 11-12.

Figure 11-12
Scatterplot of % Change in
the Dow versus Year with a
ten-month moving average

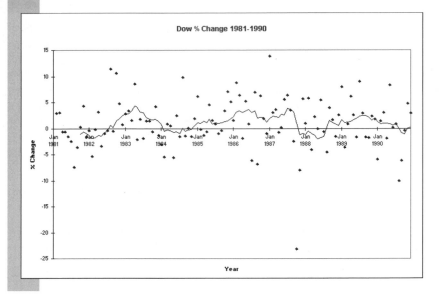

The addition of the ten-month moving average to the scatterplot smooths the ups and downs of the data. It's clear that for much of the '80s the ten-month moving average of the percentage change in the Dow remained above zero. Moreover, there appear to be three periods or phases of sustained above-average growth: from July of 1982 through January 1984, January 1985 to September 1987, and from July 1988 to July 1990. In each of these three periods the ten-month moving average remained above zero. The only period of sustained decline, as characterized by the moving average, is from October 1987 to October 1988, in which the moving average was constantly below zero. This is because you included the October 1987 value in the average.

....

Adding a Moving Average to a Chart

- Right-click the data series in the chart to which you want to add a moving average.
- Click Insert Trendline in the shortcut menu.
- Click Moving Average as the trendline type and enter the length of the period.

....

Simple Exponential Smoothing and Forecasting

Why choose ten as a moving average for the Dow? A strategy frequently used by some stock market investors is to buy if the current price of the stock exceeds the average of its ten most recent prices. Conversely, if the present price is lower than the average of the last ten, it is time to sell. In his book *Stock Market Logic*, Norman Fosback devotes Chapter 41 to the topic of moving averages. Using ten-week moving averages applied to the Standard and Poor's 500 Index, he found that the market did better after the price was above the moving average than when it was below the moving average. He compared prices three months later and also six months later. Although the differences were not great, this does lend support to the strategy of buying when the market goes above its moving average.

Although Fosback's research showed some effectiveness for ten-period moving averages, he was also critical of this method. An important point to remember with regard to using moving averages is that all prior observations within the moving-average interval are weighted equally in the calculation. That is a large part of the reason why there was a negative value for the moving average of the percentage change in the Dow after October 1987. Even in August 1988, the October '87 drop was weighted as heavily in calculating the average as the more recent July '88 values.

Many analysts advocate an average that gives greater weight to more recent prices; in particular, an exponentially weighted moving average, which uses weights that drop off exponentially. This moving average does not only include the ten most recent observations, but it gives some weight to all observations in the past. In general, the most recent observation gets weight w, the one before that gets weight $w(1 - w)$, the one before that gets weight $w(1 - w)^2$, etc. Here w is called a **smoothing factor** or **smoothing constant**. This technique is called **exponential smoothing** or, specifically, **one-parameter exponential smoothing**, because the weight assigned to an observation k units prior to the current observation is equal to $w (1 - w)^{k-1}$. The value of w ranges from 0 to 1. Table 11-3 gives the weights for prior observations under different values of w.

w	y_{n-1}	y_{n-2}	y_{n-3}	y_{n-4}	y_{n-5}	y_{n-6}	y_{n-7}	y_{n-8}	y_{n-9}	y_{n-10}
0.01	0.0100	0.0099	0.0098	0.0097	0.0096	0.0095	0.0094	0.0093	0.0092	0.0091
0.15	0.1500	0.1275	0.1084	0.0921	0.0783	0.0666	0.0566	0.0481	0.0409	0.0347
0.45	0.4500	0.2475	0.1361	0.0749	0.0412	0.0226	0.0125	0.0069	0.0038	0.0021
0.75	0.7500	0.1875	0.0469	0.0117	0.0029	0.0007	0.0002	0.0000	0.0000	0.0000

As the table indicates, different values of w cause the weights assigned to previous observations to change. For example, when w equals 0.01, approximately equal weight is given to a value from the most recent observation and to values observed ten units earlier. On the other hand, when w has the value of 0.75, the weight assigned to previous observations quickly drops so that values collected more than six units prior to the current time receive essentially no weight. In a sense you could say that as the value of w approaches zero, the smoothed average has a longer memory, whereas as w approaches 1, the memory of prior values becomes shorter and shorter.

Exponential smoothing is often used to forecast the value of the next observation, given the current and prior values. In this situation, you already know the value of y_n and are trying to give a forecast of the value of y_{n+1}. This forecasted value is called S_n. Using exponential smoothing, the value of S_n is

$$S_n = wy_n + w(1-w)y_{n-1} + w(1-w)^2 y_{n-2} + \dots + w(1-w)^{n-1} y_1$$

The value of S_n is more commonly written in an equivalent recursive formula where

$$S_n = wy_n + (1-w)S_{n-1}$$

so that S_n is equal to a weighted average of the current observation and the previous forecasted value.

To create a series of exponentially smoothed values, an initial value S_0 is required. Some statisticians use $S_0 = y_1$, the initial observation, whereas others recommend using the average of all values in the time series as the initial value for S_0. The examples in this chapter use the initial value of the time series. Once you determine the value of S_0 you can generate the exponentially smoothed values as follows:

$$S_1 = wy_1 + (1-w)S_0$$
$$S_2 = wy_2 + (1-w)S_1$$
$$\vdots$$
$$S_n = wy_n + (1-w)S_{n-1}$$

The One-parameter Exponential Smoothing Template

To help you get a visual image of the impact that differing values of *w* have on smoothing the data and forecasting the next value in the series, an instructional template, SMOOTH1.XLT, has been provided for you.

To save your work in 11DOW.XLS and then open the SMOOTH1.XLT template:

1   Click the **Save button** 🖫 to save your file as 11DOW.XLS (the drive containing your Student Disk should already be selected).

2   Open **SMOOTH1.XLT** (be sure you select the drive and directory containing your Student files).

The template, shown in Figure 11-13, opens to display the percentage change in the Dow Jones average per month in the 1980s.

Figure 11-13
The SMOOTH1.XLT template

observed line

forecasted line

area curve indicates weights

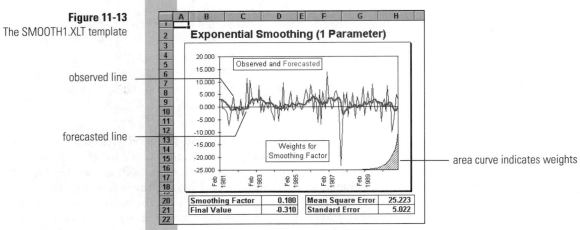

This template shows the observed values overlaid with a one-parameter exponentially smoothed curve (really a series of connected line segments). As you can see from cell D20, the smoothing factor *w* is set at 0.18. In the lower-right corner, the template contains an area curve that indicates the relative weight assigned to prior observations, as shown in Figure 11-14.

Figure 11-14
Close-up of the area curve
from the chart

most recent observation has
most weight

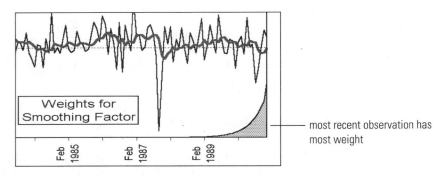

The most recent observation has the most weight, with observations exponentially decreasing in importance so that in this case an observation more than two years old becomes negligible in calculating the smoothed average. Comparing the curve to the time series tells you that the large drop the Dow experienced in October 1987 has little weight in estimating the percentage change in the Dow for the final month of the series. In fact, the memory of the final forecasted value is negligible beyond February 1989.

The chosen smoothing factor results in a forecast curve that is much less variable than the observed percentage changes in the Dow. The smoothed values are also less susceptible to the influence of large outlying values. Using this value for *w*, the final value is –0.310 (cell D21), estimating a drop in the Dow of 0.31%.

A measure of the success of the exponential smoothing in predicting the value of the percentage change in the Dow is indicated by the mean-square error value, which is the average squared difference between the observed values in the series and the forecasted values. The standard error is the square root of the mean square error and indicates the typical error in forecasting. In this example, the mean square error is 25.223 (cell H20) and the standard error is 5.022 (cell H21), showing that if you had used exponential smoothing on these data throughout the '80s, your typical error in forecasting would have been about five percentage points. One way of choosing a value for the smoothing constant is to pick the value that results in the lowest mean-square error. Let's see what happens to the mean-square error when you decrease the value of the smoothing constant.

To decrease the value of the smoothing constant:

1 Click cell **D20**, type **0.03**, then press [**Enter**]. The forecasted curve and the area curve change dramatically, as shown in Figure 11-15.

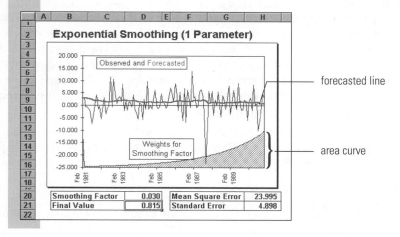

With such a small value for w, the smoothed value has a long memory. In fact, the final forecasted value, 0.815, is based in some part on observations spanning the entire 1980s. A consequence of having such a small value for w is that individual events, such as the crash of '87, have a minor impact on the smoothed values. The line of forecasted values is practically straight. Note as well that the standard error has declined from 5.022 to 4.898. This decline shouldn't be too surprising. Recall that earlier in the chapter you showed that there were no significant autocorrelations for the percentage change data, and that in fact you could regard the time series values as independent. In that case, the percentage change is best estimated by the overall average or smoothed value that has a long memory.

Now increase the value of the smoothing factor to make the forecasts *more* susceptible to month-by-month change.

To increase the smoothing factor:

1 Click cell **D20**, type **0.60**, then press [**Enter**]. Now the forecasted line more closely resembles the shape of the observed values, as shown in Figure 11-16.

Figure 11-16

Template with a smoothing
factor equal to 0.60

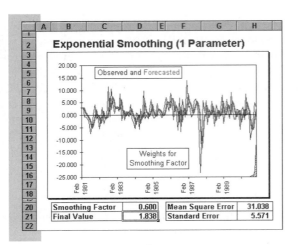

With a larger value for w, the forecasted values shown in Figure 11-16 are much more variable—almost as variable as the data. This is because the forecasted values are more like the data, and observations taken more than a few months earlier receive hardly any weight at all in the calculation. If an observation has a large upward swing, then the forecasted value for the next observation tends to be high, whereas the actual value might revert to a lower value. So even though the smoothed line better resembles the shape of the data, it does not necessarily forecast the data values better. The standard error of the forecasts has increased with this larger value for w up to 5.571.

Continue trying differing values for w to see how w affects the smoothed curve and the standard error of the forecasts. Can you find a value for w that results in forecasts with the smallest standard error? (Later in this chapter you'll learn how to estimate this value.)

To close the template:

1 Click **File > Close**, then click **No**. Because this is an instructional template, you do not need to save the changes.

Choosing a Value for w

As you saw in the instructional template, you have to choose the value of w with care. When choosing a value, keep several factors in mind. Generally you want the standard error of the forecasts to be low, but this is not the only consideration. The value of w that gives the lowest standard error might be very high (such as 0.9) so that the exponential smoothing does not result in very smooth forecasts. If your goal is to simplify the appearance of the data or to spot general trends, you would not want to use such a high value for w even if it produced forecasts with a low standard error. Analysts generally favor values for w ranging from 0.01 to 0.3. Fosback advocates using a value of 0.18 for w. However, for a particular set of time series data, choosing the appropriate value for w might be a trial-and-error process.

Choosing appropriate parameter values for exponential smoothing is often based on intuition and experience. Nevertheless, exponential smoothing has proven to be valuable in forecasting time series data.

Using Forecasting to Guide Decisions

Recall that a strategy sometimes used by investors is to purchase a stock if the current price exceeds the average of its ten most recent prices. Let's recast this idea using exponential smoothing so that it is time to buy if the stock's price is higher than was forecasted in the previous month, and time to sell if its price is lower than was forecasted. The strength of the buy or sell signal is based on how large a difference exists between the stock's observed value and its forecasted value. Note that the buy signal is not a forecasted value, but an indicator of whether the stock's value is expected to increase in the next month. A large positive difference would strongly indicate (under this strategy) that you should buy, whereas a large negative difference indicates you should sell. Let's see how well this strategy would have performed in the 1980s.

The ability to perform exponential smoothing on time series data has been provided for you with the CTI Statistical Add-Ins. Using the add-ins, let's smooth the Dow Jones average from the '80s.

To create exponentially smoothed forecasts of the Dow Jones data:

1   If you are not already in the **11DOW.XLS** workbook, click **Window > 11DOW.XLS**.

2   Click the **Dow Jones Avg sheet tab** to go to the Dow Jones data (you might have to scroll to see it).

3   Click **CTI > Exponential Smoothing**.

4   Type **Dow_Jones** in the Input Range text box and verify that the **Selection Includes Header Row check box** is selected.

5   Click the **New Sheet option button**, then type **Smoothed Dow** in the corresponding text box.

6   Double-click the **General Options Weight text box** and type **0.18** (recall that this is the value that Fosback recommends). Your dialog box should look like Figure 11-17.

Figure 11-17
Completed Perform
Exponential Smoothing
dialog box

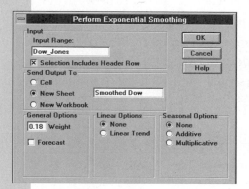

7   Click **OK**. The Smoothed Dow worksheet appears, as shown in Figure 11-18.

Figure 11-18
Exponential smoothing
output

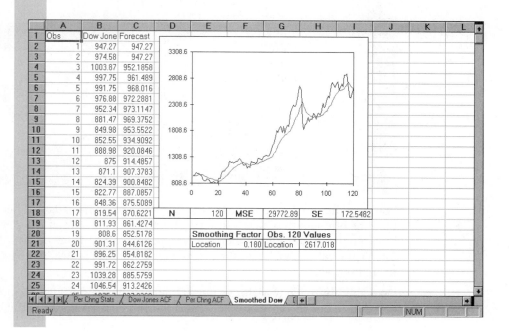

The output shown in Figure 11-18 consists of three columns: the observation number, the observed Dow Jones average, and the Dow Jones value forecasted for each particular observation. The chart indicates the exponentially smoothed values plotted with observed values. It appears that the forecasted values generally underestimated the value of the Dow Jones throughout the 1980s. Of course this is caused by the extraordinary growth. The standard error of the forecasts is 172.5482 (cell I18), indicating that the typical forecasting error was about 170 points in the '80s with this forecasting method and w value.

Now let's use the forecasted values to ascertain the reliability of our stock-purchasing rule of thumb. First you'll have to copy the forecasted values into the Dow Jones data worksheet.

To copy the data:

1 Highlight the range **C1:C121** on the Smoothed Dow sheet, then click the **Copy button** 🗎.

2 Click the **Dow Jones Avg sheet tab** (you might have to scroll to see it).

3 Click cell **G1**, then click the **Paste button** 🗎.

To calculate the buy or sell signal for each month, calculate the difference between the observed value and the forecasted value. Express this difference in terms of the percentage difference between the forecasted value and the observed value.

To calculate the percentage difference between the forecasted and observed values:

1 Click cell **H1**, type **Buy Signal**, then press **[Enter]**.

2 Highlight the range **H3:H121** (note, *not* H2:H121).

3 Type **=(D2–G2)/G2*100** in cell H3, then press **[Enter]**.

4 Click **Edit > Fill > Down** or press **[Ctrl]+D** to enter the rest of the values in the column. Your values should appear as in Figure 11-19.

Figure 11-19
Buy Signal values in column H

	A	B	C	D	E	F	G	H	I	J	K	L
1	Year	Month	Lag1 Dow	Dow Jones	Diff	Perc Diff	Forecast	Buy Signal				
2	1981	Jan		947.27			947.27					
3	1981	Feb	947.27	974.58	27.31	2.883022	947.27	-2.1E-06				
4	1981	Mar	974.58	1003.87	29.29	3.005397	952.1858	2.88302				
5	1981	Apr	1003.87	997.75	-6.12	-0.60964	961.489	5.427951				
6	1981	May	997.75	991.75	-6.00	-0.60135	968.016	3.771342				
7	1981	Jun	991.75	976.88	-14.87	-1.49937	972.2881	2.451823				
8	1981	Jul	976.88	952.34	-24.54	-2.51208	973.1147	0.472274				
9	1981	Aug	952.34	881.47	-70.87	-7.44167	969.3752	-2.13486				
10	1981	Sep	881.47	849.98	-31.49	-3.57244	953.5522	-9.06823				
11	1981	Oct	849.98	852.55	2.57	0.30236	934.9092	-10.8617				
12	1981	Nov	852.55	888.98	36.43	4.273063	920.0846	-8.80933				
13	1981	Dec	888.98	875.00	-13.98	-1.57259	914.4857	-3.38062				
14	1982	Jan	875.00	871.10	-3.90	-0.44571	907.3783	-4.3178				
15	1982	Feb	871.10	824.39	-46.71	-5.36219	900.8482	-3.99815				
16	1982	Mar	824.39	822.77	-1.62	-0.19651	887.0857	-8.48735				
17	1982	Apr	822.77	848.36	25.59	3.110225	875.5089	-7.25022				
18	1982	May	848.36	819.54	-28.82	-3.39714	870.6221	-3.10093				
19	1982	Jun	819.54	811.93	-7.61	-0.92857	861.4274	-5.86731				
20	1982	Jul	811.93	808.60	-3.33	-0.41013	852.5178	-5.74597				
21	1982	Aug	808.60	901.31	92.71	11.4655	844.6126	-5.15154				
22	1982	Sep	901.31	896.25	-5.06	-0.56141	854.8182	6.712824				
23	1982	Oct	896.25	991.72	95.47	10.65216	862.2759	4.846855				
24	1982	Nov	991.72	1039.28	47.56	4.795708	885.5759	15.01191				
25	1982	Dec	1039.28	1046.54	7.26	0.698561	913.2426	17.3564				

Dow Jones ACF / Per Chng ACF / Smoothed Dow \ **Dow Jones Avg** /

Ready   NUM

You start the Buy Signal values in the third row of the spreadsheet because you want to compare the buy signals with the resulting behavior of the market. To do this you have to lag the buy signal one row. Remember, a buy signal occurs if the observed value is greater than the forecasted value. The larger the percentage difference between the two, the stronger the signal to buy.

Does this approach work for values of the Dow during the '80s? To find out, calculate the correlation between the buy signal and the percentage change in the Dow Jones average.

To calculate the correlation:

1 Click **CTI > Correlation Matrix.**

2 Type the range **F3:F121,H3:H121** in the Input Range text box (you can also select this range by dragging your mouse over the cells).

3 Because this range does not include a header row, click the **Selection Includes Header Row check box** to deselect it.

4 Click the **Spearman Rank Correlation check box** to deselect it.

5 Click the **New Sheet option button**, type **Buy Sig Corr** in the corresponding text box, then click **OK**. The correlation matrix appears as in Figure 11-20.

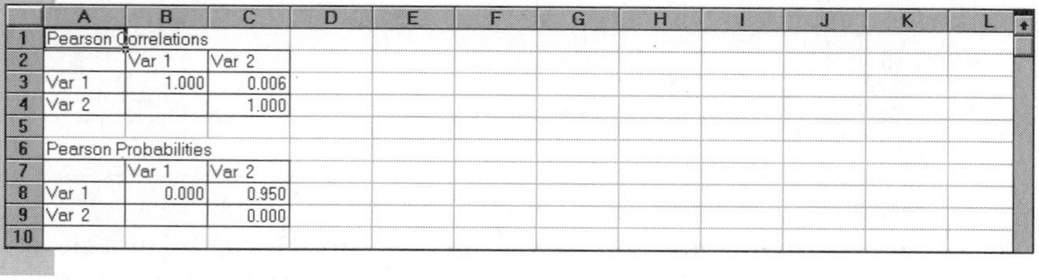

	A	B	C
1	Pearson Correlations		
2		Var 1	Var 2
3	Var 1	1.000	0.006
4	Var 2		1.000
5			
6	Pearson Probabilities		
7		Var 1	Var 2
8	Var 1	0.000	0.950
9	Var 2		0.000
10			

Figure 11-20
Correlation matrix output

The output as shown in the correlation matrix in Figure 11-20 indicates that there is no relationship between the buy signal and the subsequent behavior of the market in the following month. The correlation is 0.006 with a *p*-value of 0.95. The lack of significance is not surprising considering the low value for the lag 1 autocorrelation shown in Figure 11-10. Because there is no indication of significant correlation between the previous month's percentage change value and the current percentage change in the Dow, it is reasonable that a previous month's buy signal would not correlate well with the percentage change either.

To verify that this is the case, create a scatterplot of the percentage change in the Dow versus the buy signal.

To create a scatterplot of percentage change versus the buy signal:

1 Click the **Dow Jones Avg sheet tab** to return to the data worksheet.

Unfortunately the variable you want to plot for the y-axis (the percentage change in the Dow) is located to the left of the variable you want to plot on the x-axis. You will have to move the percentage difference column before creating the plot. For the sake of order and clarity in your worksheet, move the difference column as well.

2 Highlight columns E and F (use the column headers).

3 Right-click the selection and click **Cut** in the shortcut menu.

4 Highlight columns **I** and **J** (use the column headers).

5 Right-click the selection and click **Paste** in the shortcut menu.

Now you can create the scatterplot of percentage change versus buy signal.

6 Highlight the range **H3:H121, J3:J121** (press and hold **[Ctrl]** as you select the second range).

7 Click **Insert > Chart > As New Sheet.**

8 Using the Chart Wizard, create a **XY (Scatter)** chart, with format **1** (points but no lines), the data series in **Columns**, using the first **1** column for X data and the first **0** row for legend text. Do not add a legend to the plot. Use a Chart Title of **Percent Change in the Dow vs. Buying Signal**, a Category (X) axis title of **Buying Signal**, and **Percent Change** as the Value (Y) axis title. Your scatterplot should look like Figure 11-21.

Figure 11-21
Scatterplot of Percent
Change versus Buying
Signal

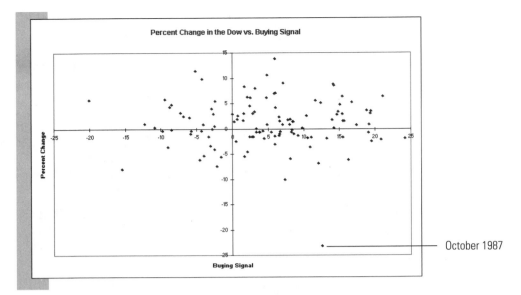

In agreement with the correlation of Figure 11-20, the plot shows no trend at all. The correlation says that there is no apparent linear relationship, and the plot says that there is no apparent relationship of any kind. Notice the outlying point at the lower right, which comes from the October 1987 market crash.

The crash illustrates that buy signals based on forecasted values can be bad advice. The two most negative signals were given by the moving average in October and November of 1987, after the market had plunged. At the end of October the Dow was 15.5% below its moving average, and at the end of November the Dow was 20.1% below its moving average. Had an investor followed these sell signals and sold all holdings, it would have been a big mistake. As shown in Figure 11-1, the market gained rapidly after November 1987.

Given the failure of the ACF and exponential smoothing to predict short-term market activity, you might expect that no simple investment formula exists. That might indeed be the case. If it were a simple matter to predict short-term market activity, we would all be doing it!

To close the 11DOW.XLS workbook and save it:

1 Click **File > Close**.

2 Click **Yes** when prompted to save changes.

Two-parameter Exponential Smoothing

The two-parameter exponential smoothing method builds upon simple exponential smoothing by adding a new factor—the trend, or slope, of the time series. To see how this works, let's express simple exponential smoothing in terms of the following equation for y_t—the value of y at time t:

$$y_t = \beta_0 + \varepsilon_t$$

where β_0 is the **location parameter** that changes slowly over time and ε_t is the random error at time t. If β_0 were constant throughout time you could estimate its value by taking the average of all the observations. Using that estimate, you would forecast values that would always be equal to your estimate of β_0. However, because β_0 varies with time, you weight the more recent observations more heavily than distant observations in any forecasts you make. Such a weighting scheme could involve exponential smoothing. How could such a situation occur in real life? Consider tracking crop yields over time. The average yield could slowly change over time as equipment or soil science technology improves. An additional factor in changing the average yield would be the weather, because a region of the country might go through several years of drought or good weather.

Now suppose the values in the time series follow a linear trend so that the series is better represented by

$$y_t = \beta_0 + \beta_1 t + \varepsilon_t$$

where β_1 is the **trend parameter** whose value might also change over time. If β_0 and β_1 were constant throughout time, you could estimate their values using simple linear regression. However, when the values of these parameters change, you can apply **two-parameter exponential smoothing**, also known as **Holt's Method**, to estimate the values of the parameters at a particular time t. This type of smoothing estimates a line fitting the time series with more weight given to recent data and less weight given to distant data. A county planner might use this method to forecast the growth of a suburb. The planner would not expect the rate of growth to be constant over time. When the suburb is new, it could have a very high growth rate, which might change as the area becomes saturated with people, as property taxes change, or as new community services are added. In forecasting the probable growth of the community, the planner tends to weight recent growth rates much higher than older ones.

Calculating the Smoothed Values

The formulas for two-parameter smoothing are very similar in form to the simple one-parameter equations. Define S_t to be the value of the location parameter forecasted at time t and T_t to be the forecast of the trend parameter. Using the same recursive form as was discussed with single parameter exponential smoothing:

$$S_t = \alpha y_t + (1-\alpha)(S_{t-1} + T_{t-1})$$

$$T_t = \gamma(S_t - S_{t-1}) + (1-\gamma)T_{t-1}$$

and the forecasted value of y_{t+1} is equal to

$$y_{t+1} = S_t + T_t$$

Note that in these equations there are two smoothing parameters, α—the smoothing factor for location—and γ—the smoothing factor for trend. The values of the parameters need not be equal. Although the equations might seem complicated, the idea is fairly straightforward. The value of S_t is a weighted average of the current observation and the previous forecast. The value of T_t is a weighted average of the change in S_t and the previous estimate of trend. As with simple exponential smoothing, you must determine values for the initial estimates S_0 and T_0. One method is to fit a linear regression line to the entire series and use the slope and intercept of the regression as initial estimates.

The Two-parameter Exponential Smoothing Template

An instructional template SMOOTH2.XLT is provided to help you explore two-parameter exponential smoothing.

To open the SMOOTH2.XLT template:

1 Open **SMOOTH2.XLT** (be sure you select the drive and directory containing your Student files). The template appears as in Figure 11-22.

Figure 11-22
The SMOOTH2.XLT template

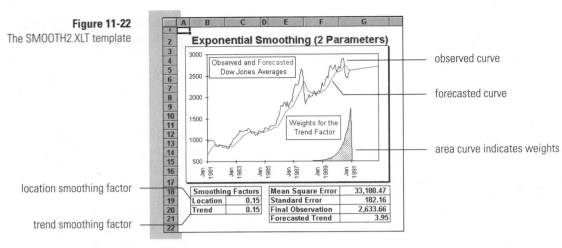

The template shows the now-familiar Dow Jones average from the 1980s. The smoothing factor for the location is equal to 0.15 (cell C19), as is the smoothing factor for trend (cell C20). The area curve indicates the relative weights assigned to the slope values in calculating the smoothed value. For γ equal to 0.15, the smoothed trend value encompasses estimates for a trend from January 1991 back to January 1990. Weights for observations prior to 1990 are too small to be visible on the chart. Based on the two-parameter exponential smoothing, a forecast of the trend for the next 12 months is shown on the chart. For this particular example, the forecasted trend is 3.95 (cell G21), or a projected increase of 3.95 points per month.

The value chosen for γ is important in determining what the forecasted value for the Dow Jones will be. An investor going into the 1990s might assume that the market will behave pretty much as it did during the '80s. Under such an assumption the investor would use a small value for the trend smoothing factor because that would result in an estimate for trend that has a longer "memory" of previous trend values.

To change the smoothing factor for trend:

1   Click cell **C20**, type **0.05**, then press [**Enter**]. The template now looks like Figure 11-23.

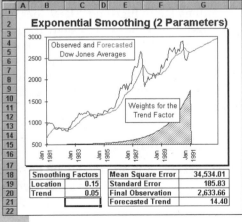

The forecasts for future values of the Dow increase 14.40 points per month, reflecting the belief that there will be an increase in the Dow increase in the '90s similar to what was observed throughout the '80s. Note that the weights for the trend factor as shown by the area curve indicate that observations as far back as January 1985 are well represented in the forecast.

To increase the smoothing factor for trend:

1   Click cell **C20**, type **0.40**, then press [**Enter**]. See Figure 11-24.

Figure 11-24
Smoothed values for the
trend using 0.40 as the
trend smoothing factor

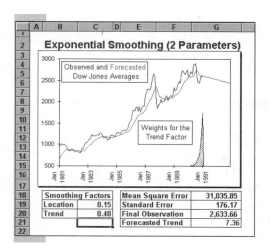

With a higher smoothing factor, the forecasted trend of the Dow declines 7.36 points per month. The area curve indicates that the smoothed trend estimate has a shorter memory—only the previous 12 months are used in estimating the trend line.

Which smoothing factor value you use might depend on what type of investment you're considering. If you are a long-term investor, then averaging trend estimates over a longer period of time might be more indicative of future performance. On the other hand, for short-term investments, the behavior of the market eight years earlier might not be indicative of future short-term prospects. Of course, as you've seen in this chapter, it is very difficult to forecast short-term changes in any case.

Using the instructional template, you can change the values of the smoothing constant for the location and trend factor. What combinations result in the lowest values for the standard error? When you are finished with your investigations, close the workbook.

To close the template:

1 Click **File > Close**, then click **No**. Because this is an instructional template, you do not need to save the changes.

Seasonality

Often time series are measured on a seasonal basis, such as monthly or quarterly. If the data are sales of ice cream, toys, or electric power, there is a pattern that repeats each year. Ice cream and electric power sales are high in the summer, and toy sales are high in December.

Multiplicative Seasonality

If the sales of some of your products are seasonal, you might want to adjust your sales for the season in order to compare figures from month to month. To compare November and December, you need to know the usual difference between November and December sales. Should this difference be expressed in multiplicative or additive terms? Recall that the Dow Jones data had a month-to-month variation that grew as the series itself grew, but the variation of the percentage change per month was relatively stable. Similarly, in many cases seasonal changes are best expressed in percentage terms, especially if there is substantive growth in yearly sales.

As annual sales increase, the difference between the November and December values should also increase, but the ratio of sales between the two months might remain nearly constant. This is called **multiplicative seasonality**. If the sales history shows September usually has high sales, this month would have a multiplicative factor greater than one, but a normally slow month would have a factor of less than one. To compare two months, you divide each by its multiplicative factor.

As an example, consider Table 11-4, which shows seasonal sales and multiplicative factors:

Table 11-4: Seasonal sales and multiplicative factors

	Jan	Feb	Mar	Apr	May	Jun	Jul	Aug	Sep	Oct	Nov	Dec
Factor	0.48	0.58	0.60	0.69	0.59	1.00	1.48	1.69	1.99	1.29	1.02	0.59
Sales	298	378	373	443	374	660	1004	1153	1388	904	715	441
Adjusted Sales	615.58	646.49	619.40	641.01	638.33	662.32	676.46	681.16	698.58	700.94	709.75	741.68

Sales figures from previous years have given you the multiplicative factors in the first row of the table; the second row contains the sales figures from the most recent year. Dividing the sales in each month by the multiplicative factor yields the adjusted sales. Although it is not readily apparent in the unadjusted sales figures, the sales for December were highest in the sense that in December you performed above what was expected for that time of the year. The adjusted sales figures allow you to make such comparisons. On the other hand, the sales figures for January were the lowest, after being adjusted for the time of the year.

Additive Seasonality

Sometimes the seasonal variation is expressed in additive terms, especially if there is not much growth. If the highest annual sales total is no more than twice the lowest annual sales total, it probably does not matter whether you use differences or ratios. If you can express the month-to-month changes in additive terms, the seasonal variation is called **additive seasonality**. For example, the November to December difference might be expected to be about the same every year. Additive seasonality is expressed in terms of differences from the mean for the year. For example, in Table 11-5 the seasonal adjustment for December sales is -240, resulting in an adjusted sales for that month of 681. After adjusting for the time of the year, December turned out to be one of the most successful months, at least in terms of exceeding goals.

Table 11-5: Seasonal sales and additive adjustments

	Jan	Feb	Mar	Apr	May	Jun	Jul	Aug	Sep	Oct	Nov	Dec
Adjustment	-325	-270	-270	-200	-280	-55	350	450	550	220	70	-240
Sales	298	378	373	443	374	660	1004	1153	1388	904	715	441
Adjusted Sales	623	648	643	643	654	715	654	703	838	684	645	681

In this chapter you'll work with multiplicative seasonality only, but you should be aware of the principles of additive seasonality.

Seasonal Example: Beer Production

Is beer production seasonal? The BEER.XLS workbook has U.S. production figures for each month, 1980 through 1991, in thousands of barrels. The workbook contains the following variables and reference names, shown in Table 11-6:

Table 11-6
BEER.XLS range names

Range Name	Range	Description
Year	A1:A145	The year
Month	B1:B145	The month
Barrels	C1:C145	The monthly production of beer in thousands of barrels

To open the BEER.XLS data:

1   Open **BEER.XLS** (be sure you select the drive and directory containing your Student files).

2   Click **File > Save As**, select the drive containing your Student Disk, then save your workbook on your Student Disk as **11BEER.XLS**. See Figure 11-25.

Figure 11-25
11BEER.XLS data

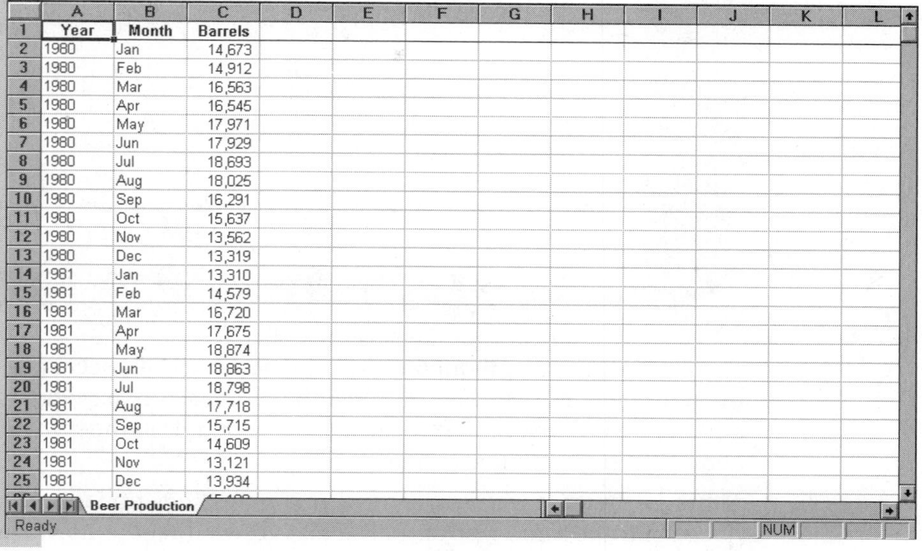

	A	B	C	D	E	F	G	H	I	J	K	L
1	Year	Month	Barrels									
2	1980	Jan	14,673									
3	1980	Feb	14,912									
4	1980	Mar	16,563									
5	1980	Apr	16,545									
6	1980	May	17,971									
7	1980	Jun	17,929									
8	1980	Jul	18,693									
9	1980	Aug	18,025									
10	1980	Sep	16,291									
11	1980	Oct	15,637									
12	1980	Nov	13,562									
13	1980	Dec	13,319									
14	1981	Jan	13,310									
15	1981	Feb	14,579									
16	1981	Mar	16,720									
17	1981	Apr	17,675									
18	1981	May	18,874									
19	1981	Jun	18,863									
20	1981	Jul	18,798									
21	1981	Aug	17,718									
22	1981	Sep	15,715									
23	1981	Oct	14,609									
24	1981	Nov	13,121									
25	1981	Dec	13,934									

Beer Production

Ready   NUM

The figures for the number of barrels produced as shown in Figure 11-25 come from pages 15–17 of the *Brewers Almanac 1992*, published by the Beer Institute in Washington, D.C. As a first step to analyzing these data, create a line plot of barrels versus year and month.

To create a time series plot:

1   Select the range **A1:C145**.

2   Click **Insert > Chart > As New Sheet** to open the Chart Wizard.

3   Verify that the Range text box contains **A1:C145** in Step 1, then click **Next**.

4   Double-click the **Line** chart type in Step 2.

5   Double-click format **2** (lines without symbols) in Step 3.

6   Verify that the **Data Series in Columns option button** is selected, that the first **2** columns are used for Category (X) labels, and that the first **1** rows are used for Legend Text, then click **Next**.

7   Click the **No option button** to omit a legend, type **Beer Production 1980-1991** in the Chart Title text box, press [**Tab**], type **Year** in the Category (X) text box, press [**Tab**], type **Barrels x 1,000** in the Value (Y) text box, then click **Finish**.

To make the chart easier to read, rescale the y-axis so that the number of barrels produced ranges from 12,000 to 20,000.

To rescale the y-axis:

1   Double-click the **y-axis** to open the Format Axis dialog box, then click the **Scale tab**.

2   Type **12000** in the Minimum text box, then click **OK**. The plot of beer production by year appears as in Figure 11-26.

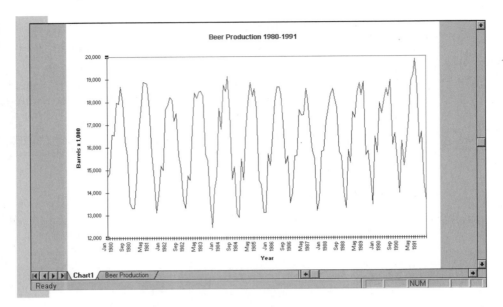

Figure 11-26
Beer production in the 1980s

The plot shows that production is seasonal, with peaks occurring each summer. Winter production is much lower. Usually, the minimum each year is about 13,000 barrels in the winter and the maximum is about 18,000 or 19,000 barrels in the summer. That is, production is about 40% higher in the summer.

Although there is a small amount of growth in the period from 1980 to 1992, the figures do not display the same type of spectacular growth that the stock market experienced, as shown in Figure 11-1.

Showing Seasonality with a Boxplot

One way to see the seasonal variation is to make a boxplot, with a box for each of the 12 months. This gives you a picture of the month-to-month variation in beer production. The shape of each box tells you how production for that month has varied throughout the 1980s.

To create the boxplot you have to restructure the data using the Make Two Way table command introduced in the previous chapter.

To create a boxplot of barrel production:

1 Click the **Beer Production sheet tab** to return to the worksheet containing the data.

2 Click **CTI > Make Two Way Table**.

3 Type **Month** in the Factor 1 Range text box, **Year** in the Factor 2 Range text box, and **Barrels** in the Response Values Range text box.

4 Click the **Selection Includes Header Row check box** to select it.

5 Click the **New Sheet option button**, type **Two Way Table** in the corresponding text box, then click **OK**. The data are restructured as shown in Figure 11-27.

Figure 11-27
Two-way table of beer production

	A	B	C	D	E	F	G	H	I	J	K	L
1	Barrels						Month					
2	Year	Jan	Feb	Mar	Apr	May	Jun	Jul	Aug	Sep	Oct	Nov
3	1980	14673	14912	16563	16545	17971	17929	18693	18025	16291	15637	1356
4	1981	13310	14579	16720	17675	18874	18863	18798	17718	15715	14609	131
5	1982	15188	14999	17654	17860	18216	18092	17174	17502	15635	15071	136
6	1983	14767	14562	16777	18420	18165	18467	18497	18273	15708	15407	136
7	1984	14148	14746	17722	16814	18745	18468	19116	17588	14581	15140	130
8	1985	15495	14551	16767	17974	18858	18232	18586	17713	14534	14358	131
9	1986	15714	15206	16506	17991	18670	18648	18327	17057	15264	15620	135
10	1987	15601	15633	17656	17422	17436	18584	18091	16807	15824	15497	131
11	1988	15801	15850	17125	17728	18310	18584	18172	17725	15777	15610	140
12	1989	15877	15292	17569	17298	18409	18821	18283	18885	15625	15825	147
13	1990	16459	15745	17968	17477	18101	18579	18246	18963	16086	16621	154
14	1991	16275	15169	16085	17228	18900	19164	19882	18627	16115	16654	144
15												

With the data in the format shown in Figure 11-27, you can now use the CTI Boxplots command.

To create a boxplot of beer production by month:

1   Select the range **B2:M14**, then click **CTI > Boxplots**.

2   Verify that the Input Range text box contains **B2:M14** and that the **Selection Includes Header Row check box** is selected.

3   Click the **New Sheet option button**, then type **Boxplot** in the corresponding text box.

4   Click the **Connect Medians Between Boxes check box** to select it, then click **OK**.

    The scale for the y-axis is not optimal, so you should rescale it to cover the range 12,000 to 20,000 barrels.

5   Double-click the **chart** to activate it.

6   Double-click the **y-axis** to open the Format Axis dialog box, click the **Scale tab** if necessary, type **12000** in the Minimum text box, then click **OK**.

7   Click the **worksheet** outside the chart to deactivate the chart, which appears in Figure 11-28.

Figure 11-28
Boxplot of barrel production versus month

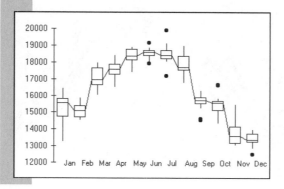

The boxplots in Figure 11-28 clearly show the seasonality. The only month that seems to depart from trend is February, which appears to be about 10% lower relative to the trend from January to March. On the other hand, February has about 10% fewer days than January and March, so it is reasonable for February to be low. The boxplot also indicates the range of production levels for each month. January, for example, is one of the most variable months in terms of beer production. How might this fact affect any forecasts you would make regarding beer production in January?

Showing Seasonality with Line Plots

You can also take advantage of the two-way table to create a line plot of barrels versus month for each year of the data set. This is another way to get insight into the monthly distribution of barrel production during the 1980s.

To create a line plot of beer production:

1   Click the **Two Way Table sheet tab** to select the worksheet containing the two-way table.

2   Highlight the range **A2:M14** and click **Insert > Chart > As New Sheet**.

3   Verify that the Range text box contains **A2:M14**, then click **Next**.

4   Double-click the **Line** chart type, then double-click format **2** (lines without points) for the line chart format.

5 Click the **Use First Column(s) up spin arrow** to set the spin box to **1** column for legend text.

At this point you can plot the time series in two ways. If you click the Data Series in Rows option button in the Step 4 dialog box, Excel creates a line chart of barrels versus month with each line representing a different year. If you select the Data Series in Columns option button, Excel shows barrels versus year with each line representing a different month. Both plots can be of interest in analyzing time series data. For now look at the barrels versus month line plot.

6 Click the **Data Series in Rows option button** if it is not already selected, then click **Next**.

7 Leave the **Yes option button** selected so that your chart will include a legend, then type **Beer Production vs. Months** in the Chart Title text box, **Month** in the Category (X) text box, and **Barrels x 1,000** in the Value (Y) text box. Click **Finish**.

8 Rescale the **y-axis** so that it ranges from 12,000 to 20,000. Your plot should look like Figure 11-29.

Figure 11-29
Line plot of barrels versus month

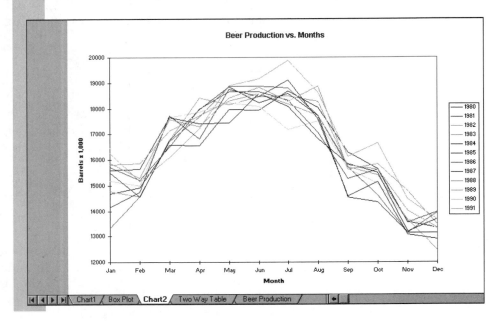

The line plot demonstrates the seasonal nature of the data and also allows you to observe individual values. Plots like this are sometimes called **spaghetti plots** for obvious reasons.

Note: You might want to enlarge the plot, or on a monochrome monitor you might need to adjust it to view the values.

Applying the ACF to Seasonal Data

You can also use the autocorrelation function to display the seasonality of the data. For a seasonal monthly series, the ACF should be very high at lag 12, because the current value should be strongly correlated with the value from a year ago.

To calculate the ACF:

1 Click the **Beer Production sheet tab** to return to the worksheet containing the production data.

2 Click **CTI > ACF Plot**.

3 Type **Barrels** in the Input Range text box and verify that the **Selection Includes Header Row check box** is selected.

4 Click the **up spin arrow** to calculate the ACF up through lag **24**.

5 Click the **New Sheet option button**, type **ACF Plot** in the corresponding text box, then click **OK**. The ACF plot appears as shown in Figure 11-30.

Figure 11-30
ACF for the beer production data

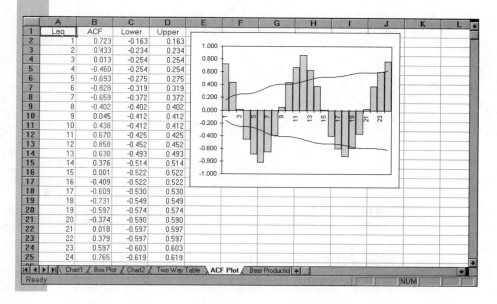

The plot shown in Figure 11-30 indicates strong seasonal, because the lag 6 autocorrelation is strongly negative and the lag 12 autocorrelation is strongly positive. Notice that these correlations are both bigger in absolute value than the lag 1 autocorrelation. So to predict this month's figure, you would do better to use the value from a year ago rather than last month's value.

Adjusting for the Effect of Seasonality

Because the beer production data have a strong seasonal component, it would be useful to adjust the values for the seasonal effect. In this way you can determine whether a drop in production during one month is due to seasonal effects or is a true decline. Adjusting the production data for seasonality also gives you a better indication of whether a trend exists for the data. You can use the CTI Statistical Add-Ins to adjust time series data for multiplicative seasonality.

To adjust the beer production data:

1 Click the **Beer Production sheet tab** to return to the worksheet containing the production data.

2 Click **CTI > Seasonal Adjustment**.

3 Type **Barrels** in the Input Range text box and verify that the **Selection Includes Header Row check box** is selected.

4 If necessary, click the **up spin arrow** to set the length of the period to **12** (months).

5 Click the **New Sheet option button**, type **Adjusted Production** in the corresponding text box, then click **OK**. The output of the seasonal adjustment is shown in Figure 11-31.

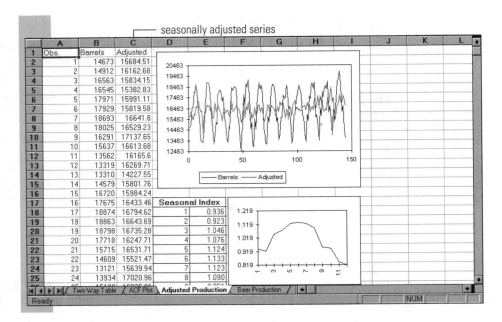

Figure 11-31
Seasonally adjusted data

The observed production levels are shown in column B, and the seasonally adjusted values are shown in column C. Using the adjusted values can give you some insight on the production levels. For example, between observations 8 and 9 (corresponding to August and September 1980) the production level drops 1,734 units from 18,025 units to 16,291. However, when you adjust these figures for the time of the year, the production level actually increases 608 units from 16,529.23 units to 17,137.65. From this you conclude that although production dropped in September 1980, the decline in production was less than would have been predicted for September based on the usual seasonal levels.

Seasonal indices for beer production data are located in cells D17:E29. From these data you can calculate the relative production in each month. For example, the seasonal index for June is 1.133 and for May it is 1.124. This indicates that you can expect a percentage increase in beer production of (1.133–1.124) / 1.124 = .00848, or 0.8%, going from May to June each year. Seasonal indices for the multiplicative model must add up to the length of the period—in this case, 12. You can use this information to tell you that 9.45% of the beer produced each year is produced in June (because 1.133 / 12 = 0.0945). A line plot of the seasonal indices is provided and shows a profile very similar to the one in the boxplot.

***

Adjusting a Time Series for Seasonal Effects

- Select Seasonal Adjustment from the CTI menu.
- Enter the input range and indicate whether the range includes a header row.
- Enter the length of the seasonal period.
- Specify where the output should be sent.

***

The scatterplot in Figure 11-31 shows both the production data and the adjusted production values. The adjusted values do not show the same seasonal trends as the unadjusted production values. It is difficult to determine whether a trend exists for the production data after adjusting for the time of the year. To try to ascertain whether a trend effect exists for the data, you can use three-parameter exponential smoothing.

Three-parameter Exponential Smoothing

You perform exponential smoothing on seasonal data using three smoothing factors. This process is known as **three-parameter exponential smoothing** or **Winters' method for exponential smoothing**. The factors in the Winters' method involve location, trend, and seasonality. The seasonality factor can either be multiplicative or additive. The form of the equation for a time series variable y_t with a multiplicative seasonality adjustment is

$$y_t = (\beta_0 + \beta_1 t) \times Season_t + \varepsilon_t$$

and for additive seasonality adjustment the equation is

$$y_t = (\beta_0 + \beta_1 t) + Season_t + \varepsilon_t$$

In these equations β_0, β_1, and ε_t represent the location, trend, and error portions of the model respectively, and $Season_t$ is the seasonal adjustment factor. Once again, these parameters are not considered to be constant but can vary with time. The beer production data might be an example of such a series. The production values are seasonal, but there might also be a time trend to the data such that production levels increase from year to year after adjusting for seasonality.

Let's concentrate on smoothing with a multiplicative seasonality factor. The smoothing equations used in three-parameter exponential smoothing are similar to equations you've already seen. For the smoothed location value S_t and the smoothed trend value T_t, from data where the length of the period is p, the recursive equations for smoothing with a multiplicative seasonality factor are

$$S_t = \alpha \frac{y_t}{I_{t-p}} + (1 - \alpha)(S_{t-1} + T_{t-1})$$

$$T_t = \gamma(S_t - S_{t-1}) + (1 - \gamma)T_{t-1}$$

Note that the recursive equation for S_t is identical to the equation used in two-parameter exponential smoothing except that the current observation y_t must be seasonally adjusted using the seasonal adjustment factor I_{t-p}. Here I_{t-p} adjusts the value of S_t based on the estimate of the seasonal adjustment factor from p units, or one period earlier. In the beer production data, for example, p equals 12 for a period of 12 months. The recursive equation for T_t is identical to the recursive equation for the two-parameter model.

As you would expect, three-parameter smoothing also smooths out the values of the seasonal adjustment factors (these might change over time). The recursive equation for I_t is

$$I_t = \delta \frac{y_t}{S_t} + (1 - \delta)I_{t-p}$$

The value of the smoothed seasonal adjustment factor I_t is a weighted average of the current seasonal adjustment and the smoothed seasonal adjustment estimate from one period earlier. Calculating initial estimates S_0, T_0, and $I_1 - I_p$ is laborious and is not covered here.

Forecasting Beer Production

Let's use exponential smoothing to predict beer production. For the purposes of demonstration, we'll assume a multiplicative model. You will have to decide upon values for each of the three smoothing constants. The values need not be the same. For example, seasonal adjustments often change more slowly than the trend and location factors, so you might want to choose a low value for the seasonal smoothing constant, say about 0.05. On the other hand, if you feel that the trend factor or location factor will change more rapidly over the course of time, you will want a higher value for the smoothing constant, such as 0.15. As you have seen in this chapter, the values you choose for these smoothing constants depend in some part upon your experience with the data. Excel does not provide a feature to do smoothing with Winters' method. One has been provided for you with the Exponential Smoothing command found in the CTI Statistical Add-Ins.

To forecast values of beer production:

1. Click the **Beer Production sheet tab** to return to the worksheet containing the production data.

2. Click **CTI > Exponential Smoothing**.

3. Type **Barrels** in the Input Range text box, then verify that the **Selection Includes Header Row check box** is selected.

4. Click the **New Sheet option button**, then type **Forecasted Values** in the corresponding text box.

5. Verify that **0.15** is entered in the General Options Weight text box to use a value of 0.15 for the location smoothing constant.

6. Click the **Forecast check box** to select it, type **12** in the Units ahead text box that appears (this forecasts beer production 12 months into the future), then verify that the Confidence Interval text box contains **0.95** (this produces a 95% confidence region around your forecast).

7. Click the **Linear Trend option button** to add a trend factor to the forecast. A Linear Weight text box appears; verify that **0.15** is entered for the trend smoothing constant.

8. Click the **Multiplicative option button** and verify that the Period text box contains **12.** Type **0.05** in the Seasonal Weight text box to use 0.05 for the seasonal adjustment smoothing constant. Your dialog box should look like Figure 11-32.

Figure 11-32
Completed Perform Exponential Smoothing dialog box

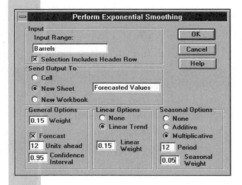

9. Click **OK**.

Output from the Exponential Smoothing command appears on the Forecasted Values worksheet. To view the forecasted values, drag the vertical scroll bar down to display cells A146:D158, as shown in Figure 11-33.

Figure 11-33
Cells A146:D158 in the exponential smoothing output

	Obs	Forecast	Lower	Upper
146				
147	145	15936.88	14843.41	17030.34
148	146	15732.28	14627.57	16836.99
149	147	17794.88	16678.3	18911.47
150	148	18263.5	17134.42	19392.58
151	149	19164.04	18021.87	20306.2
152	150	19337.61	18181.8	20493.43
153	151	19291.22	18121.19	20461.25
154	152	18695.2	17510.43	19879.97
155	153	16278.77	15078.74	17478.79
156	154	16207.34	14991.57	17423.12
157	155	14423.54	13191.54	15655.54
158	156	14020.09	12771.41	15268.78

- Select Exponential Smoothing from the CTI menu.

- Enter the input range, and indicate whether the range includes a header row.

- Specify whether you want location, linear, or seasonal smoothing.

- Enter the smoothing constants for each of the smoothing factors.

- Click the Forecast check box to forecast future values; enter the number of units into the future you want to forecast and the confidence interval of the forecast.

- Specify where the output should be sent.

The output does not give the month for each forecast, but you can easily confirm that observation 145 in column A is January 1992 because observation 144 on the Beer Production worksheet is December 1991. Based on the values in column B, you forecast that in the next year beer production will reach a peak in June 1992 (observation 150) with about 19,338 units (a unit being 1,000 barrels). The 95% prediction interval for this estimate is about 18,182 to 20,493 units. That means in June 1992 you would not expect to have less than 18,182 units nor more than 20,493 units produced. You could use these estimates to plan your production strategy for the upcoming year. Before putting much faith in the prediction intervals, you should verify the assumptions for the smoothed forecasts. If the smoothing model is correct, the residuals should be independent (show no discernible ACF pattern) and follow a normal distribution with mean 0. You would find for the beer data that these assumptions are met.

Scrolling back up the worksheet, you can view how well the smoothing method forecasted the beer production in the previous 12 years, as shown in Figure 11-34.

Figure 11-34
Cells A1:I23 in the exponential smoothing output

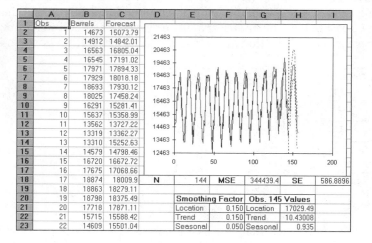

The standard error of the forecast is 586.8896 (cell I18), indicating that the average forecasting error in the time series is about 600 units. Because the average monthly production is about 16,400 units, this does not represent a large error as compared to the mean.

The values for observation 145 (January 1992) in cells G20:H23 use the final estimates for the location, trend, and seasonal adjustment. The value of the seasonal adjustment, 0.935, is the adjustment given to production estimates in January. The estimate of the trend factor (cell H22) indicates that production is increasing at a rate of about 10.43 units per month—hardly a large increase given the magnitude of the monthly production. You can use these values to calculate the forecasted value for January 1992, which would be equal to (17,029.49 + 10.43008) * 0.935 = 15,936.88. Note that this is indeed the forecasted value for January 1992 as indicated in cell B147 from Figure 11-33.

The output also includes a scatterplot comparing the observed, smoothed, and forecasted production values. Due to the number of points in the time series, the seasonal curves are close together and difficult to interpret. To make it easier to view the comparison between the observed and forecasted values, rescale the x-axis to show only the last year and the forecasted year's data.

To rescale the x-axis:

1 Double-click the **scatterplot** in the range D1:I17 to activate it.

2 Double-click the **x-axis** to open the Format Axis dialog box.

3 Click the **Scale tab** if it is not in the forefront of the dialog box, type **130** in the Minimum text box, then click **OK**.

4 Click outside the chart to deactivate it. The chart appears as in Figure 11-35.

Figure 11-35
Observed, smoothed, and forecasted beer production values with x-axis rescaled

smoothed values

observed values

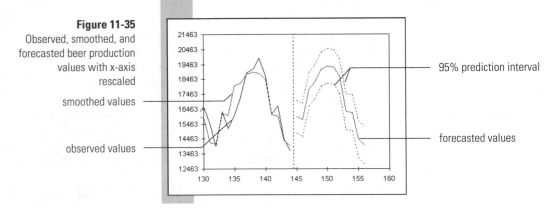

95% prediction interval

forecasted values

Observing the rescaled plot in Figure 11-35, you would conclude that exponential smoothing has done a good job of modeling the production data and that the resulting forecasts appear reasonable. The final part of the exponential smoothing output is the seasonal indices for beer production data located in cells E25:J37, as shown in Figure 11-36.

Figure 11-36
Seasonal indices and scatterplot of seasonal indices

Seasonal Index	
1	0.935
2	0.923
3	1.043
4	1.070
5	1.122
6	1.131
7	1.128
8	1.092
9	0.951
10	0.946
11	0.841
12	0.817

The values for the seasonal indices are very similar to those you calculated using the seasonal adjustment command. The difference is due to the fact that these seasonal indices are calculated using a smoothed average whereas the earlier indices were calculated using an unweighted average.

To save and close the 11BEER workbook:

1 Click the **Save button** 🖫 to save your file as 11BEER.XLS (the drive containing your Student Disk should already be selected).

2 Click **File > Close**.

Optimizing the Exponential Smoothing Constant (Optional)

As you've seen in this chapter, the choice for the value of the exponential smoothing constant depends partly upon the analyst's experience and intuition. Many analysts advocate using the value that minimizes the mean squared error. You can use an Excel Add-In called the Solver to calculate this value. To demonstrate how this technique works, open the file EXPSOLVE.XLS, which contains the monthly percentage changes in the Dow that you worked with earlier in this chapter.

To open the EXPSOLVE.XLS data:

1   Open **EXPSOLVE.XLS** (be sure you select the drive and directory containing your Student files).

2   Click **File > Save As**, select the drive containing your Student Disk, then save your workbook on your Student Disk as **11EXP.XLS**.

The workbook displays the column of percentage differences. Let's create a column of exponentially smoothed forecasts. First you must decide on an initial estimate for the smoothing constant w; you can start with any value you want, so let's start with 0.15.

To minimize the mean squared error:

1   Click cell **F1**, type **0.15**, then press [**Enter**].

Next determine an initial value for the smoothed forecasts to be put in cell C2. The initial value of the time series is usually used.

2   Click cell **C2**, type **=B2**, then press [**Enter**].

Now create a column of smoothed forecasts using the recursive smoothing equation.

3   Select the range **C3:C120**.

4   Type **=$F$1*B2+(1-$F$1)*C2**, then press [**Enter**].

5   Click **Edit > Fill > Down** or press [**Ctrl**]+**D** to fill in the rest of the forecasts.

Now create a column of squared errors [(forecast – observed)^2].

6   Select the range **D2:D120**.

7   Type **=(C2-B2)^2**, then press [**Enter**].

8   Click **Edit > Fill > Down** or press [**Ctrl**]+**D** to fill in the rest of the squared errors.

Finally, calculate the mean square error for this particular value of w.

9   Click cell **F2**, type **=SUM(D2:D120)/119**, then press [**Enter**].

Verify that the values in your spreadsheet match the values in Figure 11-37.

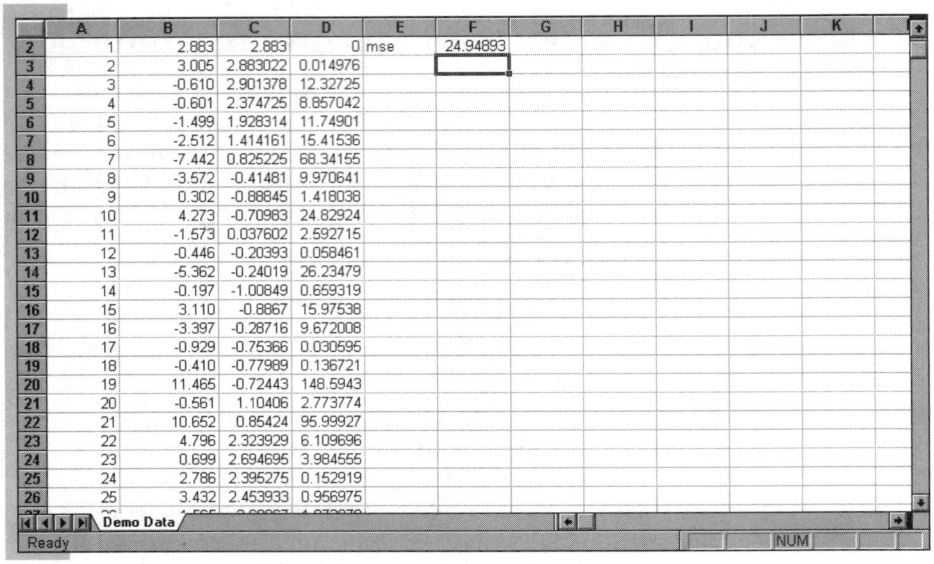

Figure 11-37
Worksheet values

	A	B	C	D	E	F	G	H	I	J	K
2	1	2.883	2.883	0	mse	24.94893					
3	2	3.005	2.883022	0.014976							
4	3	-0.610	2.901378	12.32725							
5	4	-0.601	2.374725	8.857042							
6	5	-1.499	1.928314	11.74901							
7	6	-2.512	1.414161	15.41536							
8	7	-7.442	0.825225	68.34155							
9	8	-3.572	-0.41481	9.970641							
10	9	0.302	-0.88845	1.418038							
11	10	4.273	-0.70983	24.82924							
12	11	-1.573	0.037602	2.592715							
13	12	-0.446	-0.20393	0.058461							
14	13	-5.362	-0.24019	26.23479							
15	14	-0.197	-1.00849	0.659319							
16	15	3.110	-0.8867	15.97538							
17	16	-3.397	-0.28716	9.672008							
18	17	-0.929	-0.75366	0.030595							
19	18	-0.410	-0.77989	0.136721							
20	19	11.465	-0.72443	148.5943							
21	20	-0.561	1.10406	2.773774							
22	21	10.652	0.85424	95.99927							
23	22	4.796	2.323929	6.109696							
24	23	0.699	2.694695	3.984555							
25	24	2.786	2.395275	0.152919							
26	25	3.432	2.453933	0.956975							

Demo Data
Ready · NUM

You now have everything you need to use the Solver.

To open the Solver:

1 Click **Tools > Add Ins**.

2 If the **Solver Add-In check box** appears in the list box, click it to select it if necessary, then click **OK**. If the check box does not appear, then you must install the Solver. Check your *Excel User's Guide* for information on installing the Solver from your original Excel installation disks.

Once the Solver is installed you can determine the optimal value for the smoothing constant.

To determine the optimal value for the smoothing constant:

1 Click **Tools > Solver**.

2 Type **F2** in the Set Target Cell text box. This is the cell that you will use as a target for the Solver.

3 Click the **Min option button** to indicate that you want to minimize the value of the mean square error (cell F2).

4 Type **F1** in the By Changing Cells text box to indicate that you want to change the value of F1, the smoothing constant, in order to minimize cell F2.

Because the exponential smoothing constant can only take on values between 0 and 1, you have to add some constraints to the values that the Solver will investigate.

5 Click the **Add button**.

6 Type **F1** in the Cell Reference text box, select <= from the Constraint drop-down list, type **1** in the Constraint text box, then click **Add**.

7 Type **F1** in the Cell Reference text box, select >= from the Constraint drop-down list, type **0** in the Constraint text box, then click **Add**.

8 Click **Cancel** to return to the Solver Parameters dialog text box. The completed Solver Parameters dialog box should look like Figure 11-38.

Figure 11-38
Completed Solver
Parameters dialog box

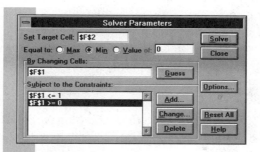

9   Click **Solve**.

The Solver now determines the optimal value for the smoothing constant (at least in terms of minimizing the mean square error). When the Solver is finished it will prompt you to either keep the Solver solution or restore the original values.

10   Click **OK** to keep the solution.

The Solver returns a value of 0.028792 (cell F1) for the smoothing constant, resulting in a mean square error of 23.99456 (cell F2). This is the optimal value for the smoothing constant.

It's possible to set up similar spreadsheets for two-parameter and three-parameter exponential smoothing, but that will not be demonstrated here. The main difficulty in setting up the spreadsheet to do these calculations is in determining the initial estimates : S_0 , T_0 , and the seasonal indices.

In the case of two-parameter exponential smoothing, you would use linear regression on the entire time series to derive initial estimates for the location and trend values. Once this is done, you derive the forecasted values using the recursive equations described earlier in the chapter. You would then apply the Solver to minimize the mean square error of the forecasts by modifying *both* the location and trend smoothing constants. Using the Solver to derive the best smoothing constants for the three-parameter model is more complicated, because you have to come up with initial estimates for all of the seasonal indices. The interested student can refer to more advanced texts for techniques to calculate the initial estimates.

To save and close the 11EXP workbook:

1   Click the **Save button** 🖫 to save your file as 11EXP.XLS (the drive containing your Student Disk should already be selected).

2   Click **File > Close**.

E XERCISES

1. The BBAVER.XLS workbook contains data on the leading major league baseball batting averages for the years 1901 to 1991.

   a. Create a line chart of the batting average versus year. Do you see any apparent trends? Do you see any outliers? Does George Brett's average of 0.390 in 1980 stand out compared with other observations?

   b. Insert a trendline smoothing the batting average using a ten-year moving average.

2. For the BBAVER.XLS workbook of Exercise 1, calculate the ACF and state your conclusions (notice that the ACF does not drop off to zero right away, which suggests a trend component).

3. Calculate the difference of the BBAVER.XLS averages from one year to the next. Plot the different series and also compute its ACF.

   a. Does the plot show that the variance of the original series is reasonably stable? That is, are the changes roughly the same size at the beginning, middle, and end of the series?

b. Looking at the ACF of the differenced series, do you see much correlation after the first few lags? If not, it suggests that the differenced series does not have a trend, and this is what you would expect. Interpret any lags that are significantly correlated.

4. Perform simple exponential smoothing for the BBAVER data, forecasting one year ahead. This means to smooth the Average variable based only on the location parameter with no linear or seasonal effect.

   a. Use the values 0.2, 0.3, 0.4, and 0.5 as the smoothing parameters. In each case, notice the value predicted for 1992 (observation 92). Which parameter gives the minimum prediction standard error?

   b. An almanac shows that the actual highest batting average of 1992 is 0.343. Compare the predictions with the actual value. Does the parameter with the minimum standard error also give the best prediction for 1992?

5. How can you tell if a series is seasonal? Mention plots, including the ACF. What is the difference between additive and multiplicative seasonality?

6. The ELECTRIC.XLS workbook has monthly data on U.S. electric power production, 1978 through 1990. The variable Power is measured in billions of kilowatt hours. The figures come from the *1992 CRB Commodity Year Book*, published by the Commodity Research Bureau in New York.

   a. Create a line chart of the power data. Is there any seasonality to the data?

   b. Fit an exponential model with location, linear, and seasonal parameters. Use a smoothing constant of 0.05 for the location parameter, 0.15 for the linear parameter, and 0.05 for the seasonal parameter. What level of power production do you forecast for the next 12 months?

   c. Using the seasonal index, which are the three months of highest power production? Is this in accord with the plots that you have seen? Does it make sense to you as a consumer? By what percentage does the busiest month exceed the slowest month?

   d. Repeat the exponential smooth of part 6b with the smoothing constants shown in Table 11-7:

Table 11-7
Smoothing constraints

Location	Linear	Seasonal
0.05	0.30	0.05
0.15	0.15	0.05
0.15	0.30	0.05
0.30	0.15	0.05
0.30	0.30	0.05

Which forecasts give the smallest standard error?

7. The VISIT.XLS workbook contains monthly visitation data for two sites at the Kenai Fjords National Park in Alaska from January 1990 to June 1994. You'll analyze the visitation data for the Exit Glacier site.

   a. Create a line plot of visitation for Exit Glacier versus year and month. Summarize the pattern of visitation at Exit Glacier between 1990 and mid-1994.

   b. Create two line plots, one showing the visitation at Exit Glacier plotted against year with different lines for different months, and the second showing visitation plotted against month with different lines for different years (you will have to create a two-way table for this). Are there any unusual values? How might the June 1994 data influence future visitation forecasts?

   c. Calculate the seasonally adjusted values for visits to the park. Is there a particular month in which visits to the park jump to a new and higher level?

   d. Smooth the visitation data using exponential smoothing. Use smoothing constants of 0.15 for both the location and the linear parameter, and 0.05 for the seasonal parameter. Forecast the visitation 12 months into the future. What are projected values for the next 12 months?

e. A lot of weight of the projected visitations for 1994–1995 in part 7d is based on the jump in visitation in June 1994. Assume that this jump was an aberration, and refit two exponential smoothing models with 0.05 and 0.01 for the location parameter (to reduce the effect of the June 1994 increase), 0.15 for the linear parameter, and 0.05 for the seasonal parameter. Compare your values with the values from part 7d. How do the standard errors compare? Which projections would you work with and why? What further information would you need to decide between these three projections?

f. What problems do you see with either forecasted value? (Hint: Look at the confidence intervals for the forecasts.)

8. The visitation data in the VISIT.XLS workbook cover a wide range of values. It might be appropriate to analyze the $\log_{10}$ of the visitation counts instead of the raw counts. Create a new column in the workbook of the $\log_{10}$ counts of the Exit Glacier data (use the Excel function LOG_{10}).

a. Create a line plot of $\log_{10}$ (visitation) for the Exit Glacier site from 1990 to mid-1994. What seasonal values does this chart reveal that were hidden when you charted the raw counts?

b. Use exponential smoothing to smooth the $\log_{10}$ (visitation) data. Use a value of 0.15 for the location and linear effects, and 0.05 for the seasonal effect. Project $\log_{10}$ (visitation) 12 months into the future. Untransform the projections and the prediction intervals by raising 10 to the power of $\log_{10}$ (visitation) (that is, if $\log_{10}$ (visitation) = 1.6 then visitation = $10^{1.6}$ = 39.8). What do you project for the next year at Exit Glacier? What are the 95% prediction intervals? Are the upper and lower limits reasonable?

c. Repeat part 8b twice, using 0.01 and then 0.05 for the location parameter, 0.15 for the linear parameter and 0.05 for the seasonal parameter. Which of the three projections results in the smallest standard error?

d. Compare your chosen projections from Exercise 7, using the raw counts, with your chosen projections from this exercise, using the $\log_{10}$-transformed counts. Which would you use to project the 1994–1995 visitations? Which would you use to determine the amount of personnel you will need in the winter months and why?

9. The workbook NFP.XLS contains daily body temperature data for 239 consecutive days for a woman in her twenties. Daily temperature readings is one component of natural family planning (NFP) in which a woman uses her monthly cycle with a number of biological signs to determine the onset of ovulation. The file has four columns: Observation, Period (the menstrual period), Day (the day of the menstrual period), and Waking Temperature. Day 1 is the first day of menstruation.

a. Create a line plot of the daily body temperature values. Do you see any evidence of seasonality in the data?

b. Create a boxplot of temperature versus day (you will have to create a two-way table for this). What can you determine about the relationship between body temperature and the onset of menstruation?

c. Calculate the ACF for the temperature data up through lag 70. Based on the shape of the ACF what would you estimate as the length of the period in days?

d. Smooth the data using exponential smoothing. Use 0.15 as the location parameter, 0.01 for the linear parameter (it will not be important in this model), and 0.05 for the seasonal parameter. Use the period length that you estimated in part 9c of this exercise. What body temperature values do you forecast for the next cycle?

e. Repeat part 9d with values of 0.15 and 0.25 for the seasonal parameters. Which model has the lowest standard error?

10. A politician citing the latest raw monthly unemployment figures claimed that unemployment had fallen by 88,000 workers. The Bureau of Labor Statistics, however, using seasonally adjusted totals, claimed that unemployment had increased by 98,000. Discuss the two interpretations of the data. Which number gives a better indication of the state of the economy?

CHAPTER 12

QUALITY CONTROL

O B J E C T I V E S

In this chapter you will learn to:

- Distinguish between controlled and uncontrolled variation

- Distinguish between variables and attributes

- Determine control limits for several types of control charts

- Use graphics to create statistical control charts with Excel

- Interpret control charts

- Create a Pareto chart

Statistical Quality Control

The immediately preceding chapters have been dedicated to the identification of relationships and patterns among variables. Such relationships are not immediately obvious, mainly because they are never exact for individual observations. There is always some sort of variation that obscures the true association. In some instances, once the relationship has been identified, an understanding of the types and sources of variation also becomes critical. This is especially true in business where people are interested in controlling the variation of a process. A **process** is any activity that takes a set of inputs and creates a product. The process for an industrial plant takes raw materials and creates a finished product. A process need not be industrial; for example, the process for an office worker might be to take unorganized information and produce an organized analysis. Teaching could even be considered a process because the teacher takes uninformed students and produces students capable of understanding a subject such as statistics. In all such processes, people are interested in controlling the procedure so as to improve quality. The analysis of processes for this purpose is called **statistical quality control** (SQC) or **statistical process control** (SPC).

Statistical process control originated in 1924 with Walter A. Shewhart, a researcher for Bell Telephone. A certain Bell product was being manufactured with great variation in quality, and the production managers could not seem to reduce the variation to an acceptable level. Dr. Shewhart developed the rudimentary tools of statistical process control to improve the homogeneity of Bell's output. Shewhart's ideas were later championed and refined by W. Edwards Deming, who tried unsuccessfully to persuade American firms to implement SPC as a methodology underlying all production processes. Having failed to convince U.S. executives of the merits of SPC, Deming took his cause to Japan, which, before World War II, was renowned for its shoddy goods. The Japanese adopted SPC wholeheartedly, and Japanese production became synonymous with high and uniform quality. American firms have jumped on the SPC bandwagon only in the last decade, and many of their products have regained market share.

Two Types of Variation

The reduction of variation in any process is beneficial. However, you can never eliminate all variation, even in the simplest process, because there are bound to be many small, unobservable, chance effects that influence the process outcome. Variation of this kind is called **controlled variation** and is analogous to the random error effects in the ANOVA and regression models you studied earlier. As in those statistical models, many individually insignificant random factors interact to have some net effect on the process output. In quality control terminology, this random variation is said to be "in control," not because the process operator is able to control the factors absolutely, but rather because the variation is the result of normal disturbances, called **common causes**, within the process. This type of variation can be predicted. In other words, given the limitations of the process, each of these common causes is controlled to the extent possible.

The other type of variation that can occur within a process is called **uncontrolled variation**. Uncontrolled variation is due to **special causes**, which arise sporadically and for reasons outside the normally functioning process. Variation induced by a special cause is usually significant in magnitude and occurs only occasionally. Examples of special causes include differences between machines, different skill or concentration levels of workers, changes in atmospheric conditions, and variation in the quality of inputs.

Because controlled variation is the result of small variations in the normally functioning process, it cannot be reduced unless the entire process is redesigned. Furthermore, any attempts to reduce the controlled variation without redesigning the process will create more, not less, variation in the process. Endeavoring to reduce controlled variation is called **tampering**, which increases costs and must be avoided. Tampering might occur, for instance, when operators adjust machinery in response to normal variations in the production process. Because normal variations will always occur, adjusting the machine is more likely to harm the process than to help it. Uncontrolled variation, on the other hand, can be reduced by eliminating its special cause. The failure to bring uncontrolled variation into control is costly.

SPC is a methodology for distinguishing whether variation is controlled or uncontrolled. If variation is controlled, then only improvements in the process itself can reduce it. If variation is uncontrolled, further analysis designed to identify and eliminate the special cause and its associated variation is required.

Control Charts

The principal tool of SPC is the control chart. A **control chart** is simply a graph on which points representing some feature of a process are plotted. The control chart contains **upper** and **lower control limits** (UCL and LCL, respectively), which appear as dotted horizontal lines in standard control charts. The solid line between the upper and lower control limits is the center line and acts as another check of whether a process is in control. Figure 12-1 shows the principal features of a control chart.

Figure 12-1
Control chart basics

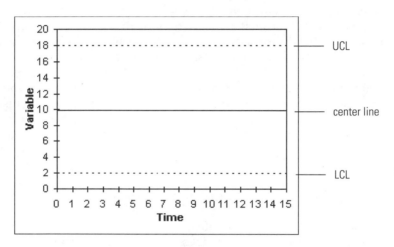

A variable representing time is plotted on the horizontal axis, while the variable or attribute of interest is plotted on the vertical axis. Control limits mark the boundaries of a process that is in control. With the few exceptions discussed below (see Figures 12-4 and 12-5), provided that all observations fall between the upper and lower control limits, the process is in control, all variation is controlled variation due to common causes, and there is no need to be concerned that the variation is due to special causes. On the other hand, if an observation falls outside the control limits, that observation represents a point at which the process is out of control. In that case, the special cause associated with the observation must be identified and corrected.

It is important to note that control limits do not represent specification limits or maximum variation targets. Rather, control limits illustrate the limits of normal process variation.

Figure 12-2 shows a process that is in statistical control, because all data points fall between the two control limits. Consecutive points are usually connected in control charts to improve their readability.

Figure 12-2
Process in statistical control

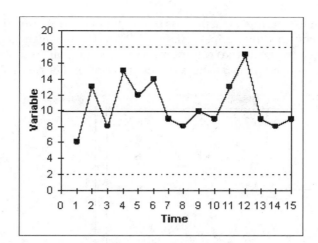

In contrast, the process depicted in Figure 12-3 is out of control. The fourth observation is too large to be the outcome of a normally functioning process. Similarly, observation 12 is abnormally small relative to the typical process outcome.

Figure 12-3
Out-of-control process

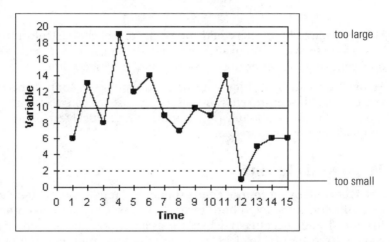

Even control charts in which all points lie between the control limits might suggest that a process is out of control. In particular, the existence of an obvious pattern in eight or more consecutive points should be interpreted as a process that is out of control. In Figure 12-4, for example, the last eight observations depict a steady upward trend. Even though all of the points lie within the control limits, you must conclude that this process is out of control because of the trend.

Figure 12-4
Process out of statistical control because of an upward trend

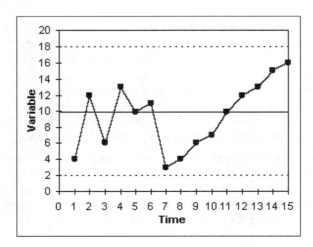

Another common example of a process that is out of control, even though all points lie between the control limits, appears in Figure 12-5. The first eight observations are all less than the center line while the second seven observations all lie above the center line. Because of prolonged periods where values are small and large, this process is out of control.

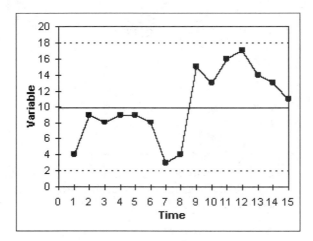

Other suspicious patterns could appear in control charts. Unfortunately, there is insufficient space here to discuss them all. In general, though, any clear pattern that appears in eight consecutive points indicates a process that is not in control.

Statisticians usually highlight out-of-control points in control charts by circling them. As you can see, the control chart makes it very easy for you to identify points and processes that are out of control without using complicated statistical tests. This makes the control chart an ideal tool for the shop floor.

Control Charts and Hypothesis Testing

The idea underlying control charts should be familiar to you. It is closely related to confidence intervals and hypothesis testing. The associated null hypothesis is that the process is in control; you reject this null hypothesis if any point lies outside the control limits or if any clear pattern appears in eight consecutive points. Another insight from this analogy is that the possibility of making errors exists, just as errors can occur in standard hypothesis testing. In other words, occasionally a point that lies outside the control limits does not have any special cause, but occurs because of normal process variation. On the other hand, there could exist a special cause that is not big enough to move the point outside of the control limits. Statistical analysis can never be one hundred percent certain.

Variables and Attributes

There are two categories of control charts: those that monitor variables and those that monitor attributes. Variables are continuous measures, such as weight, diameter, thickness, purity, and temperature. As you have probably already noticed, much statistical analysis focuses on the mean values of such measures. In a process that is in control, you expect the mean output of the process to be stable over time. The **mean chart**, or $\bar{x}$-**chart**, is designed to determine whether this criterion is satisfied. A closely associated chart is the **range chart**, which analyzes whether the variation in a process is under control.

Most texts discuss mean charts before tackling range charts, because the interpretation of mean charts is usually easier to grasp. However, it is important to understand that in practice the range chart must often be drawn first, because the range must be in control in order for the mean chart to be useful. This problem is discussed further in this chapter in the sections "The Mean Chart When Sigma Is Unknown" and "The Range Chart."

Attributes differ from variables in that they describe a feature of the process rather than a continuous variable such as a weight or volume. Attributes can be either discrete quantities, such as the number of defects in a sample, or proportions, such as the percentage of defects per lot. Accident and safety rates are also typical examples of attributes. SPC includes two charts, the P- and C-charts, that you can use to determine whether process attributes are in control over time. The control charts for attributes are discussed after the variable charts.

Mean Charts and General Theory of Control Charts

In order to compare mean output levels at various points in time, several individual observations are averaged during each of several short subintervals, called **subgroups**, throughout a given time period. The best procedure is to take n observations during each of these k subgroups. You can then average the n values for each subinterval.

Each point in the $\bar{x}$-chart plots the subgroup average against the subgroup number. Because observations usually are taken at regular time intervals, the subgroup number is typically a variable that measures time, with subgroup two occurring after subgroup one and before subgroup three. In any case, the variable plotted on the horizontal axis in all control charts must be a sequential measure. As an example consider a clothing store in which the owner monitors the length of time customers wait to be served. He decides to calculate the average wait-time in half-hour increments. The first half-hour (for instance, customers who were served between 9 AM and 9:30 AM) forms the first subgroup and the owner records the average wait-time during this interval. The second subgroup covers the time from 9:30 AM to 10:00 AM and so forth.

The mean chart, like all SPC charts, is based upon the standard normal distribution. The standard normal distribution underlies the mean chart, because the Central Limit Theorem (see Chapter 5) states that averaged values have approximately normal distributions even when the underlying observations are not normally distributed.

The applicability of the normal distribution allows the control limits to be calculated very easily when the standard deviation of the process is known. The control limits used most commonly in mean charts are three-σ limits. You might recall from Chapter 5 that 99.74% of the observations in a normal distribution fall within three standard deviations (i.e., 3σ) of the mean (μ). In SPC, this means that points outside the three-σ limits occur by chance only 0.26% of the time. Because this probability is so small, points outside the control limits are assumed to be the result of uncontrolled special causes. Why not narrow the control limits to two-σ limits? The problem with this approach is that you might increase the **false-alarm rate**, that is, the number of times you stop a process that you incorrectly believed was out of control. Stopping a process can be expensive, and adjusting a process that doesn't need adjusting might increase the variability through tampering. For this reason a three-σ control limit was chosen as a balance between running an out-of-control process and incorrectly stopping a process when it doesn't need to be stopped.

You might also recall that the mean tests you learned earlier in the book differed slightly depending on whether the population standard deviation was known or unknown. An analogous situation occurs with control charts. The two possibilities are considered in separate sections below.

The Mean Chart When Sigma Is Known

If the true standard deviation of the process (σ) is known, then the control limits are:

$$\mathrm{LCL} = \mu - \frac{3\sigma}{\sqrt{n}}$$

$$\mathrm{UCL} = \mu + \frac{3\sigma}{\sqrt{n}}$$

and 99.74% of the points should lie between the control limits if the process is in control. If σ is known, it usually derives from historical values.

The value for μ might also be known from past values. Alternatively, μ might represent the target mean of the process, rather than the actual mean attained. In practice, though, μ might also be unknown. In that case the mean of the subgroup averages replaces μ in the formulas as follows:

$$LCL = \bar{x} - \frac{3\sigma}{\sqrt{n}}$$

$$UCL = \bar{x} + \frac{3\sigma}{\sqrt{n}}$$

The interpretation of the mean chart is the same whether the true process mean is known or unknown.

Here is an example to help you understand the basic mean chart. Students are often concerned about getting into courses with "good" professors and staying out of courses taught by "bad" ones. In order to provide students with information about the quality of instruction provided by different instructors, many universities use end-of-semester surveys in which students rate various professors on a numerical scale. At some schools, such results are even posted and used by students to help them decide in which section of a course to enroll. Many faculty members object to such rankings on the grounds that, although there is always some apparent variation among faculty members, there are seldom any significant differences. However, students often believe that variations in scores reflect the professors' relative aptitudes for teaching and are not simply random variations due to chance effects.

One way to shed some light on the value of student evaluations of teaching is to examine the scores for one instructor over time. The workbook TEACH.XLS provides data ratings of one professor who has taught principles of economics at the same university for 20 consecutive semesters. The instruction in this course can be considered a process because the instructor has used the same teaching methods and covered the same material over the entire period. Five student evaluation scores were recorded for each of the 20 courses. The five scores for each semester constitute a subgroup. Possible teacher scores run from 0 (terrible) to 100 (outstanding). The following range names have been defined for the workbook in Table 12-1:

Table 12-1
TEACH.XLS range names

Range Name	Range	Description
Semester	A1:A21	The semester of the evaluation
Score_1	B1:B21	First student evaluation
Score_2	C1:C21	Second student evaluation
Score_3	D1:D21	Third student evaluation
Score_4	E1:E21	Fourth student evaluation
Score_5	F1:F21	Fifth student evaluation

To open TEACH.XLS, be sure that Excel is started and the windows are maximized, and then:

1 Open **TEACH.XLS** (be sure you select the drive and directory containing your Student files).

2 Click **File > Save As**, select the drive containing your Student Disk, then save your workbook on your Student Disk as **12TEACH.XLS**.

The data in TEACH.XLS are shown in Figure 12-6. There is obviously some variation between scores across semesters, with scores varying from a low of 54.0 to a high of 100. Without further analysis, you and your friends might think that such a spread indicates that the professor's classroom performance has fluctuated widely over the course of 20 semesters. Is this interpretation valid?

Figure 12-6
The 12TEACH.XLS workbook

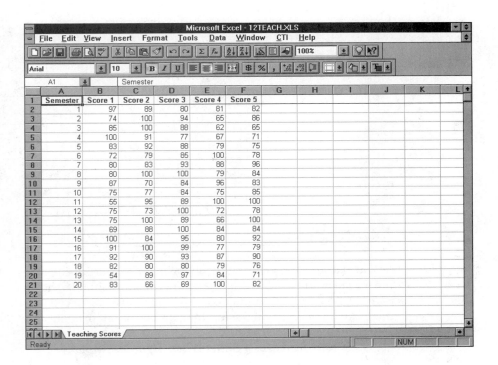

	Semester	Score 1	Score 2	Score 3	Score 4	Score 5
1	Semester	Score 1	Score 2	Score 3	Score 4	Score 5
2	1	97	89	80	81	82
3	2	74	100	94	65	86
4	3	85	100	88	62	65
5	4	100	91	77	67	71
6	5	83	92	88	79	75
7	6	72	79	85	100	78
8	7	80	83	93	88	96
9	8	80	100	100	79	84
10	9	87	70	84	96	83
11	10	75	77	84	75	85
12	11	55	95	89	100	100
13	12	75	73	100	72	78
14	13	75	100	89	66	100
15	14	69	88	100	84	84
16	15	100	84	95	80	92
17	16	91	100	99	77	79
18	17	92	90	93	87	90
19	18	82	80	80	79	76
20	19	54	89	97	84	71
21	20	83	66	69	100	82

If you consider teaching to be a process with student evaluation scores as one of its products, you can use SPC to determine whether the process is in control. In other words, you can use SPC techniques to determine whether the variation in scores is due to identifiable differences in the quality of instruction that can be attributed to a particular semester's course (i.e., special causes) or due merely to chance (common causes).

Historical data from other sources show that σ for this professor is 5.0. Because there are five observations from which each mean is calculated, $n = 5$. You can use the CTI Statistical Add-Ins to calculate the mean scores for each semester and then the average of all 20 mean scores.

To create an $\bar{x}$-chart of the teacher's scores:

1   Click **CTI > XBAR Chart**.

2   Type **B1:F21** in the Input Range text box and then verify that the **Selection Includes Header Row check box** is selected.

3   Click the **Sigma Known option button**, then type **5** in the corresponding text box.

4   Click the **New Sheet option button**, then type **XBAR Chart** in the corresponding text box. Your dialog box should look like Figure 12-7.

Figure 12-7
Completed Create XBAR
Chart dialog box

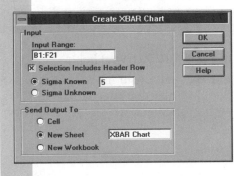

5   Click **OK**. The $\bar{x}$-chart looks like Figure 12-8.

Figure 12-8

The x̄-chart for the teacher score data

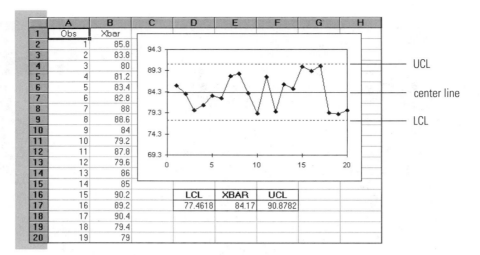

	A	B
1	Obs	Xbar
2	1	85.8
3	2	83.8
4	3	80
5	4	81.2
6	5	83.4
7	6	82.8
8	7	88
9	8	88.6
10	9	84
11	10	79.2
12	11	87.8
13	12	79.6
14	13	86
15	14	85
16	15	90.2
17	16	89.2
18	17	90.4
19	18	79.4
20	19	79

LCL	XBAR	UCL
77.4618	84.17	90.8782

As you can see from Figure 12-8, no mean score falls outside the control limits. The semester means are given in cells B1:B21. The lower control limit as shown in cell D17 is 77.4618, x̄ is 84.17 (cell E17) and the upper control limit is 90.8782 (in cell F17). You conclude that there is no reason to believe that the teaching process is out of control.

Creating a Mean Chart When Sigma Is Known

- Select the range of cells containing the process values. The range must be contiguous.

- Click CTI > XBAR Chart.

- Verify the input range and specify whether the selection includes a header row.

- Click the Sigma Known button and type the value of sigma in the text box that appears.

- Specify the output destination, then click OK.

The fact that no mean score falls outside the control limits implies that no score should be interpreted as differing from the mean score for any reason other than common causes. In other words, in contrast to what the typical student might conclude from the data, there is no evidence that this professor's performance was better or worse in one semester than in another. The raw scores are misleading. A student might claim that using a historical value for σ is also misleading because a smaller value for σ could lead one to conclude that the scores were not in control after all. Exercise 2 at the end of this chapter will deal with this issue by redoing the control chart with an unknown value for σ.

One corollary to the above analysis should be stated: Because even a single professor experiences wide fluctuations in student evaluations over time, apparent differences among various faculty members can also be deceptive. You should use all such statistics with caution.

To save and close the 12TEACH.XLS workbook:

1   Click the **Save button** 🖫 to save your file as 12TEACH.XLS (the drive containing your Student Disk should already be selected).

2   Click **File > Close**.

The Mean Chart When Sigma Is Unknown

In many instances, the value of σ is not known. You learned in Chapter 6 that the normal distribution does not strictly apply when σ is unknown and must be estimated. In that chapter, the *t* distribution was used instead of the standard normal distribution. Because SPC is often implemented on the shop floor by workers who have had little or no formal statistical training (and might not have handy access to Excel), the method for estimating σ is simplified and the normal approximation is used to construct the control chart. The difference is that when σ is unknown, the control limits are estimated using the average range of observations within a subgroup as the measure of the variability of the process. The control limits are:

$$LCL = \mu - A_2\overline{R}$$
$$UCL = \mu + A_2\overline{R}$$

$\overline{R}$ represents the average range in the sample. It is computed by first calculating the range for each subgroup and then averaging those ranges. As before, $\bar{x}$ can replace μ in practice. A_2 is a correction factor from Table 12-2:

Table 12-2

Factors for computing three-σ control limits

Number of Observations in Each Sample	Mean Charts	Range Charts (LCL)	Range Charts (UCL)
n	A_2	D_3	D_4
2	1.88	0	3.268
3	1.023	0	2.574
4	0.729	0	2.282
5	0.577	0	2.114
6	0.483	0	2.004
7	0.419	0.076	1.924
8	0.373	0.136	1.864
9	0.337	0.184	1.816
10	0.308	0.223	1.777
11	0.285	0.256	1.744
12	0.266	0.284	1.717
13	0.249	0.308	1.692
14	0.235	0.329	1.671
15	0.223	0.348	1.652
16	0.212	0.363	1.637
17	0.203	0.378	1.622
18	0.194	0.391	1.608
19	0.187	0.403	1.597
20	0.180	0.415	1.585
21	0.173	0.425	1.575
22	0.167	0.434	1.566
23	0.162	0.443	1.557
24	0.157	0.451	1.548
25	0.153	0.459	1.541

Source: Adapted from "1950 ASTM Manual on Quality Control of Materials," American Society for Testing Materials, in J.M. Juran, ed., Quality Control Handbook *(New York: McGraw-Hill Book Company, 1974), Appendix II, p. 39. Reprinted with permission of McGraw-Hill.*

A_2 accounts for both the factor of three from the earlier equations (used when σ was known) and for the fact that the average range represents a proxy for the common cause variation. (There are other alternative methods for calculating control limits when σ is unknown.) As you can see from the table, A_2 depends only on the number of observations in each subgroup. Furthermore, the control limits become tighter when the subgroup sample size increases. The most typical sample size is 5 because this usually ensures normality of sample means. You will learn to use the values of D_3 and D_4 located in the last two columns of the table in the next section.

The data in the workbook COATS.XLS come from a manufacturing firm that sprays one of its metal products with a special coating to prevent corrosion. Because this company has just begun to implement SPC, σ is unknown for the coating process.

To open COATS.XLS:

1 Open **COATS.XLS** (be sure you select the drive and directory containing your Student files).

2 Click **File > Save As**, select the drive containing your Student Disk, then save your workbook on your Student Disk as **12COATS.XLS**. See Figure 12-9.

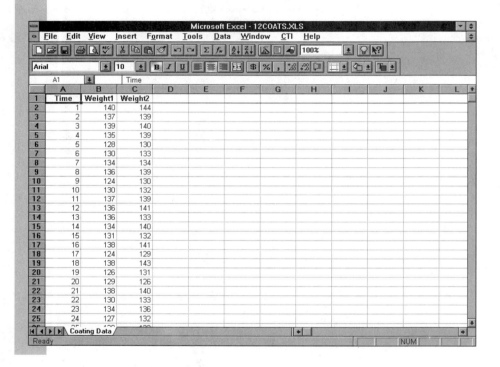

The weight of the spray in milligrams is recorded, with two observations taken at each of 28 times each day, as shown in Figure 12-9. The following range names have been defined for the workbook in Table 12-3:

Table 12-3
12COATS.XLS range names

Range Name	Range	Description
Time	A1:A29	The subgroup
Weight1	B1:B29	The first weight observation
Weight2	C1:C29	The second weight observation

As before, you can use the CTI Statistical Add-Ins to create the $\bar{x}$-chart. Note that because $n = 2$ (there are two observations per subgroup), $A_2 = 1.880$.

To create an $\bar{x}$-chart of the weight values:

1. Click **CTI > XBAR Chart**.

2. Type **B1:C29** in the Input Range text box and then verify that the **Selection Includes Header Row check box** is selected.

3. Click the **Sigma Unknown option button**.

4. Click the **New Sheet option button**, type **XBAR Chart** in the corresponding text box, and click **OK**. The $\bar{x}$-chart appears as in Figure 12-10.

Figure 12-10
The $\bar{x}$-chart output

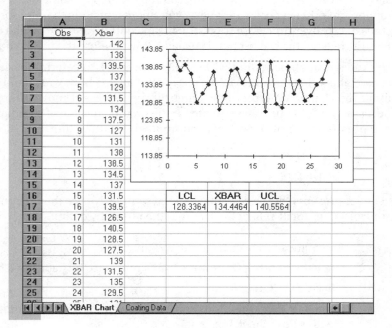

	A	B	C	D	E	F	G	H
1	Obs	Xbar						
2	1	142						
3	2	138						
4	3	139.5						
5	4	137						
6	5	129						
7	6	131.5						
8	7	134						
9	8	137.5						
10	9	127						
11	10	131						
12	11	138						
13	12	138.5						
14	13	134.5						
15	14	137						
16	15	131.5		LCL	XBAR	UCL		
17	16	139.5		128.3364	134.4464	140.5564		
18	17	126.5						
19	18	140.5						
20	19	128.5						
21	20	127.5						
22	21	139						
23	22	131.5						
24	23	135						
25	24	129.5						

XBAR Chart / Coating Data /

Creating a Mean Chart When Sigma Is Unknown

- Select the range of cells containing the process values. The range must be contiguous.

- Click CTI > XBAR Chart.

- Verify the input range and specify whether the selection includes a header row.

- Click the Sigma Unknown button.

- Specify the output destination, then click OK.

Column B contains the mean weight values for each subgroup. The lower control limit is 128.3364, $\bar{x}$ is 134.4464, and the upper control limit is 140.5564. Note that while most of the points in the mean chart lie between the control limits, there are four points (observations 1, 9, 17, and 20) that lie outside the limits. This process is not in control.

Because the process is out of control, you should attempt to identify the special causes associated with each out-of-control point. Observation 1, for example, has too much coating. Perhaps the coating mechanism became stuck for an instant while applying the spray to that item. The other three observations indicate too little coating on the associated products. In talking with the operator, you might learn that he had not added coating material to the sprayer on schedule, so there was insufficient material to spray.

It is common practice in SPC to note the special causes either on the front of the control chart (if there is room) or on the back. This is a convenient way of keeping records of special causes.

In many instances, proper investigation leads to the identification of the special causes underlying out-of-control processes. However, there might be out-of-control points whose special causes cannot be identified.

The Range Chart

The $\bar{x}$-chart , or mean chart, provides information about whether variation about the mean value for each subgroup is too large. It is also important to know whether the range of values is stable from group to group. In the coating example, if some observations exhibit very large ranges and others very small ranges, you might conclude that the sprayer is not functioning consistently over time. Moreover, the mean chart is valid only when the range is in control. For this reason, the range chart is usually drawn before the mean chart in practice.

Use the information in the 12COATS.XLS workbook to determine whether the range of coating weights is in control. As in the case of the mean chart, you would use the three-σ control limits if the standard error of the range for the process is known. In practice, it is usually necessary to substitute an estimate for the unknown value of standard error. In that case, you use the values of D_3 and D_4 to compute the control limits:

$$LCL = D_3 \bar{R}$$
$$UCL = D_4 \bar{R}$$

$\bar{R}$ still represents the average range in the sample. Notice that the possible lower and upper control values for $\bar{R}$ reflect the fact that the range, by definition, can never be negative.

To create a range chart of the weight values:

1. Click the **Coating Data sheet tab** to return to the data worksheet.

2. Click **CTI > Range Chart**.

3. Type **B1:C29** in the Input Range text box and then verify that the **Selection Includes Header Row check box** is selected.

4. Click the **New Sheet option button**, type **Range Char**t in the corresponding text box, then click **OK**. The range chart appears as in Figure 12-11.

Figure 12-11
Range chart output

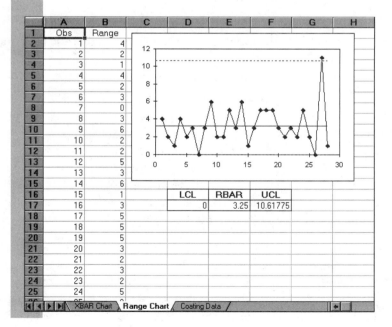

Creating a Range Chart

- Select the range of cells containing the process values. The range must be contiguous.
- Click CTI > Range Chart.
- Verify the input range and specify whether the selection includes a header row.
- Specify the output destination, then click OK.

Each point on the range chart represents the difference between the maximum value and the minimum value in each corresponding time period. The center line of the range is 3.25 with the control limits going from 0 to 10.61775. According to the range chart shown in Figure 12-11, only the twenty-seventh observation has an out-of-control value. As always, the special cause should be identified if possible. However, in discussing the problem with the operator, sometimes you might not be able to determine a special cause. This does not necessarily mean that no special cause exists, but could mean instead that you are unable to determine what the cause is in this instance. It is also possible that there really is no special cause. However, because you are constructing control charts with the width of about three standard deviations, an out-of-control value is unlikely unless there is something wrong with the process (that is, a special cause).

You might have noticed that the range chart identifies as out-of-control a point that was apparently in control in the mean chart, but does not identify as out-of-control any of the four observations that are out-of-control in the mean chart. This is a common occurrence. For this reason, the mean and range charts are often used in conjunction to determine whether a process is in control. In fact, the mean and range charts often appear on the same page in practice, because viewing both charts simultaneously improves the overall picture of the process. In this example you would judge that the process is out of control with both charts but based on different observations.

Finally, the fact that the range is out of control implies that you must be cautious in evaluating the mean chart. Remember that use of the mean chart requires that the range be in control.

To save and close the 12COATS.XLS workbook:

1 Click the **Save button** 💾 to save your file as 12COATS.XLS (the drive containing your Student Disk should already be selected).

2 Click **File > Close**.

The C-Chart

Some processes can be described by counting a certain feature, or attribute, of the process. Such counts are often used to track the number of flaws in a standardized section of continuous sheet metal or the number of defects in lots of a certain size. The number of accidents in a plant might also be counted in this manner.

The ACCID.XLS workbook contains the number of accidents that occurred each month during a period of a few years at a production site. Let's create control charts of the number of accidents per month to determine whether the process is in control.

To open ACCID.XLS:

1 Open **ACCID.XLS** (be sure you select the drive and directory containing your Student files).

2 Click **File > Save As**, select the drive containing your Student Disk, then save your workbook on your Student Disk as **12ACCID.XLS**. See Figure 12-12.

Figure 12-12

The 12ACCID.XLS workbook

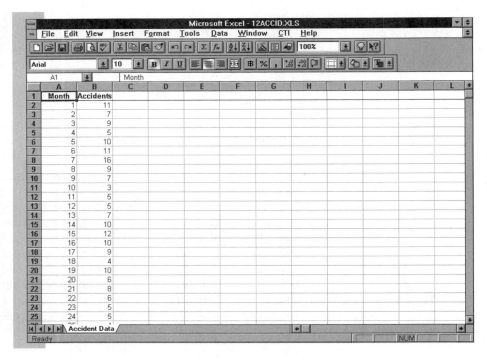

The following range names have been defined for the workbook in Table 12-4:

Table 12-4

12ACCID.XLS range names

Range Name	Range	Description
Month	A1:A45	The month
Accidents	B1:B45	The number of accidents that month

The form of the control limits is very similar to the form used for the $\bar{x}$-charts. Let $\bar{c}$ denote the average monthly number of accidents over the entire data set. Then the control limits are:

$$LCL = \bar{c} - 3\sqrt{\bar{c}}$$
$$UCL = \bar{c} + 3\sqrt{\bar{c}}$$

In the 12ACCID.XLS workbook, the average number of accidents per month is $\bar{c} = 7.114$ (which you can verify using the average function). Therefore:

$$LCL = 7.114 - 3\sqrt{7.114} = 0$$
$$UCL = 7.114 + 3\sqrt{7.114} = 15.115$$

Note that LCL has been set to zero because the formula yields a negative LCL, which is non-sensical. There can never be fewer than zero accidents in a month.

To create a C-chart for accidents at this firm:

1. Click **CTI > C-Chart**.

2. Type **Accidents** in the Input Range text box and then verify that the **Selection Includes Header Row check box** is selected.

3. Click the **New Sheet option button**, type **C Chart** in the corresponding text box, and click **OK**. The C-chart shown in Figure 12-13 appears.

Figure 12-13
C-chart output

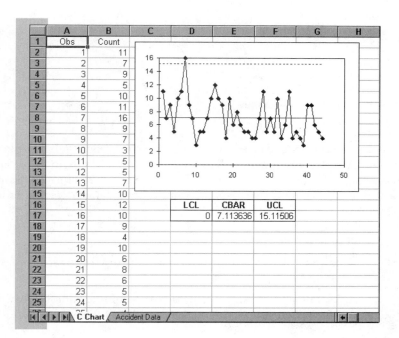

	A	B	C	D	E	F	G	H
1	Obs	Count						
2	1	11						
3	2	7						
4	3	9						
5	4	5						
6	5	10						
7	6	11						
8	7	16						
9	8	9						
10	9	7						
11	10	3						
12	11	5						
13	12	5						
14	13	7						
15	14	10			LCL	CBAR	UCL	
16	15	12			0	7.113636	15.11506	
17	16	10						
18	17	9						
19	18	4						
20	19	10						
21	20	6						
22	21	8						
23	22	6						
24	23	5						
25	24	5						

C Chart / Accident Data /

Creating a C-Chart

- Select the range of cells containing the process values. The range must be contiguous.
- Click CTI > C-Chart.
- Verify the input range and specify whether the selection includes a header row.
- Specify the output destination, then click OK.

Each point on the C-chart in Figure 12-13 represents the number of accidents (the vertical axis) per month (horizontal axis). Only in the seventh month did the number of accidents exceed the upper control limit of 15.115. Since then, the process appears to have been in control. Of course, it is appropriate to determine the special causes associated with the large number of accidents in the seventh month. In the case of this firm, the workload was particularly heavy during that month and a substantial amount of overtime was required. Because employees put in longer shifts than they were accustomed to working, fatigue is likely to have been the source of the extra accidents.

To save and close the 12ACCID.XLS workbook:

1 Click the **Save button** 🖫 to save your file as 12ACCID.XLS (the drive containing your Student Disk should already be selected).

2 Click **File > Close**.

The P-Chart

Closely related to the C-chart is the **P-chart**, which depicts the proportion of items with a particular attribute, such as defects. The P-chart is often used to analyze the proportion of defects in lots of a given size n.

Let $\overline{p}$ denote the average proportion of the sample that is defective. The distribution of the proportions can be approximated by the normal distribution, provided that both $\overline{p}*n$ and $(1-\overline{p})*n$ are both at least five. If $\overline{p}$ is very small or very large, a very large sample size might be required for the approximation to be legitimate.

The three-σ control limits are:

$$LCL = \overline{p} - 3\sqrt{\frac{\overline{p}(1-\overline{p})}{n}}$$

$$UCL = \overline{p} + 3\sqrt{\frac{\overline{p}(1-\overline{p})}{n}}$$

For example, suppose that a manufacturer of steel rods regularly tests whether the rods will withstand 50% more pressure than the company claims them to be capable of withstanding. A rod that fails this test is a defect. Twenty samples of 200 rods each were obtained over a period of time, and the number and proportion of defects were recorded in the STEEL.XLS workbook.

To open STEEL.XLS:

1   Open **STEEL.XLS** (be sure you select the drive and directory containing your Student files).

2   Click **File > Save As**, select the drive containing your Student Disk, then save your workbook on your Student Disk as **12STEEL.XLS**. See Figure 12-4.

Figure 12-14
The 12STEEL.XLS workbook

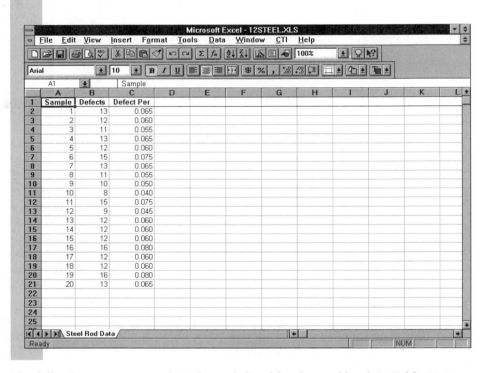

The following range names have been defined for the workbook in Table 12-5:

Table 12-5
12STEEL.XLS range names

Range Name	Range	Description
Sample	A1:A21	The sample number
Defects	B1:B21	The number of defects in the sample
Defect_Per	C1:C21	The proportion of defects in the sample (the number of defects divided by 200)

To create a P-chart for accidents at this firm:

1   Click **CTI > P-Chart**.

2   Type **Defect_Per** in the Input Range text box and then verify that the **Selection Includes Header Row check box** is selected.

3   Type **200** in the Average size of subgroup text box because this is the size of each sample.

4   Click the **New Sheet option button**, type **P Chart** in the corresponding text box, then click **OK**. The P-chart appears as in Figure 12-15.

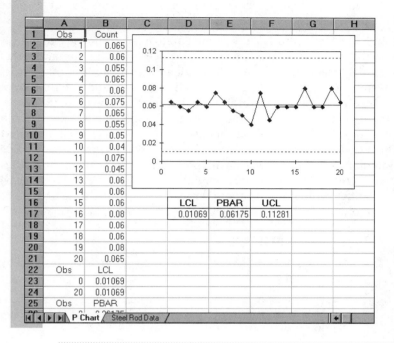

Figure 12-15
P-chart output

Creating a P-Chart

- Select the range of cells containing the process values. The range must be contiguous.
- Click CTI > P-Chart.
- Verify the input range and specify whether the selection includes a header row.
- Enter the number of parts or subgroups that the proportion comes from.
- Specify the output destination, then click OK.

As shown in Figure 12-15, the lower control limit is 0.01069 or a defect percentage of about 1%. The upper control limit is 0.11281, or about 11%. The average defect percentage is 0.06175—about 6%. The control chart clearly demonstrates that no point is anywhere near the three-σ limits.

Note that not all out-of-control points indicate the existence of a problem. For example, suppose that another sample of 200 rods was taken and that only one rod failed the stress test. In other words, only one-half of one percent of the sample was defective. In this case, the proportion is 0.005, which falls below the lower control limit, so technically it is out of control. Yet you would not be concerned about the process being out of control in this case, because the proportion of defects is so low.

To save and close the 12STEEL.XLS workbook:

1   Click the **Save button** 🖫 to save your file as 12STEEL.XLS (the drive containing your Student Disk should already be selected).

2   Click **File > Close**.

The Pareto Chart

After you have determined that your process is resulting in an unusual number of defective parts, the next natural step is to determine what component in the process is causing the defects. This investigation can be aided by a **Pareto chart**, which creates a bar chart of the causes of the defects in order from most to least frequent so that you can focus attention on the most important problems. The chart also includes the cumulative percentage of these components so that you can determine what combination of factors cause, for example, 75% of the defects.

Excel does not automatically create Pareto charts, but you can usually manipulate the workbook to create your own. The workbook POWDER.XLS contains data from a company that manufactures baby powder. Part of the process includes a machine called a filler that pours the powder into bottles to a specified limit. The quantity of powder placed in the bottle varies due to uncontrolled variation, but the final weight of the bottle filled with powder cannot be less than 368.6 grams. Any bottle weighing less than this amount is rejected and must be refilled manually (at a considerable cost in terms of time and labor). Bottles are filled from a filler that has 24 valve heads so that 24 bottles can be filled at one time. Sometimes a head is clogged with powder, and this causes the bottles being filled on that head to receive less than the minimal amount of powder. To gauge whether the machine is operating within limits, random samples of 24 bottles (one from each head) are selected at about one-minute intervals over the nighttime shift at the factory.

To open POWDER.XLS:

1   Open **POWDER.XLS** (be sure you select the drive and directory containing your Student files).

2   Click **File > Save As**, select the drive containing your Student Disk, then save your workbook on your Student Disk as **12POWDER.XLS**. See Figure 12-16.

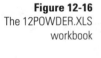

Figure 12-16
The 12POWDER.XLS
workbook

	Time	Head 1	Head 2	Head 3	Head 4	Head 5	Head 6	Head 7	Head 8	Head 9	Head 10	Head 11
2	12:18 AM	439.3	435.5	430.6	429.2	428.6	440.2	433.2	444.4	437.7	441.3	424.9
3	12:19 AM	446.8	445.9	438.9	409.7	448.9	445.6	445.9	448.8	439.1	445.3	434.0
4	12:20 AM	443.4	449.7	439.7	433.0	449.2	445.6	445.7	442.2	447.7	440.9	442.1
5	12:21 AM	451.1	444.2	447.7	433.6	455.1	441.6	439.2	452.6	457.1	440.3	439.3
6	12:22 AM	445.2	446.9	448.0	435.5	446.9	445.5	443.2	445.4	452.9	449.5	451.8
7	12:23 AM	453.3	446.9	439.9	430.3	445.5	441.1	455.7	451.3	447.8	450.1	448.1
8	12:25 AM	450.7	446.1	439.8	422.1	452.2	443.1	452.4	447.6	449.3	449.4	443.7
9	12:26 AM	455.2	467.4	438.1	441.5	470.4	432.2	434.5	439.5	435.7	445.4	449.3
10	12:27 AM	449.8	445.5	438.9	440.2	443.4	446.5	449.4	449.9	453.7	446.8	455.2
11	12:28 AM	460.6	441.1	444.2	426.8	441.1	442.3	454.3	458.2	451.1	447.3	456.4
12	12:29 AM	455.4	455.1	435.1	428.6	428.5	448.2	437.4	455.1	441.9	451.6	430.2
13	12:30 AM	440.6	435.3	431.0	425.3	440.7	435.3	446.2	442.0	440.4	450.6	408.9
14	12:31 AM	428.9	455.0	438.1	424.9	438.1	422.3	447.0	434.0	444.0	372.7	432.7
15	12:34 AM	442.3	432.3	420.7	102.6	424.0	460.1	452.7	430.9	410.8	427.9	428.7
16	12:34 AM	433.2	452.5	398.1	400.2	449.0	436.2	445.1	427.9	443.4	432.0	437.1
17	12:42 AM	434.4	431.8	425.0	428.1	426.2	438.7	441.1	441.0	440.4	440.2	427.4
18	12:43 AM	460.6	442.6	411.5	428.1	419.9	396.3	404.3	415.1	439.9	429.5	438.8
19	12:44 AM	414.9	430.6	421.2	427.9	440.9	424.6	434.8	437.3	434.4	435.6	437.4
20	12:45 AM	426.4	421.6	402.4	422.7	423.9	440.4	426.8	426.9	427.0	414.9	409.5
21	12:46 AM	408.0	396.6	387.0	411.6	406.7	391.5	401.3	415.8	414.4	405.5	410.8
22	12:47 AM	375.8	386.4	382.4	365.2	383.7	394.3	398.6	396.1	394.5	389.2	397.4
23	12:48 AM	358.6	368.8	371.6	364.9	396.8	391.3	393.3	392.5	387.0	403.6	395.1
24	12:49 AM	386.2	390.2	391.2	394.8	401.1	399.4	400.4	402.8	409.6	402.3	401.4
25	12:50 AM	397.7	382.4	376.8	389.5	408.5	390.0	399.9	395.2	400.4	398.4	407.0

The following range names have been defined for the workbook in Table 12-6:

Table 12-6

12POWDER.XLS range names

Range Name	Range	Description
Time	A1:A352	The time of the sample
Head_1	B1:B352	Quantity of powder from head #1
Head_2	C1:C352	Quantity of powder from head #2
...	...	...
Head_24	Y1:Y352	Quantity of powder from head #24

You want to determine which heads are the major source of under-filled bottles. The first step is to create a list of the filler heads.

To create a list of filler heads:

1 Click **Insert > Worksheet**.

2 Double-click the **Sheet1 sheet tab**, type **Reject Bottles** in the Name text box, then click **OK**.

3 Select the range **A1:B1**, type **Head** in cell A1, press [**Tab**], type **Rejects** in cell B1, then press [**Enter**].

4 Select the range **A2:A3**, type **Head_1** (include the underscore) in cell A2, press [**Enter**], type **Head_2** in cell A3, then press [**Enter**].

5 Drag the **fill handle** of the selection down to cell **A25**. The values Head_1 ... Head_24 should now be filled in the range A2:A25.

Notice that the labels you entered into column A correspond to the range reference names already defined for this workbook. That is, the text entered in cell A2—"Head_1"—is also the range name for the reference B1:B352 on the Powder Data worksheet. Similarly the text entered into cell A3—"Head_2"—is also the range name for the range C1:C352. This is not by accident, as you'll make use of these range names in calculating the number of rejects per filler head.

To do this you use Excel's COUNTIF function, which counts the number of cells in a given range whose values fulfill a criterion you specify. In this case, the criterion is the number of bottles for each filler head that weigh less than 368.6 grams. You can refer to the range of cells containing bottle weights for each head using Excel's INDIRECT function. The INDIRECT function allows you to use text entered in one cell to make reference to a range of cells.

To calculate the number of rejects for each head:

1 Select the range **B2:B25**.

2 Type **=COUNTIF(INDIRECT(A2),"<368.6")**, then press [**Enter**].

3 Click **Edit > Fill > Down** or press [**Ctrl**]+**D**.

The number of bottles weighing less than 368.6 grams (that is, the rejects) is displayed in cells B2:B25 for each head. Note that the use of the INDIRECT function allows you to indirectly refer to the proper ranges in the Powder Data sheet by referring to the range reference names in column A. Using the INDIRECT function saved you a lot of typing. The COUNTIF function counts the number of bottles for each filler head that weigh less than 368.6 grams and are, therefore, rejects. Because the Pareto chart is ordered by the number of defects, you should now sort the table by decreasing number of rejects.

To sort the table:

1   Select the range **A1:B25** and click **Data > Sort**.

2   Click the **Sort By list arrow**, then click **Rejects**.

3   Click the **Descending button**, then click **OK**.

Head 18 has the most rejects with 87, while the 14th head is second with 50.

Now calculate the cumulative percentage for the number of rejects so that you can include these on your Pareto chart.

To calculate the cumulative percentage:

1   Click cell **C1**, type **Percent**, then press [**Enter**].

2   Type **=B2/SUM($B$2:$B$25)** in cell **C2** and press [**Enter**].

3   Select the range **C3:C25**, type **=B3/SUM($B$2:$B$25)+C2** in cell **C3**, then press [**Enter**].

4   Click **Edit > Fill > Down** or press [**Ctrl**]+**D**. Click any cell to deselect the selection and compare your worksheet to Figure 12-17.

Figure 12-17
Number of rejects and cumulative percents in cells A1:C25

	A	B	C	D	E	F	G	H	I	J	K	L
1	Head	Rejects	Percent									
2	Head_18	87	0.183158									
3	Head_14	50	0.288421									
4	Head_23	46	0.385263									
5	Head_6	29	0.446316									
6	Head_3	27	0.503158									
7	Head_4	26	0.557895									
8	Head_17	26	0.612632									
9	Head_12	24	0.663158									
10	Head_5	22	0.709474									
11	Head_15	20	0.751579									
12	Head_20	17	0.787368									
13	Head_16	14	0.816842									
14	Head_19	13	0.844211									
15	Head_21	12	0.869474									
16	Head_13	11	0.892632									
17	Head_2	10	0.913684									
18	Head_22	8	0.930526									
19	Head_24	7	0.945263									
20	Head_1	6	0.957895									
21	Head_10	6	0.970526									
22	Head_11	6	0.983158									
23	Head_7	5	0.993684									
24	Head_9	2	0.997895									
25	Head_8	1	1									

Reject Bottles / Powder Data /

As shown in Figure 12-17, the cumulative percentage in cell C6, 0.503158, indicates that more than 50% of the rejects come from five heads: 18, 14, 23, 6 and 3. With the data in this format you can create a Pareto chart graphically showing the distribution of rejects versus the filler heads.

To create the Pareto chart:

1   Select the range **A1:C25** and click **Insert > Chart > As New Sheet**.

2   In the first step of the Chart Wizard verify that the range **A1:C25** is entered in the Range text box, then click **Next**.

3   Double-click the **Combination** chart type to create a chart that shows both a bar chart and a line chart.

4   Double-click **Format 2** in the Combination format list to plot the line chart on a second y-axis scale. This tells Excel to create two separate axes, one for the bars and one for the points on the line. You need to do this because the range for the number of rejects extends from 1 to 87, whereas the cumulative percentage ranges only from 0 to 1.

5 Verify that the data series are organized by **Columns**, that the first **1** column is used for Category (X) Axis Labels, and the first **1** row is used for Legend Text, then click **Next**.

6 Verify that the **Yes option button** is selected to add a legend, then type **Pareto Chart** in the Chart Title text box, **Heads** in the Category (X) text box, **Rejects** in the Value (Y) text box, and **Percentage** in the Second Y text box.

7 Click **Finish**. Excel produces the Pareto chart shown in Figure 12-18.

Figure 12-18
Pareto chart for the rejected baby powder bottles

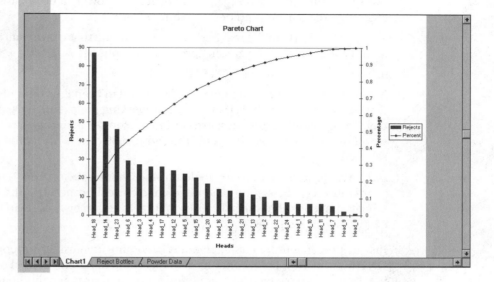

Creating a Pareto Chart

- Create a table listing the sources of defects in the first column.
- In the second column calculate the total number of defects per source. Use Excel's COUNTIF and INDIRECT functions if appropriate for your data.
- Sort the table by the total number of defects in descending order.
- In the third column calculate the cumulative percentage for each row in the table.
- Create a chart of the three columns with the Chart Wizard using the first column as the X variable, the second column as the first Y variable (shown as a bar), and the third column as the second Y variable (shown as a line).

The Pareto chart displayed in Figure 12-18 shows that a majority of the rejects come from a few heads. This might indicate some structural problems with those heads. There might be something physically wrong with the heads that made them more liable to clogging up with powder. If rejects were being produced randomly from the filler heads, you would expect that each filler head would produce $\frac{1}{24}$ or about 4% of the total rejects. Using the information from the Pareto chart shown in Figure 12-18, you might want to investigate whether you have to make fundamental changes in the process to reduce clogging.

To save and close the 12POWDER.XLS workbook:

1 Click the **Save button** 🖫 to save your file as 12POWDER.XLS (the drive containing your Student Disk should already be selected).

2 Click **File > Close**.

E X E R C I S E S

1. Use the information in the TEACH.XLS workbook to determine whether the range of the teaching process is in control. What would you conclude about the instructor's performance from the range chart? If the process is not in control, what special causes might be to blame?

2. One student complained that using historical data in the TEACH.XLS workbook to determine the value of σ wasn't really fair because the lower value of σ might make one conclude that the teacher's grades were out of control.

   a. Fit an $\bar{x}$-chart to the teacher's scores assuming that σ is actually unknown. Do you still conclude that the grading process is in control?

   b. Save your workbook as E12TEACH.XLS.

3. A can manufacturing company must be careful to keep the width of its cans consistent. One associated problem is that the metal-working tools tend to wear down during the day. To compensate, the pressure behind the tools is increased as the blades become worn. In the CANS.XLS workbook, the width of 39 cans is measured at four randomly selected points.

   a. Use range and mean charts to determine whether the process is in control. Does the pressure-compensation scheme seem to correct properly for the tool wear? If not, suggest some possible special causes that seem still to be present in the process.

   b. Save your workbook as E12CANS.XLS.

4. Just-in-time inventory management is becoming increasingly important. The ONTIME.XLS workbook contains data regarding the proportion of on-time deliveries during each month over a two-year period for each of several paperboard products (cartons, sheets, and total). The Total column includes cartons, sheets, and other products. Because sheets were not produced for the entire two-year period, a few data points are missing for that variable. Assume that 1,044 deliveries occurred during each month.

   a. For each of these products, use a P-chart to determine whether the delivery process is in control. If not, suggest some possible special causes that might exist.

   b. Save your workbook as E12ONTIM.XLS.

5. A steel sheet manufacturer is concerned about the number of defects, such as scratches and dents, that occur as the sheet is made. In order to track defects, 10-foot lengths of sheet are examined at regular intervals. For each length, the number of defects is counted. Data have been recorded in the SHEET.XLS workbook.

   a. Determine whether the process is in control. If it is not, suggest some possible special causes that might exist.

   b. Save your workbook as E12SHEET.XLS.

6. A firm is concerned about safety in its workplace. This company does not consider all accidents to be identical. Instead, it calculates a safety index, which assigns more importance to more serious accidents.

   a. Construct a C-chart for the SAFETY.XLS workbook to determine whether safety is in control at this firm.

   b. Save your workbook as E12SAFE.XLS.

7. A manufacturer subjects its steel bars to stress tests to be sure that they are up to standard. Three bars were tested in each of 23 subgroups. The amount of stress applied before the bar breaks is recorded in the STRESS.XLS workbook.

   a. Use SPC to determine whether the production process is in control. If not, what factors might be contributing to the lack of control?

   b. Save your workbook as E12STRES.XLS.

8. A steel rod manufacturer has contracted to supply rods 180 millimeters in length to one of its customers. Because the cutting process varies somewhat, not all rods are exactly the desired length. Five rods were measured from each of 33 subgroups during a week.

    a. Is the cutting process in the ROD.XLS workbook in statistical control? If not, why not?

    b. Save your workbook as E12ROD.XLS.

9. SPC can also be applied to services. An amusement park sampled customers leaving the park over an 18-day period. The total number of customers and the number of customers who indicated they were satisfied with their experience in the park were recorded in the SATISFY.XLS workbook.

    a. Determine whether the percentage of people satisfied is in statistical control. Use the average number of customers as the subgroup size for the P-chart.

    b. What factors, if any, might have contributed to the lack of control?

    c. Save your workbook as E12SATIS.XLS.

10. The number of flaws on the surfaces of a particular model of automobile leaving the plant was recorded in the AUTOS.XLS workbook for each of 40 automobiles during a one-week period.

    a. Determine whether this attribute is in control. If so, is the process perfect?

    b. Save your workbook as E12AUTOS.XLS.

11. Open the POWDER.XLS workbook discussed in this chapter.

    a. Create an $\bar{x}$-chart for bottles filled between 4 AM and 5 AM, assuming that the value of σ is known to be 18. Was the process in control during this time period?

    b. Create an $\bar{x}$-chart for bottles filled between 5 AM and 6 AM Is this process in control during this hour?

    c. Look at the times the filler was working. Is there anything in the time log that would support your conclusions in parts 11a and 11b?

    d. Save your workbook as E12POWD.XLS.

EXCEL REFERENCE

C ONTENTS

The Excel Reference contains the following:

- Summary of Data Sets
- Math and Statistical Functions
- CTI Functions
- Analysis ToolPak Add-Ins
- CTI Statistical Add-Ins
- Instructional Templates
- Bibliography

Summary of Data Sets

ACCID.XLS has the average number of accidents each month for several years (used in Chapter 12).

ALUM.XLS has mass and volume for eight chunks of aluminum, as measured in a high school chemistry class (used in Chapter 2 Exercises, Chapter 4 Exercises, and Chapter 8 Exercises).

AUTOS.XLS has data on surface flaws for 40 cars (used in Chapter 12).

BASE.XLS is part of a data set collected by Lorraine Denby of AT&T Bell Labs for a graphics exposition at the 1988 Joint Statistical Meetings. BASE.XLS has batting and salary data for 263 major league baseball players (used in Chapter 3 Exercises, Chapter 4 Exercises, Chapter 5, and Chapter 5 Exercises).

BASE26.XLS gives batting data for the 26 major league teams on Tuesday, July 29, 1992, as reported the next day in the Bloomington, Illinois *Pantagraph* (used in Chapter 9 Exercises).

BASEINFD.XLS is a subset of BASE.XLS with only the infielders (used in Chapter 10 Exercises).

BBAVER.XLS has the leading major league batting average for each of the years 1901-1991 (used in Chapter 11 Exercises).

BEER.XLS gives monthly United States beer production for 1980-1991, as given by the *Brewers Almanac 1992*, published by the Beer Institute in Washington, D.C (used in Chapter 11).

BIGTEN.XLS has data on graduation rates for ten universities in the Big Ten Conference. The data are from the *Chronicle of Higher Education*, March 27, 1991, pages A39-A44, *U.S. News America's Best Colleges 1992*, and the *Computer Industry Almanac 1991* (used in Chapter 4, Chapter 4 Exercises, Chapter 6 Exercises, and Chapter 8).

BOOTH.XLS has data on the largest banks in the U.S.A. in the early 1970's, taken from Booth (1985) (used in Chapter 8 Exercises).

BREWER.TXT has data on American beer consumption for the major brands. The data are from the *Beverage Industry Annual Manual 91/92*, page 37 (used in Chapter 2 Exercises).

CALC.XLS has the final grade in first semester calculus and several predictor variables for 154 students (used in Chapter 4 Exercises, Chapter 8, Chapter 8 Exercises, and Chapter 9).

CALCFM.XLS contains a subset of the data from CALC.XLS: just the calculus scores for the males and females, broken into two columns, one for each gender (used in Chapter 6 Exercises).

CANS.XLS has width of 39 cans (used in Chapter 12).

CARS.XLS has data on 392 car models. It is based on material collected by David Donoho and Ernesto Ramos (1982), and has been used by statisticians to compare the abilities of different statistics packages (used in Chapter 9 Exercises).

COATS.XLS has data on coating weight for sprayed coatings on metals (used in Chapter 12).

COLA.XLS has data on foam volume, type (diet or regular), and brand for 48 cans of cola (used in Chapter 10).

DISCRIM.XLS has a subset of the data from JRCOL.XLS, reorganized to facilitate using the Analysis ToolPak (used in Chapter 9).

DOW.XLS has the closing Dow Jones average for each month, 1981-1990 (used in Chapter 11).

ECON.XLS has data related to the economy of the United States from 1947 to 1962, as given by Longley (1967). The Deflator variable is a measure of the inflation of the dollar, arbitrarily set to 100 for 1954. Total contains total employment. Population shows increase in population each year (used in Chapter 2 Exercises, and Chapter 4).

ELECTRIC.XLS has monthly data on U.S. electric power production, 1978-1990. The source is the *1992 CRB Commodity Year Book*, published by the Commodity Research Bureau in New York (used in Chapter 11 Exercises).

EXPSOLVE.XLS has the monthly percent change in the Dow from DOW.XLS, set up to facilitate use by the Solver (used in Chapter 11).

FIDELITY.XLS has figures from 1989, 1990, and 1991 for 35 Fidelity sector funds. The source is the *Morningstar Mutual Fund Sourcebook 1992, Equity Mutual Funds* (used in Chapter 8 Exercises).

FOURYR.XLS has information on 24 colleges from the 1992 edition of *U.S. News America's Best Colleges.* They list 140 national liberal arts colleges, 35 in each of four quartiles. A random sample of six colleges was taken from each quartile. An additional variable on computer availability is from the *Computer Industry Almanac 1991* (used in Chapter 10 Exercises).

GAS.XLS contains average daily gasoline sales and other (non-gasoline) sales for each of ten service station/convenience franchises in a store chain in a western city (used in Chapter 2).

HONDA12.XLS contains a subset of the HONDA25.XLS workbook in which the age variable is categorical with the values 1-3, 4-5, and 6+. Some observations of HONDA25.XLS have been removed to balance the data (used in Chapter 10 Exercises).

HONDA25.XLS has price, age, and transmission for 25 used Hondas advertised in the *San Francisco Examiner-Chronicle*, November 25, 1990 (used in Chapter 10 Exercises).

HONDACIV.XLS includes price, age, and miles for used Honda Civics advertised in the *San Francisco Examiner-Chronicle*, November 25, 1990 (used in Chapter 9 Exercises).

HOTEL.XLS has data on 32 hotels, eight chosen randomly from each of four cities, as given in the *1992 Mobil Travel Guide to Major Cities* (used in Chapter 10).

HOTEL2.XLS is taken from the same source as the HOTEL.XLS data. However, the random sample has only two hotels in each of twelve cells, which are defined by four levels of city and three levels of stars (used in Chapter 10 Exercises).

JRCOL.XLS has employment data for 81 faculty members at a junior college (used in Chapter 4, Chapter 4 Exercises, and Chapter 7 Exercises).

LONGLEY.XLS has seven variables related to the economy of the United States 1947-1962, as given by Longley (1967) (used in Chapter 4 Exercises).

MATH.XLS has data from students in a low-level math course (used in Chapter 6 Exercises).

MUSTANG.XLS includes prices and ages of used Mustangs from the *San Francisco Examiner-Chronicle*, November 25, 1990 (used in Chapter 8 Exercises).

NFP.XLS contains daily temperature data taken for 239 consecutive days for a woman in her twenties as one of the measures used in natural family planning (NFP) in which a woman uses her monthly cycle in combination with a number of biological signs to determine the onset of ovulation. The file has four columns: Observation, Period (the menstrual period), Day (the day of the menstrual period; day 1 is the first day of menstruation), and Waking Temperature (used in Chapter 11 Exercises).

ONTIME.XLS has the percentage of on-time deliveries each month for a two-year period (used in Chapter 12 Exercises).

PARK.XLS contains monthly public-use data (the number of visitors) for Kenai Fjords National Park in Alaska in 1993. Data courtesy of Maria Gillett, Chief of Interpretation, and Glenn Hart, Park Ranger (used in Chapter 1).

PCINFO.XLS contains data resulting from tests of 191 prominent 486 PCs, published in *PC Magazine*, Vol. 12, #21, pp. 148-149. Used by permission of Ziff-Davis Publishing (used in Chapter 3, and Chapter 7 Exercises).

PCPAIR.XLS contains a subset of the data from PCINFO.XLS, arranged in two columns, for pairs of computers sharing the same characteristics except for CPU (used in Chapter 6 and Chapter 6 Exercises).

PCSURV.XLS has survey results on 35 makes of personal computers, taken from *PC Magazine*, February 9, 1993. There were 17,000 questionnaires sent to randomly chosen subscribers, and the data are based on 8,176 responses. There are five variables. Used by permission of Ziff-Davis Publishing (used in Chapter 9 Exercises).

PCUNPAIR.XLS contains a subset of the data from PCINFO.XLS, with price and processor information for different BUS types for 486 DX2/66 machines (used in Chapter 6 Exercises).

POLU.XLS gives the number of unhealthy days in each of 14 cities, for six years from the 80's. The figures are from *Environmental Protection Agency National Air Quality and Emissions Trends Report*, 1989, as quoted in the *Universal Almanac 1992*, p. 534 (used in Chapter 2 Exercises, Chapter 3 Exercises, and Chapter 6 Exercises).

POWDER.XLS contains data from a company that manufactures baby powder, with random samples of bottle weights selected at one-minute intervals over the nighttime shift. Data courtesy of Kemp Wills, Johnson & Johnson (used in Chapter 12 and Chapter 12 Exercises).

ROD.XLS has data on the lengths of steel rods (used in Chapter 12 Exercises).

SAFETY.XLS has values for the seriousness of accidents at a firm (used in Chapter 12 Exercises).

SATISFY.XLS has satisfaction data for customers leaving an amusement park (used in Chapter 12 Exercises).

SHEET.XLS has data on defects in steel sheets (used in Chapter 12 Exercises).

SPACE.XLS has measurements on the electrical resistance at the calf (to get an idea of the loss of blood from the legs to the upper part of the body) over a period of 24 days, for seven men and eight women. The study was financed by NASA (used in Chapter 6).

STATE.XLS has the death rate from cardiovascular problems and the death rate from pulmonary problems, for each of the 50 states. The data are from the *1986 Metropolitan Area Data Book* of the U.S. Census Bureau (used in Chapter 8 Exercises).

STEEL.XLS has the proportion of defects in each of 20 samples of steel rods (used in Chapter 12).

STRESS.XLS has strength data for steel bars (used in Chapter 12 Exercises).

SURVEY.XLS has data from 390 professors who responded to a questionnaire designed to help plan this book. There are 16 variables (used in Chapter 7 and Chapter 7 Exercises)

TEACH.XLS has five evaluations for each of 20 instructors (used in Chapter 12 and Chapter 12 Exercises).

VISIT.XLS contains monthly visitation data for two sites at the Kenai Fjords National Park in Alaska, from January 1990 through June 1994. Data courtesy of Maria Gillett, Chief of Interpretation, and Glenn Hart, Park Ranger (used in Chapter 11).

WBUS.XLS contains data on the top 50 women-owned businesses from the Wisconsin State Department of Development (used in Chapter 3, Chapter 3 Exercises, and Chapter 5 Exercises).

WHEAT.TXT, WHEAT.XLS have nutrition information on ten wheat products, as given on the labels (used in Chapter 2, Chapter 2 Exercises, Chapter 4 Exercises, Chapter 8 Exercises, and Chapter 9 Exercises).

WHEATDAN.XLS is the same as WHEAT.XLS with an additional case, the Apple Danish from McDonald's (used in Chapter 9 Exercises).

Math and Statistical Functions

This section documents all the functions provided with Excel that are relevant to statistics. So that you can more easily find the function you need, similar functions are grouped together in six categories: Descriptive Statistics for One Variable, Descriptive Statistics for Two or More Variables, Distributions, Mathematical Formulas, Statistical Analysis, and Trigonometric Functions.

Descriptive Statistics for One Variable

Function Name	Description
AVEDEV	AVEDEV(*number1, number2, . . .*) returns the average of the (absolute value of the) deviations of the points from their mean.
AVERAGE	AVERAGE(*number1, number2, . . .*) returns the average of the *numbers* (up to 30).
CONFIDENCE	CONFIDENCE(*alpha, standarddev, n*) returns a confidence interval for the mean.
COUNT	COUNT(*value1, value2, . . .*) returns how many numbers are in the *value(s)*.
COUNTA	COUNTA(*value1, value2, . . .*) returns the count of non-blank values in the list of arguments.
COUNTBLANK	COUNTBLANK(*range*) returns the count of blank cells in the *range*.
COUNTIF	COUNTIF(*range, criteria*) returns the count of non-blank cells in the *range* that meet the *criteria*.
DEVSQ	DEVSQ(*number1, number2, . . .*) returns the sum of squared deviations from the mean of the *numbers*.
FREQUENCY	FREQUENCY(*data-array, bins-array*) returns the frequency distribution of *data-array* as a vertical array, based on *bins-array*).
GEOMEAN	GEOMEAN(*number1, number2, . . .*) returns the geometric mean of up to 30 *numbers*.
HARMEAN	HARMEAN(*number1, number2, . . .*) returns the harmonic mean of up to 30 *numbers*.
KURT	KURT(*number1, number2, . . .*) returns the kurtosis of up to 30 *numbers*.
LARGE	LARGE(*array, n*) returns the *n*th largest value in *array*.
MAX	MAX(*number1, number2, . . .*) returns the largest of up to 30 *numbers*.
MEDIAN	MEDIAN(*number1, number2, . . .*) returns the median of up to 30 *numbers*.
MIN	MIN(*number1, number2, . . .*) returns the smallest of up to 30 *numbers*.
MODE	MODE(*number1, number2, . . .*) returns the value most frequently occurring in up to 30 *numbers*, or in a specified array or reference.
PERCENTILE	PERCENTILE(*array, n*) returns the *n*th percentile of the values in *array*.
PERCENTRANK	PERCENTRANK(*array, value, significant digits*) returns the percent rank of the *value* in the *array*, with the specified number or *significant digits* (optional).
PRODUCT	PRODUCT(*number1, number2, . . .*) returns the product of up to 30 *numbers*.
RANK	RANK(*number, range, order*) returns the rank of the *number* in the *range*. If *order* = 0 then the range is ranked from largest to smallest; if *order* = 1 the range is ranked from smallest to largest.
QUOTIENT	QUOTIENT(*dividend, divisor*) returns the quotient of the numbers, truncated to integers.

SKEW	SKEW(*number1, number2,* . . .) returns the skewness of up to 30 *numbers* (or a reference to numbers).
SMALL	SMALL(*array, n*) returns the *n*th smallest number in *array*.
STANDARDIZE	STANDARDIZE(*x, mean, standard-deviation*) normalizes a distribution and returns the z-score of *x*.
STDEV	STDEV(*number1, number2,* . . .) returns the sample standard deviation of up to 30 *numbers*, or of an array of numbers.
STDEVP	STDEVP(*number1, number2,* . . .) returns the population standard deviation of up to 30 *numbers*, or of an array of numbers.
SUM	SUM(*number1, number2,* . . .) returns the sum of up to 30 *numbers*, or of an array of numbers.
SUMIF	SUMIF(*range, criteria, sum-range*) returns the sum of the numbers in *range* (optionally in *sum-range*) according to *criteria*.
SUMSQ	SUMSQ(*number1, number2,* . . .) returns the sum of the squares of up to 30 *numbers*, or of an array of numbers.
TRIMMEAN	TRIMMEAN(*array, percent*) returns the mean of a set of values in an *array*, excluding *percent* of the values, half from the top and half from the bottom.
VAR	VAR(*number1, number2,* . . .) returns the sample variance of up to 30 *numbers* (or an array or reference).
VARP	VARP(*number1, number2,* . . .) returns the population variance of up to 30 *numbers* (or an array or reference).

Descriptive Statistics for Two or More Variables

Function Name	Description
CORREL	CORREL(*array1, array2*) returns the coefficient of correlation between *array1* and *array2*.
COVAR	COVAR(*array1, array2*) returns the covariance of *array1* and *array2*.
PEARSON	PEARSON(*array1, array2*) returns the Pearson correlation coefficient between *array1* and *array2*.
RSQ	RSQ(*known-y's, known-x's*) returns the square of Pearson's product moment correlation coefficient.
SUMPRODUCT	SUMPRODUCT(*array1, array2,* . . .) returns the sum of the products of corresponding entries in up to 30 *arrays*.
SUMX2MY2	SUMX2MY2(*array1, array2*) returns the sum of the differences of squares of corresponding entries in two *arrays*.
SUMX2PY2	SUMX2PY2(*array1, array2*) returns the sum of the sums of squares of corresponding entries in two *arrays*.
SUMXMY2	SUMXMY2(*array1, array2*) returns the sum of the squares of differences of corresponding entries in two *arrays*.

Distributions

Function Name	Description
BETADIST	BETADIST(*x, alpha, beta, a, b*) returns the value of the cumulative beta probability density function.
BETAINV	BETAINV(*p, alpha, beta, a, b*) returns the value of the inverse of the cumulative beta probability density function.
BINOMDIST	BINOMDIST(*successes, trials, p, type*) returns the probability for the binomial distribution (*type* is TRUE for the cumulative distribution function, FALSE for the probability mass function).

CHIDIST	CHIDIST(x, df) returns the probability for the chi-squared distribution.
CHIINV	CHIINV(p, df) returns the inverse of the chi-squared distribution.
CRITBINOM	CRITBINOM(*trials*, p, *alpha*) returns the smallest value so that the cumulative binomial distribution is less than or equal to the criterion value, *alpha*.
EXPONDIST	EXPONDIST(x, *lambda*, *type*) returns the probability for the exponential distribution (*type* is TRUE for cumulative distribution function, FALSE for probability density function).
FDIST	FDIST(x, *df1*, *df2*) returns the probability for the F distribution.
FINV	FINV(p, *df1*, *df2*) returns the inverse of the F distribution.
GAMMADIST	GAMMADIST(x, *alpha*, *beta*, *type*) returns the probability for the Gamma distribution with parameters *alpha* and *beta* (*type* is TRUE for cumulative distribution function, FALSE for probability mass function).
GAMMAINV	GAMMAINV(p, *alpha*, *beta*) returns the inverse of the Gamma distribution.
GAMMALN	GAMMALN(x) returns the natural log of the Gamma function evaluated at x.
HYPGEOMDIST	HYPGEOMDIST(*sample-successes*, *sample-size*, *population-successes*, *population-size*) returns the probability for the hypergeometric distribution.
LOGINV	LOGINV(p, *mean*, *sd*) returns the inverse of the lognormal distribution, where the natural logarithm of the distribution is normally distributed with mean *mean* and standard deviation *sd*.
LOGNORMDIST	LOGNORMDIST(x, *mean*, *sd*) returns the probability for the lognormal distribution, where the natural logarithm of the distribution is normally distributed with mean *mean* and standard deviation *sd*.
NEGBINOMDIST	NEGBINOMDIST(*failures*, *threshold-successes*, *probability*) returns the probability for the negative binomial distribution.
NORMDIST	NORMDIST(x, *mean*, *sd*, *type*) returns the probability for the normal distribution with mean *mean* and standard deviation *sd* (type is TRUE for the cumulative distribution function, and FALSE for the probability mass function).
NORMINV	NORMINV(p, *mean*, *sd*) returns the inverse of the normal distribution with mean *mean* and standard deviation *sd*.
NORMSDIST	NORMSDIST(*number*) returns the probability for the standard normal distribution.
NORMSINV	NORMSINV(*probability*) returns the inverse of the standard normal distribution.
POISSON	POISSON(x, *mean*, *type*) returns the probability for the Poisson distribution (*type* is TRUE for the cumulative distribution, and FALSE for probability mass function).
TDIST	TDIST(x, df, *number-of-tails*) returns the probability for the T distribution.
TINV	TINV(p, df) returns the inverse of the T distribution.
WEIBULL	WEIBULL(x, *alpha*, *beta*, *type*) returns the probability for the Weibull distribution (*type* is TRUE for cumulative distribution function, FALSE for probability mass function).

Mathematical Formulas

Function Name	Description
ABS	ABS(*number*) returns the absolute value of the *number* to the point specified.
COMBIN	COMBIN(x, n) returns the number of combinations of x objects taken n at a time.
EVEN	EVEN(*number*) returns the *number* rounded up to the nearest even integer.
EXP	EXP(*number*) returns the exponential function of the *number* with base e.
FACT	FACT(*number*) returns the factorial of *number*.
FACTDOUBLE	FACTDOUBLE(*number*) returns the double factorial of *number*.

FLOOR	FLOOR(*number, significance*) returns the *number* down to the nearest multiple of the *significance* value.
GCD	GCD(*number1, number2,* . . .) returns the greatest common divisor of up to 29 *numbers.*
GESTEP	GESTEP(*number, step*) returns 1 if *number* is greater than or equal to *step*, and 0 if not.
LCM	LCM(*number1, number2,* . . .) returns the least common multiple of up to 29 *numbers.*
INT	INT(*number*) truncates a *number* to the units place.
LN	LN(*number*) returns the natural logarithm of the *number.*
LOG	LOG(*number, base*) returns the logarithm of the *number*, with the specified (optional; default is 10) *base.*
LOG10	LOG10(*number*) returns the common logarithm of a *number.*
MOD	MOD(*number, divisor*) returns the remainder of the division of *number* by *divisor.*
MULTINOMIAL	MULTINOMIAL(*number1, number2,* . . .) returns the quotient of the factorial of the sum of *numbers* and the product of the factorials of the *numbers.*
ODD	ODD(*number*) returns the *number* rounded up to the nearest odd integer.
POWER	POWER(*number, power*) returns the *number* raised to the *power.*
PERMUT	PERMUT(*x, n*) returns the number of permutations of *x* items taken *n* at a time.
RAND	RAND( ) returns a randomly chosen number from 0 to but not including 1.
ROUND	ROUND(*number, places*) rounds *number* to a certain number of decimal *places* (if *places* is positive), or to an integer (if *places* is 0), or to the left of the decimal point (if *places* is positive).
ROUNDDOWN	ROUNDDOWN(*number, places*) rounds like ROUND, except always toward 0.
ROUNDUP	ROUNDUP(*number, places*) rounds like ROUND, except always away from 0.
SERIESSUM	SERIESSUM(*x, n, m, coefficients*) returns the sum of the power series: $a_1 x^n + a_2 x^{n+m} + \ldots + a_i x^{n+(i-1)m}$, where $a_1, a_2, \ldots a_i$ are the *coefficients.*
SIGN	SIGN(*number*) returns 0, 1, or -1, the sign of *number.*
SQRT	SQRT(*number*) returns the square root of *number.*
SQRTPI	SQRTPI(*number*) returns the square root of $number * \pi$.
TRIM	TRIM(*text*) returns *text* with spaces removed, except for single spaces between words.
TRUNC	TRUNC(*number, digits*) truncates *number* to an integer (optionally, to a number of *digits*).

Statistical Analysis

Function Name	Description
CHITEST	CHITEST(*actual-range, expected-range*) returns the value of the chi-squared distribution for the independence test statistic.
GROWTH	GROWTH(*known -y's, known-x's, new-x's, constant*) returns the predicted (y) values for the *new-x's*, based on exponential regression of the *known-y's* on the *known-x's.*
FISHER	FISHER(*x*) returns the value of the Fisher transformation evaluated at *x.*
FISHERINV	FISHERINV(*y*) returns the value of the inverse Fisher transformation evaluated at *y.*
FORECAST	FORECAST(*x, known-y's, known-x's*) returns a predicted (y) value for *x*, based on linear regression of the *known-y's* on the *known-x's.*
FTEST	FTEST(*array1, array2*) returns the *p*-value of the one-tailed F statistic, based on the hypothesis that *array1* and *array2* have the same variance (which is rejected for low *p*-values).

INTERCEPT	INTERCEPT(*known-y's, known-x's*) returns the y-intercept of the linear regression of *known-y's* on *known-x's*.
LINEST	LINEST(*known-y's, known-x's, constant, stats*) returns coefficients on each x-value in linear regression of *known-y's* on *known-x's* (*constant* is TRUE if the intercept is forced to be 0, and *stats* is TRUE if regression statistics are desired).
LOGEST	LOGEST(*known-y's, known-x's, constant, stats*) returns the bases of each x-value in exponential regression of *known-y's* on *known-x's* (*constant* is TRUE if the leading coefficient is forced to be 0, and *stats* is true if regression statistics are desired).
PROB	PROB(*x-values, probabilities, value*) returns the *probability* associated with *value*, given the *probabilities* of a range of values.
PROB	PROB(*x-values, probabilities, lower-limit, upper-limit*) returns the *probability* associated with values between *lower-limit* and *upper-limit*.
SLOPE	SLOPE(*known-y's, known-x's*) returns the slope of a linear regression line.
STEYX	STEYX(*known-y's, known-x's*) returns the standard error of the linear regression.
TREND	TREND(*known-y's, known-x's, new-x's, constant-not-zero*) returns the predicted y-values for given input values (*new-x's*) based on regression of *known-y's* on *known-x's*.
TTEST	TTEST(*array1, array2, number-of-tails, type*) returns the *p*-value of a *t*-test, of *type* paired (1), two-sample equal variance (2), or two-sample unequal variance (3).
ZTEST	ZTEST(*array, x, sigma*) returns the *p*-value of a two-tailed z-test.

Trigonometric Formulas

Function Name	Description
ACOS	ACOS(*number*) returns the arccosine (inverse cosine) of the *number*.
ACOSH	ACOSH(*number*) returns the inverse hyperbolic cosine of the *number*.
ASIN	ASIN(*number*) returns the arcsine (inverse sine) of the *number*.
ASINH	ASINH(*number*) returns the inverse hyperbolic sine of the *number*.
ATAN	ATAN(*number*) returns the arctangent (inverse tangent) of the *number*.
ATAN2	ATAN2(*x,y*) returns the arctangent (inverse tangent) of the angle from the positive x-axis.
ATANH	ATANH(*number*) returns the inverse hyperbolic tangent of the *number*.
COS	COS(*angle*) returns the cosine of *angle*.
COSH	COSH(*number*) returns the hyperbolic cosine of the *number*.
DEGREES	DEGREES(*angle*) returns the radian measurement into degrees.
PI	PI( ) returns π accurate to 15 digits.
RADIANS	RADIANS(*angle*) returns the radian measure of an *angle* given in degrees.
SIN	SIN(*angle*) returns the sine of *angle*.
SINH	SINH(*number*) returns the hyperbolic sine of *number*.
TAN	TAN(*angle*) returns the tangent of *angle*.
TANH	TANH(*number*) returns the hyperbolic tangent of *number*.

CTI Functions

The following functions are added to Excel when the CTI Statistical Add-Ins are loaded.

Descriptive Statistics for One Variable

Function Name	Description
COUNTBETW	COUNTBETW(*range, lower, upper, boundary*) returns the count of non-blank cells in the *range* that lie between the *lower* and *upper* values. The *boundary* variable determines how the end points are used. If *boundary* = 1 the interval is > the lower value and < the upper value. If *boundary* = 2 the interval is > the lower value and < the upper value. If *boundary* =3 the interval is > the lower value and < the upper value. If *boundary* = 4 the interval is ≥ lower value and ≤ the upper value.
NSCORE	NSCORE(*number, range*) returns the normal score of the *number* (or cell reference to a *number*) from a *range* of values.
RANGEVAL	RANGEVAL(*range*) calculates the difference between the maximum and minimum value from a *range* of values.
RANKTIED	RANKTIED(*number, range, order*) returns the rank of the *number* in *range*, adjusting the rank for ties. If *order* = 0 then the range is ranked from largest to smallest; if *order* = 1 the range is ranked from smallest to largest.
SIGNRANK	SIGNRANK(*number, range*) returns the sign rank of the *number* in *range*, adjusting the rank for ties. Values of zero receive a sign rank of 0. If *order* = 0 then the range is ranked from largest to smallest in absolute value; if *order* = 1 the range is ranked from smallest to largest in absolute value.

Descriptive Statistics for Two or More Variables

Function Name	Description
CORRELP	CORRELP(*range1, range2*) returns the *p*-value for the Pearson coefficient of correlation between *range1* and *range2*. Note: *Range values must be in two columns.*
PEARSONCHISQ	PEARSONCHISQ(*range*) returns the Pearson chi-square test statistic for data in the *range*.
PEARSONP	PEARSONP(*range*) returns the *p*-value for the Pearson chi-square test statistic for data in the *range*.
SPEARMAN	SPEARMAN(*range*) returns the Spearman nonparametric rank correlation for values in *range*. Note: *Range values must be in one column only.*
SPEARMANP	SPEARMAN(*range*) returns the *p*-value for the Spearman nonparametric rank correlation for values in *range*. Note: *Range values must be in one column only.*

Distributions

Function Name	Description
NORMBETW	NORMBETW(*lower, upper, mean, stdev*) calculate the area under the curve between the *lower* and *upper* limits for a normal distribution with μ = *mean* and σ = *stdev*.

NORMCENT	NORMCENT (*center, length, mean, stdev*) calculates the area under the curve a fixed *length* around a *center* point for a normal distribution with μ = *mean* and σ = *standard deviation*.
TDF	TDF(*number, df, cumulative*) calculates the area under the curve to the left of *number* for a *t*-distribution with degrees of of freedom *df*, if *cumulative*=TRUE. If *cumulative* = FALSE, this function calculates the probability density function for *number*.

Mathematical Formulas

Function Name	Description
RANDBETA	RANDBETA(*alpha, beta, a, b*) returns a random number from the beta distribution with parameters *alpha, beta,* and (optionally) *a,* and *b*..
RANDCHISQ	RANDCHISQ(*df*) returns a random number from the chi-square distribution with degrees of freedom *df*.
RANDF	RANDF(*df1, df2*) returns a random number from the *F*-distribution with numerator degrees of freedom *df1* and denominator degrees of freedom *df2*.
RANDGAMMA	RANDGAMMA(*alpha, beta*) returns a random number from the gamma distribution with parameters *alpha* and *beta*.
RANDLOG	RANDLOG(*mean, stdev*) returns a random number from the log-normal distribution with μ = *mean* and σ = *stdev*.
RANDNORM	RANDNORM(*mean, stdev*) returns a random number from the normal distribution with μ = *mean* and σ = *stdev*.
RANDT	RANDT(*df*) returns a random number from the *t*-distribution with degrees of freedom *df*.

Statistical Analysis

Function Name	Description
ACF	ACF(*range, lag*) calculates the autocorrelation function for values in *range* for lag = *lag*. *Note: Range values must be in one column only.*

Analysis ToolPak Add-Ins

The Analysis ToolPak Add-Ins come with Excel and allow you to perform basic statistical analysis. None of the output from the Analysis ToolPak is updated for changing data, so if the source data changes, you will have to rerun the command. To use the Analysis ToolPak you must first verify that it is available to your workbook.

To check whether the Analysis ToolPak is available:

1   Click **Tools** from the menu. If the menu option **Data Analysis** appears, the Analysis ToolPak commands are available to you.

2   If the **Data Analysis** menu command does not appear, click **Tools > Add-Ins** from the menu. If the Analysis ToolPak Add-In is listed in the Add-Ins list box, click **Analysis ToolPak** and click **OK**. The Analysis ToolPak commands are now available to you.

3   If the Analysis ToolPak is not listed in the Add-Ins list box, you will have to install it from your installation disks. See your *Excel User's Guide* for details.

The rest of this section documents each Analysis ToolPak command, showing each corresponding dialog box and then listing each option in the dialog box.

Output Options

All the dialog boxes that produce output share the following output storage options:

Output Range

Click to send output to a cell in the current worksheet, then type the cell; Excel uses that cell as the upper-left corner of the range.

New Worksheet Ply

Click to send output to a new worksheet, then type the name of the worksheet.

New Workbook

Click to send output to a new workbook.

Anova: Single-Factor

The **Anova: Single-Factor** command calculates the one-way analysis of variance, testing whether means from several samples are equal.

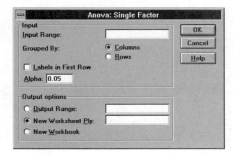

Input Range

Enter the range of worksheet data you want to analyze. The range must be contiguous.

Grouped By

Indicate whether the range of samples is grouped by columns or by rows.

Labels in First Row/Column

Indicate whether the first row (or column) includes header information.

Alpha

Enter the alpha level used to determine the critical value for the F statistic.

See "Output Options" at the beginning of this section for information on the output storage options.

Anova: Two-Factor with Replication

The **Anova: Two-Factor with Replication** command calculates the two-way analysis of variance with multiple observations for each combination of the two factors. An analysis of variance table is creating that tests for the significance of the two factors and the significance of an interaction between the two factors.

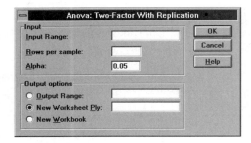

Input Range

Enter the range of worksheet data you want to analyze. The range must be rectangular with the columns representing the first factor and the rows representing the second factor. An equal number of rows are required for each level of the second factor.

Rows per Sample

Enter the number of repeated values for each combination of the two factors.

Alpha

Enter the alpha level used to determine the critical value for the F statistic.

See "Output Options" at the beginning of this section for information on the output storage options.

Anova: Two-Factor Without Replication

The **Anova: Two-Factor without Replication** command calculates the two-way analysis of variance with one observation for each combination of the two factors. An analysis of variance table is created that tests for the significance of the two factors.

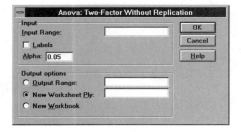

Input Range

Enter the range of worksheet data you want to analyze. The range must be contiguous, with each row and column representing a combination of the two factors.

Labels

Indicate whether the row and first column includes header information.

Alpha

Enter the alpha level used to determine the critical value for the *F* statistic.

See "Output Options" at the beginning of this section for information on the output storage options.

Correlation

The **Correlation** command creates a table of the Pearson correlation coefficient for values in rows or columns on the worksheet.

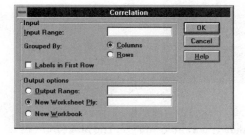

Input Range

Enter the range of worksheet data you want to analyze. The range must be contiguous.

Grouped By

Indicate whether the range of samples is grouped by columns or by rows.

Labels in First Row/Column

Indicate whether the first row (or column) includes header information.

See "Output Options" at the beginning of this section for information on the output storage options.

Covariance

The **Covariance** command creates a table of the covariance for values in rows or columns on the worksheet.

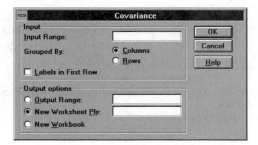

Input Range

Enter the range of worksheet data you want to analyze. The range must be contiguous.

Grouped By

Indicate whether the range of samples is grouped by columns or by rows.

Labels in First Row/Column

Indicating whether the first row (or column) includes header information.

See "Output Options" at the beginning of this section for information on the output storage options.

Descriptive Statistics

The **Descriptive Statistics** command creates a table of univariate descriptive statistics for values in rows or columns on the worksheet.

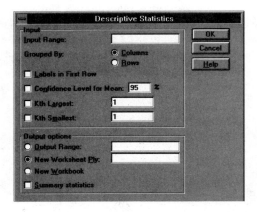

Input Range

Enter the range of worksheet data you want to analyze. The range must be contiguous.

Grouped By

Indicate whether the range of samples is grouped by columns or by rows.

Labels in First Row/Column

Indicate whether the first row (or column) includes header information.

Confidence Level for Mean

Click to print the alpha-% confidence level for the mean in each row or column of the input range where the confidence level = Z_α(standard error).

Kth Largest

Click to print the kth largest value for each row or column of the input range; enter k in the corresponding box.

Kth Smallest

Click to print the kth smallest value for each row or column of the input range; enter k in the corresponding box.

Summary Statistics

Click to print the following statistics in the output range: Mean, Standard Error (of the mean), Median, Mode, Standard Deviation, Variance, Kurtosis, Skewness, Range, Minimum, Maximum, Sum, Count, Largest (#), Smallest (#), and Confidence Level.

See "Output Options" at the beginning of this section for information on the output storage options.

Exponential Smoothing

The **Exponential Smoothing** command creates a column of smoothed averages using simple one-parameter exponential smoothing.

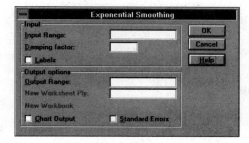

Input Range

Enter the range of worksheet data you want to analyze. The range must be a single row or a single column.

Damping Factor

Enter the value of the smoothing constant. The value 0.3 is used as a default if nothing is entered.

Labels

Indicate whether the first row or first column includes header information.

Chart Output

Click to create a chart of observed and forecasted values.

Standard Errors

Click to create a column of standard errors to the right of the forecasted column.

Output Options

You can only send output from this command to a cell on the current worksheet.

....

F-Test: Two-Sample for Variances

The **F-Test: Two-Sample for Variances** command performs an F-test to determine whether the population variance of two samples are equal.

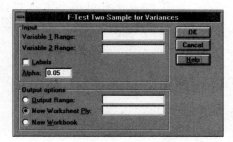

Variable 1 Range

Enter the range of the first sample, either a single row or column.

Variable 2 Range

Enter the range of the second sample, either a single row or column.

Labels

Indicate whether the first row or first column includes header information.

Alpha

Enter the alpha level used to determine the critical value for the F statistic.

See "Output Options" at the beginning of this section for information on the output storage options.

....

Histogram

The **Histogram** command creates a frequency table for data values located in a row, column or list. The frequency table can be based on default or customized bin widths. Additional output options include calculating the cumulative percentage, creating a histogram, and creating a histogram sorted in descending frequency order (also known as a *Pareto Chart*).

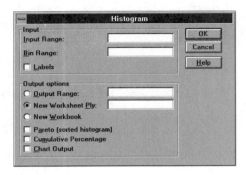

Input Range

Enter the range of worksheet data you want to analyze. The range must be either a row, column or rectangular region.

Bin Range

Enter an optional range of values that define the boundaries of the bins.

Labels

Indicate whether the first row or first column includes header information.

Pareto (sorted histogram)

Click to create a Pareto Chart sorted by descending order of frequency.

Cumulative Percentage

Click to calculate the cumulative percents.

Chart Output

Click to create a histogram of frequency versus bin values.

See "Output Options" at the beginning of this section for information on the output storage options.

Moving Average

The **Moving Average** command creates a column of moving averages over the preceding observations for an interval specified by the user.

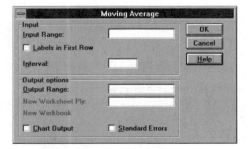

Input Range

Enter the range of worksheet data for which you want to calculate the moving average. The range must be a single row or a single column containing four or more cells of data.

Labels in First Row

Indicating whether the first row or first column includes header information.

Interval

Enter the number of cells you want to include in the moving average. The default value is three.

Chart Output

Click to create a chart of observed and forecasted values.

Standard Errors

Click to create a column of standard errors to the right of the forecasted column.

Output Options

You can only send output from this command to a cell on the current worksheet.

Random Number Generation

The **Random Number Generation** command creates columns of random numbers following a user-specified distribution.

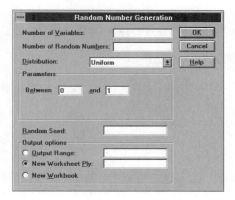

Number of Variables

Enter the number of columns of random variables you want to generate. If no value is entered, Excel fills up all available columns.

Number of Random Numbers

Enter the number of rows in each column of random variables you want to generate. If no value is entered, Excel fills up all available columns. This command is not available for the patterned distribution (see below).

Distribution

Click the down arrow to open a list of seven distributions from which you can choose to generate random numbers.

Uniform

Specify the lower and upper limits of the uniform distribution.

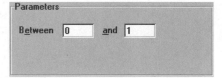

Normal

Specify the mean and standard deviation of the normal distribution.

Bernoulli

Specify the *p*-value for the Bernoulli distribution. If the random number is ≤ the *p*-value, the random variable = 1; otherwise a value of 0 is generated.

Binomial

Specify the *p*-value for the probability of success in a trial and the number of trials. A random variable is generated, which is the number of successful trials.

Poisson

Specify the lambda parameter of the Poisson distribution (used for data in which the values are counts).

Patterned

This option does not create random data; instead it creates data according to a specified pattern with values going from a lower value to an upper value in specified steps. Each value and each sequence can be repeated any number of times you specify.

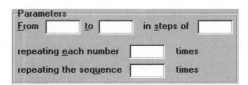

Discrete

Specify the values and the associated probability from a range on the worksheet. The range must contain two columns: the left column must contain values, and the right column must contain probabilities associated with the value in that row. The sum of the probabilities must be 1.

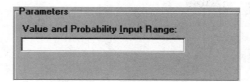

Random Seed

Enter an optional value used as a starting point, called a *random seed*, for generating a string of random numbers. You need not enter a random seed, but using the same random seed ensures the same string of random numbers will be generated. This box is not available for patterned or discrete random data.

See "Output Options" at the beginning of this section for information on the output storage options.

Rank and Percentile

The **Rank and Percentile** command produces a table with ordinal and percentile values for each cell in the input range. The four columns of the table include:

Point, the row number (if the data are grouped by columns) or column number (if the data are grouped by rows) of the input cells

Column or *Row*, the values in the input range (sorted in descending order)

Rank, the rank in descending order of the values in the input range

Percent, the percentile of the values in the input range (in descending order)

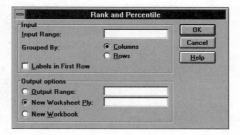

Input Range

Enter the range of worksheet data you want to analyze. The range must be contiguous.

Grouped By

Indicate whether the range of samples is grouped by columns or by rows.

Labels in First Row/Column

Indicating whether the first row (or column) includes header information.

See "Output Options" at the beginning of this section for information on the output storage options.

Regression

The **Regression** command performs multiple linear regression for a variable in an input column based on up to 16 predictor variables. The user has the option of calculating residuals and standardized residuals and producing line fit plots, residuals plots and normal probability plots.

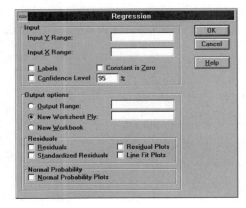

Input Y Range

Enter a single column of values that will be the response variable in the linear regression.

Input X Range

Enter up to 16 contiguous columns of values that will be the predictor variables in the regression.

Labels

Indicate whether the first row of the Y and X ranges include header information.

Constant is Zero

Click to include an intercept term in the linear regression or to assume that the intercept term is zero.

Confidence Level

Click to indicate a confidence interval for linear regression parameter estimates. A 95% confidence interval is automatically included; enter a different one in the corresponding box.

Residuals

Click to create a column of residuals (observed – predicted) values.

Residual Plots

Click to create a plot of residuals versus each of the predictor variables in the model.

Standardized Residuals

Click to create a column of residuals divided by the standard error of the regression's analysis of variance table.

Line Fit Plots

Click to create a plot of observed and predicted values against each of the predictor variables.

Normal Probability Plots

Click to create a normal probability plot of the Y variable in the Input Y Range.

See "Output Options" at the beginning of this section for information on the output storage options.

Sampling

The **Sampling** command creates a sample of an input range. The sample can either be random or periodic (sampling values a fixed number of cells apart). The sample generated is placed into a single column.

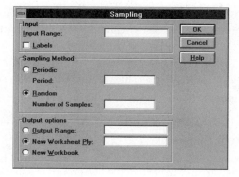

Input Range

Enter the range of worksheet data you want to sample. The range must be contiguous.

Labels

Indicate whether the first row of the Y and X ranges include header information.

Sampling Method

Click the sampling method you want.

Periodic

Click to sample values from the input range *period* cells apart; enter a value for period in the corresponding box.

Random

Click to create a random sample the size of which you enter in the corresponding box.

See "Output Options" at the beginning of this section for information on the output storage options.

t-Test: Paired Two-Sample for Means

The **t-Test: Paired Two Sample for Means** calculates the paired two sample Student's *t*-Test. The output includes both the one-tail and two-tail critical values.

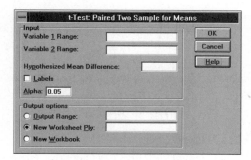

Variable 1 Range

Enter the input range of the first sample; it must be a single row or column.

Variable 2 Range

Enter the input range of second sample; it must be a single row or column.

Hypothesized Mean Difference

Enter a mean difference value with which to calculate the *t*-test. If no value is entered, a mean difference of zero is assumed.

Labels

Indicate whether the first row of the Y and X ranges includes header information.

Alpha

Enter an alpha value used to calculate the critical values of the *t* shown in the output.

See "Output Options" at the beginning of this section for information on the output storage options.

t-Test: Two-Sample Assuming Equal Variances

The **t-Test: Two Sample Assuming Equal Variances** calculates the unpaired two sample Student's *t*-Test. The tests assumes that the variance in each of the two groups is equal. The output includes both the one-tail and two-tail critical values.

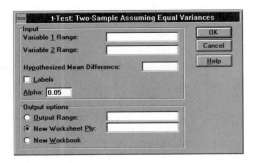

Variable 1 Range

Enter an input range for the first sample; it must be a single row or column.

Variable 2 Range

Enter an input range for the second sample; it must be a single row or column.

Hypothesized Mean Difference

Enter a mean difference with which to calculate the *t*-test. If no value is entered, a mean difference of zero is assumed.

Labels

Indicate whether the first row of the Y and X ranges include header information.

Alpha

Enter an alpha value used to calculate the critical values of the *t* shown in the output.

See "Output Options" at the beginning of this section for information on the output storage options.

t-Test: Two-Sample Assuming Unequal Variances

The **t-Test: Two Sample Assuming Unequal Variances** calculates the unpaired two sample Student's *t*-test. The test allows the variances in the two groups to be unequal. The output includes both the one-tail and two-tail critical values.

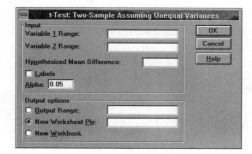

Variable 1 Range

Enter the input range of the first sample; it must be a single row or column.

Variable 2 Range

Enter the input range of the second sample; it must be a single row or column.

Hypothesized Mean Difference

Enter a mean difference with which to calculate the t-test. If no value is entered, a mean difference of zero is assumed.

Labels

Indicate whether the first row of the Y and X ranges includes header information.

Alpha

Enter an alpha value used to calculate the critical values of the t shown in the output.

See "Output Options" at the beginning of this section for information on the output storage options.

z-Test: Two-Sample for Means

The **z-Test: Two Sample for Means** calculates the unpaired two sample z-test. The test assumes that the variance in each of the two groups is known (though not necessarily equal to each other). The output includes both the one-tail and two-tail critical values.

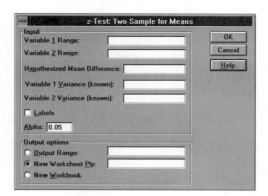

Variable 1 Range

Enter an input range of the first sample; it must be a single row or column.

Variable 2 Range

Enter an input range of the second sample; it must be a single row or column.

Hypothesized Mean Difference

Enter a mean difference with which to calculate the *t*-test. If no value is entered, a mean difference of zero is assumed.

Variable 1 Variance (known)

Enter the known variance σ_1^2 of the first sample.

Variable 2 Variance (known)

Enter the known variance σ_2^2 of the second sample.

Labels

Indicate whether the first row of the Y and X ranges includes header information.

Alpha

Enter an alpha value used to calculate the critical values of the *t* shown in the output.

See "Output Options" at the beginning of this section for information on the output storage options.

CTI Statistical Add-Ins

The CTI Statistical Add-Ins come with the textbook *Data Analysis with Microsoft Excel 5.0 for Windows* to perform basic statistical analysis not covered by the Analysis ToolPak. Some of the output from the add-ins is updated for source data that you update; the degree to which the output is dynamic is indicated in the following descriptions of the various command options. To use the CTI Statistical Add-Ins you must first verify that they are available to your workbook.

To check whether the add-ins are available:

1. Check the worksheet menu bar. If the **CTI** menu option appears on the menu bar, the CTI Statistical Add-Ins are loaded.

2. If the menu command does not appear, click **Tools > Add-Ins** from the menu. If the CTI Statistical Add-Ins option is listed in the Add-Ins list box, click the **CTI Statistical Add-Ins** option and click **OK**. The add-in commands are now available to you.

3. If the add-ins are not listed in the Add-Ins list box, you will have to install them from your instructor's CTI disk. See Chapter 1 for more information. The name of the add-in file is **CTI.XLA** and can be copied to any directory on the drive. This book assumes that the file is located in the Student subdirectory of the Excel directory on your hard drive. Once the add-ins are copied to your hard drive, click **Tools > Add-Ins**, then click the **Browse** button in the Add-Ins dialog box. Locate the copy of **CTI.XLA** and click it, then click **OK** in the Browse dialog box. The **CTI Statistical Add-Ins** option should now appear in the Add-Ins list box. It is not recommended that you run CTI.XLA from a floppy disk.

The rest of this section documents each CTI Statistical Add-Ins command, showing each corresponding dialog box and then listing each option in the dialog box.

Output Options

All the dialog boxes that produce output share the following output storage options:

Cell

Click to send the output to a specific cell in the current worksheet, then enter the cell in the corresponding box that appears. This cell becomes the upper-left corner of the output range.

New Sheet

Click to send the output to cell A1 in a new worksheet; enter the title in the corresponding box.

New Workbook

Click to send the output to cell A1 in sheet Sheet1 of a new workbook.

Stack Columns

The **Stack Columns** command takes data that lie in separate columns and stacks the values into two columns. The column to the left contains the values; the column to the right is a category column. Values for the category are found from the header rows in the input columns, or if there are no header rows, the categories are labeled as *Level 1*, *Level 2*, and so forth. The input range need not be contiguous. Values placed in the output range are static and will not update automatically if the source data in the input range changes.

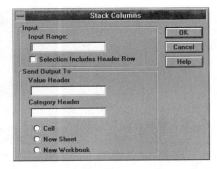

Input Range

Enter the range of worksheet columns you want to stack. The columns need not be contiguous, nor must they be of the same length.

Selection Includes Header Row

Indicate whether the first row of each column in the input range contains header information.

Value Header

Enter the text you want to place in the first row of the value column in the output range.

Category Header

Enter the text you want to place in the first row of the category column in the output range.

See "Output Options" at the beginning of this section for information on the output storage options.

Unstack Column

The **Unstack Column** command takes data found in two columns: a column of data values and a column of categories, and outputs the values into different columns–one column for each category level. The length of the values column and the length of the category column must be equal. Values in the output range are static and will not update automatically if the source data in the input range changes.

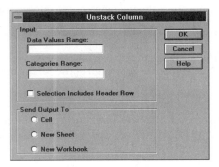

Data Values Range

Enter the worksheet column containing the data you want to unstack.

Categories Range

Enter the worksheet column containing the category levels that you want to separate into different columns. The categories column and the data values column must have the same number of rows. You do *not* have to sort the category levels.

Selection Includes Header Row

Indicate whether the first row of each column in the input range contains header information.

See "Output Options" at the beginning of this section for information on the output storage options.

Make Indicator Variables

The **Make Indicator Variables** command takes a column of category levels and creates columns of indicator variables, one for each category level in the input range. An indicator variable for a particular category = 1 if the row comes from an observation belonging to that category and 0 otherwise. Values in the output range are static and will not update automatically if the source data in the input range changes

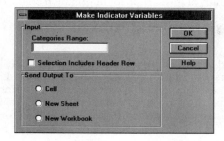

Categories Range

Enter the worksheet column containing the category levels from which you want to create indicator variables. You do *not* have to sort the category levels.

Selection Includes Header Row

Indicate whether the first row of each column in the input range contains header information.

See "Output Options" at the beginning of this section for information on the output storage options.

Make Two Way Table

The **Make Two-Way Table** command takes data arranged into three columns–a column of values, a column of category levels for one factor, a second column of category levels for a second factor–and arranges the data into a two-way table. The columns of the table consist of the different levels of the first factor; the rows of table consist of different levels of the second factor. Multiple values for each combination of the two factors show up in different rows within the table. Output from this command can be used in the Analysis ToolPak's ANOVA commands. Values in the output range are static and will not update automatically if the source data in the input range change. The number of rows in each of the three columns must be equal.

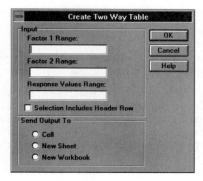

Factor 1 Range

Enter the worksheet column range containing the category levels for the first factor which will be placed into the columns of the output table.

Factor 2 Range

Enter the worksheet column range containing the category levels for the second factor which will be placed into the rows of the output table.

Response Values Range

Enter the worksheet column range containing the values for each combination of the two factors which will be placed in the cells of the output table.

Selection Includes Header Row

Indicate whether the first row of each column in the input range contains header information.

See "Output Options" at the beginning of this section for information on the output storage options.

1-Sample Wilcoxon

The **1-Sample Wilcoxon** command creates a table consisting of the 1-sample Wilcoxon test statistic for data from an input column. The first column of the table contains the titles of the descriptive statistics; the second column shows their values. Descriptive statistics include the number of values in the input range, the median, the number of positive values, the number of negative values, the number of zeroes, the sum of the positive ranks (adjusted for ties), the sum of the negative ranks (adjusted for ties), the sum of the squared ranks, the 1-tailed p-value, and the 2-tailed p-value. The p-value for the Wilcoxon test statistic will be exact if the number of rows in the input range is ≤ 16 and an approximate p-value will be used if the number of rows > 16. The test statistic is based on a null hypothesis that the median value of the input range is 0, versus an alternative hypothesis that the median is not zero. Values in the output range are static and will not update automatically if the source data in the input range changes.

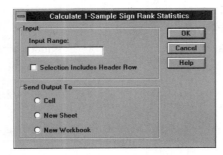

Input Range

Enter the worksheet column range containing data. Only one column is allowed.

Selection Includes Header Row

Indicate whether the first row of the column in the input range contains header information.

See "Output Options" at the beginning of this section for information on the output storage options.

Table Statistics

The **Table Statistics** command creates a table of descriptive statistics for a two-way cross-classification table. The first column of the table contains the titles of the descriptive statistics, the second column shows their values, the third column indicates the degrees of freedom, and the fourth column shows the p-value or asymptotic standard error. Test statistics included in the table include: Pearson's Chi-Square, Continuity Adjusted Chi-Square, Likelihood Ratio Chi-Square, Phi, Contingency, Cramer's V, Goodman-Kruskal Gamma, Kendall's tau-b, Stuart's tau-c, and Somer's D (C|R). Values in the output range are static and will not update automatically if the source data in the input range changes.

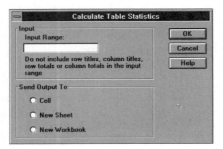

Input Range

Enter the worksheet range containing the counts of the cross classification table. The range must be contiguous and not contain row or column headers or totals.

See "Output Options" at the beginning of this section for information on the output storage options.

Means Matrix (1-way ANOVA)

The **Means Matrix (1-way ANOVA)** command creates a matrix of pairwise mean differences for the means of data arranged in different columns. The output includes a matrix of p-values with an option to adjust the p-value for the number of comparisons using the Bonferroni correction factor. Values in the output range are static and will not update automatically if the source data in the input range changes.

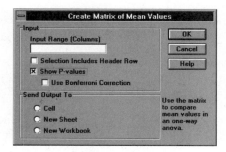

Input Range

Enter the worksheet range containing the data values arranged in different columns. The range need not be contiguous but the number of rows in each column must be equal.

Selection Includes Header Row

Indicate whether the first row of the column in the input range contains header information.

Show P-values

Click to create a matrix of p-values below the matrix of paired mean differences.

Use Bonferroni Correction

Click to correct p-values in the matrix of p-values for multiple comparisons using the Bonferroni correction factor.

See "Output Options" at the beginning of this section for information on the output storage options.

Correlation Matrix

The **Correlation Matrix** command creates a correlation matrix for data arranged in different columns. The correlation matrix can use either the Pearson correlation coefficient or the nonparametric Spearman rank correlation coefficient. You can also output a matrix of p-values for the correlation matrix. Values in the output range are dynamic and will update automatically if the source data in the input range changes.

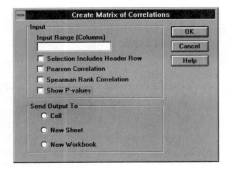

Input Range

Enter the worksheet range containing the data values arranged in different columns. The range need not be contiguous but the number of rows in each column must be equal.

Selection Includes Header Row

Indicate whether the first row of the column in the input range contains header information.

Pearson Correlation

Click to create a matrix of Pearson correlations.

Spearman Rank Correlation

Click to create a matrix of Spearman rank correlations.

Show P-values

Click to create a matrix of p-values for the correlation matrix.

See "Output Options" at the beginning of this section for information on the output storage options.

Boxplots

The **Boxplots** command creates a boxplot chart for data arranged in different columns. The user can specify whether to connect the boxes.

Input Columns

Enter the worksheet range containing the data values arranged in different columns. The range need not be contiguous and the number of rows in each column need not be equal.

Selection Includes Header Row

Indicate whether the first row of the columns in the input range contains header information.

Connect Medians Between Boxes

Click to connect the medians between the boxes with a straight line.

See "Output Options" at the beginning of this section for information on the output storage options.

Scatterplot Matrix

The **Scatterplot Matrix** command creates a matrix of scatterplots for data values arranged in columns. The scatterplot matrix is dynamic and will update automatically if the source data in the input range changes.

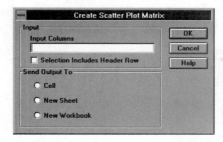

Input Columns

Enter the worksheet range containing the data values arranged in different columns. The range need not be contiguous but the number of rows in each column must be equal.

Selection Includes Header Row

Indicate whether the first row of the columns in the input range contains header information.

See "Output Options" at the beginning of this section for information on the output storage options.

Histogram

The Histogram command creates a histogram of data in a single column with a superimposed normal curve on the plot. The output includes the frequency table. You can specify the number of bins in the histogram. The histogram and frequency table are dynamic and will update automatically if the source data in the input range changes.

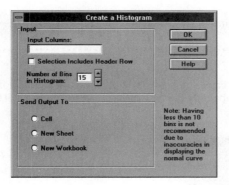

Input Columns

Enter the worksheet range containing the data values arranged in a single column.

Selection Includes Header Row

Indicate whether the first row of the column in the input range contains header information.

Number of Bins in Histogram

Enter the number of bins in the histogram. The bins will be equally spaced between the minimum and maximum value. Setting the number of bins less than 10 is not recommended due to inaccuracies in displaying the superimposed normal curve.

See "Output Options" at the beginning of this section for information on the output storage options.

Multiple Histograms

The **Multiple Histograms** command creates stacked histogram charts for data arranged in columns. The histograms have common bin values and are shown in the same vertical axis scale. A frequency table is also included in the output. The input range must be contiguous and the number of rows in each column must be the same (you can select blank rows to "fill-out" the columns if one column has fewer data values than another). The histograms and frequency tables are dynamic and will update automatically if the source data in the input range changes.

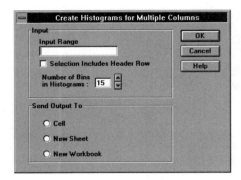

Input Range

The worksheet range containing the data values arranged in a contiguous selection. The columns may contain blank values.

Selection Includes Header Row

Indicating whether the first row of the columns in the input range contains header information.

Number of Bins in Histogram

The number of bins in the histogram. The bins will be equally spaced between the minimum and maximum value.

See "Output Options" at the beginning of this section for information on the output storage options.

Normal P-Plot

The **Normal P-Plot** command creates a normal probability plot with a table of normal scores for data in a single column. The normal probability plot and table are dynamic and will update automatically if the source data in the input range changes.

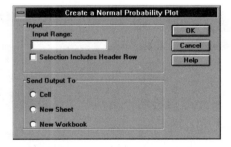

Input Range

Enter the worksheet range containing the data values arranged in a single column.

Selection Includes Header Row

Indicate whether the first row of the column in the input range contains header information.

See "Output Options" at the beginning of this section for information on the output storage options.

ACF Plot

The **ACF Plot** command creates a table of the autocorrelation function for time series data arranged in a single column and a chart of the autocorrelation function. The first column in the output table contains the lag values up to a number specified by the user, the second column contains the autocorrelation, the third column of the table contains the lower 95% confidence boundary and the fourth column contains the upper 95% confidence boundary. Autocorrelation values that lie outside the 95% confidence interval are shown in red for users with color monitors. The chart shows the autocorrelations and the 95% confidence interval. The ACF output is not dynamic and will not update if the source data in the input range changes.

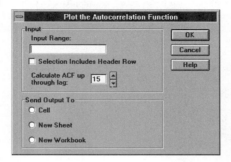

Input Range

Enter the worksheet range containing the time series values arranged in a single column.

Selection Includes Header Row

Indicate whether the first row of the column in the input range contains header information.

Calculate ACF up through lag

Click the spin arrows to specify the highest lag for which you want to calculate the ACF of the time series.

See "Output Options" at the beginning of this section for information on the output storage options.

Seasonal Adjustment

The **Seasonal Adjustment** command creates a column of seasonally-adjusted values for time series data that show periodicity and creates a plot of unadjusted and adjusted values. A plot of the seasonal indices is included in the output (multiplicative seasonality is assumed). The seasonal adjustment output is not dynamic and will not update if the source data in the input range changes.

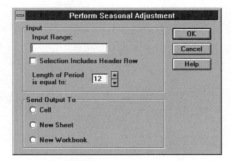

Input Range

Enter the worksheet range containing the time series values arranged in a single column.

Selection Includes Header Row

Indicating whether the first row of the column in the input range contains header information.

Length of period equal to

Specify the length of period.

See "Output Options" at the beginning of this section for information on the output storage options.

Exponential Smoothing

The **Exponential Smoothing** command calculates one, two or three parameter exponential smoothing models for a single column of time series data. You can forecast future values of the time series based on the smoothing model for a specified number of units and include a confidence interval of size α. The output includes a table of observed and forecasted values, future forecasted values, a table of descriptive statistics including the mean square error and final values of the smoothing factors. A plot of the seasonal indices (for three parameter exponential smoothing) is included. The exponential smoothing output is not dynamic and will not update if the source data in the input range changes.

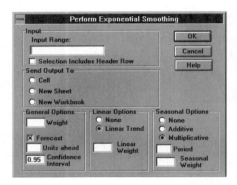

Input Range

Enter the worksheet range containing the time series values arranged in a single column.

Selection Includes Header Row

Indicate whether the first row of the column in the input range contains header information.

Weight

Enter the weight of the location smoothing parameter. It is set to 0.15 as a default. This value must be between 0 and 1.

Forecast

Click to indicate whether to forecast future values.

Units Ahead

Enter the number of observations you want to forecast ahead.

Confidence Interval

Enter the width of the confidence interval. A value of 0.95 is entered for the 95% confidence interval. The value must be between 0 and 1.

Linear Options: None

Click to omit a linear smoothing parameter from the smoothing model.

Linear Options: Linear Trend

Click to add a linear smoothing parameter to the smoothing model.

Linear Weight

Enter the weight of the linear smoothing parameter. It is set to 0.15 as a default. This value must be between 0 and 1.

Seasonal Options: None

Click to omit a seasonal smoothing parameter from the smoothing model.

Seasonal Options: Additive

Click to add an additive seasonal smoothing parameter to the smoothing model. *Not included with this release.*

Seasonal Options: Multiplicative

Click to add a multiplicative seasonal smoothing parameter to the smoothing model.

Period

Enter the length of the seasonal period.

Seasonal Weight

Enter the weight of the seasonal smoothing parameter. It is set to 0.15 as a default. This value must be between 0 and 1.

See "Output Options" at the beginning of this section for information on the output storage options.

XBAR Chart

The **XBAR Chart** command creates an xbar chart for quality control data arranged in multiple columns. The mean is the mean of the rows over the columns. The input range must be contiguous with the same number of rows in each column. The user can specify either a known sigma or an unknown sigma. The quality control chart includes a mean line and lower and upper control limits. The xbar chart output is not dynamic and will not update if the source data in the input range changes

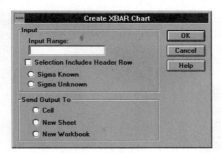

Input Range

Enter the worksheet range containing each sample of quality control data within a separate column. The columns must occupy a contiguous range with the same number of rows in each column.

Selection Includes Header Row

Indicate whether the first row of the columns in the input range contains header information.

Sigma Known

Click if the population standard deviation is known, then enter the value in the corresponding box.

Sigma Unknown

Click if the population standard deviation is not known.

See "Output Options" at the beginning of this section for information on the output storage options.

Range Chart

The **Range Chart** command creates a range chart for quality control data arranged in multiple columns. The range is the range of the rows over the columns. The input range must be contiguous with the same number of rows in each column. The quality control chart includes a mean line and lower and upper control limits. The range chart output is not dynamic and will not update if the source data in the input range changes.

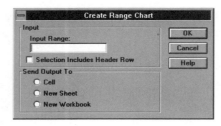

Input Range

Enter the worksheet range containing each sample of quality control data within a separate column. The columns must occupy a contiguous range with the same number of rows in each column.

Selection Includes Header Row

Indicate whether the first row of the columns in the input range contains header information.

See "Output Options" at the beginning of this section for information on the output storage options.

C-Chart

The **C-Chart** command creates a C-chart (count chart) of quality control data for a single column of counts (for example, the number of defects in an assembly line). The quality control chart includes a mean line and lower and upper control limits. The C-chart output is not dynamic and will not update if the source data in the input range changes.

Input Range

Enter the worksheet range containing a sample of counts arranged within a single column.

Selection Includes Header Row

Indicate whether the first row of the column in the input range contains header information.

See "Output Options" at the beginning of this section for information on the output storage options.

P-Chart

The **P-Chart** command creates a P-chart (proportion chart) of quality control data for a single column of proportions (for example, the proportions of defects in an assembly line). The quality control chart includes a mean line and lower and upper control limits. The P-chart output is not dynamic and will not update if the source data in the input range changes.

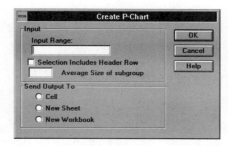

Input Range

Enter the worksheet range containing a sample of proportions arranged within a single column.

Selection Includes Header Row

Indicate whether the first row of the column in the input range contains header information.

Average Size of Subgroup

Enter the average subgroup size from which the proportions have been calculated.

See "Output Options" at the beginning of this section for information on the output storage options.

Insert ID Labels

The **Insert ID Labels** command is available from the shortcut menu when a series has been selected within a scatterplot chart. Click within the worksheet to activate it and drag the mouse over the range containing the labels to be added to the plot. The labels must be arranged within a single column. Then, specify whether the selected range includes a header row or not. A single point within the scatterplot series may also be selected and in that case only that selected point will have a label added to it. The number of rows in the labels column must equal the number of points in the chart series.

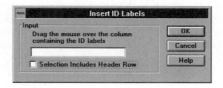

Instructional Templates

Instructional templates are special files included with the CTI disks to teach basic concepts of statistics. A template is an Excel file that is used as a pattern from which other files are created. When you open a template you do not open the source document but instead open a copy of that document. Instructional templates are self-contained and do not require the CTI Statistical Add-Ins to run properly. You do not have to save the copy of the template that you opened unless you want to preserve the final condition of the template for later use.

BOXPLOT.XLT

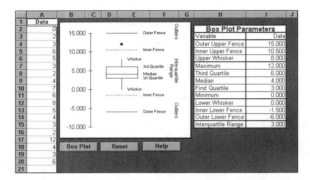

The **BOXPLOT.XLT** template provides information on the different parts of a boxplot. To run the template, enter a label for the data in cell A1 and place the data values in the rest of column A starting with cell A2. When you've completed entering the data, click the **Box Plot** button on the template sheet. The boxplot will then be generated. Changing any of the data values in column A will automatically modify the boxplot and boxplot parameters table. If you want to enter a new set of data, click the **Reset** button and reenter the new data.

CENTRAL.XLT

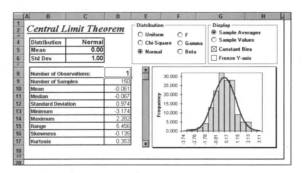

The **CENTRAL.XLT** template gives an interactive demonstration of the central limit theorem and the concept of the sampling distribution. To run the template, click one of six distribution option buttons (uniform, normal, chi-square, gamma, beta and F) and enter the parameters for the distribution in the dialog box that appears. The template then generates 150 random samples following that distribution with the sample size of each ranging from one to nine.

You can drag the scroll box in the template to change the number of observations per sample displayed from one to nine. You can specify whether to see the average of the first *n* observations for each of the 150 samples, or observation *n* from each sample (where *n* ranges from 1 to 9). Descriptive statistics of the sample averages or sample values are shown including the mean, standard deviation, minimum and maximum. You can specify whether to keep the x-axis constant for each histogram or the vary to x-axis and display each histogram on a different scale. You can freeze the y-axis scale or keep it constant from one histogram to another.

CONF.XLT

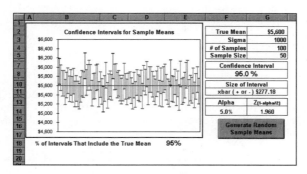

The **CONF.XLT** template gives an interactive demonstration of the principle of the confidence interval. The template shows 100 randomly generated samples from a population with $\mu = 5600$ and $\sigma = 1000$. The size of each sample is 50. The chart in the template shows the α % confidence interval for each of the 100 samples (where α is the confidence level set by the user and can vary from 0-100). Confidence intervals that cross the population mean are shown as green vertical lines; confidence intervals that do not cross the population mean are shown in red. You can regenerate the random samples by clicking the **Generate Sample Means** button on the template.

CPDF.XLT

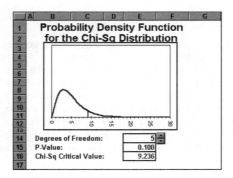

The **CPDF.XLT** template lets you change the degrees of freedom for the chi-square distribution and see how this modifies the shape of the distribution. Changing the degrees of freedom is accomplished by clicking the spin arrow on the template. The chart also displays the critical value for a user-specified *p*-value.

FPDF.XLT

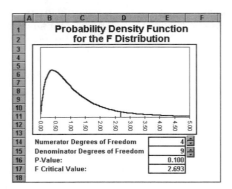

The **FPDF.XLT** template lets you change the numerator and denominator degrees of freedom for the F distribution and see how this modifies the shape of the distribution. Changing the degrees of freedom is accomplished by clicking the spin arrow on the template. The chart also displays the critical value for a user-specified p-value.

HYPOTH.XLT

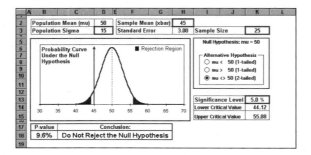

The **HYPOTH.XLT** template gives an interactive demonstration of the concept of hypothesis testing. The hypothesis test is based on a z-test where the null hypothesis mean $\mu = 50$ and the population variance is known to be $\sigma = 15$. A random sample of sample size 25 has been drawn from the population with a sample mean of 45. The null hypothesis is that $\mu = 50$, versus an alternative hypothesis that $\mu \neq 50$. The significance level is 5%. The chart in the template demonstrates the sampling distribution of the mean under the null hypothesis. Rejection regions are shown in black on the chart. You can change five parameters in the hypothesis test: the population standard deviation σ, the observed sample mean, the sample size, the significance level of the test and the alternative hypothesis (1-tailed or 2-tailed). The chart and conclusions update automatically as these five parameters are changed.

PDF.XLT

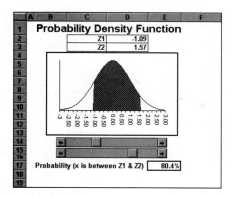

The **PDF.XLT** template gives an interactive demonstration of a probability density function and how the area under the pdf relates to probability. You can drag a scroll box to change the lower and upper boundaries of an input range. The probability of a random variable falling within that range is shown and the chart of the pdf is updated to show the input range as a shaded area.

SMOOTH1.XLT

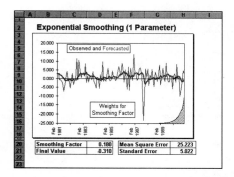

The **SMOOTH1.XLT** template interactively displays the effect of changing the smoothing parameter on the values for a one-parameter exponential smoothing model fit to the 1980's Dow Jones Average. You can change the smooth parameter from 0 to 1. The chart updates automatically, showing both the observed values and the smoothed values. A table of descriptive statistics showing the mean square error and standard error is also updated.

SMOOTH2.XLT

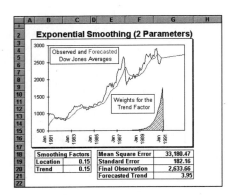

The **SMOOTH2.XLT** template interactively displays the effect of changing the smoothing parameters for location and trend on the values for a two-parameter exponential smoothing model fit to the 1980's Dow Jones Average. You can change the smooth parameters from 0 to 1. The chart updates automatically, showing both the observed values and the smoothed values and the forecasted values for the next 12 months. A table of descriptive statistics showing the mean square error, standard error and forecasted trend is also updated.

TARGET.XLT

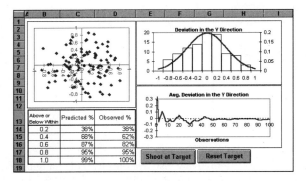

The **TARGET.XLT** template shows the relationship between a probability density function and a randomly generated sample of values following that pdf. This is using the example of a marksman shooting 100 times at a target. You can define the marksman's ability by choosing one of five levels of accuracy. The output shows the distribution of the shots around the target, a histogram of the deviation of the shots from the target in the vertical direction, a table of the distribution of shots about the target and a chart of the average vertical deviation from the target as the number of shots increases from 1 to 100.

TPDF.XLT

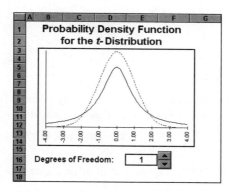

The **TPDF.XLT** template shows the relationship between the *t*-distribution and the standard normal distribution. The *t*-distribution is shown on a chart as a solid blue line; the standard normal as a dotted red line. Clicking a spin arrow on the template allows you to increase the degrees of freedom for the *t* from 1 to 30 and observe how the *t* closely follows the normal as the degrees of freedom increase.

Bibliography

Bliss, C. I. (1964). *Statistics in Biology*. New York: McGraw-Hill.

Booth, D. E. (1985). Regression methods and problem banks. *Umap Modules: Tools for Teaching 1985*, Arlington, MA: Consortium for Mathematics and Its Applications, 179–216.

Bowerman, B.L. and O'Connell, R.T. (1987). *Forecasting and Time Series, An Applied Approach*, Belmont, CA: Duxbury Press.

Cushny, A. R. and Peebles, A. R. (1905). The action of optical isomers, II: Hyoscines. *Journal of Physiology*, 32, 501–10.

D'Agostino, R. B., Chase, W., and Belanger A. (1988). The appropriateness of some common procedures for testing the equality of two independent binomial populations. *The American Statistician*, 42, 198–202.

Deming, W. E. (1982). *Quality, Productivity, and Competitive Position*. Cambridge, MA: MIT Center for Advanced Engineering Study.

Deming, W. E. (1982). *Out of the Crisis*. Cambridge, MA: MIT Center for Advanced Engineering Study.

Donoho, David and Ramos, Ernesto (1982), PRIMDATA: Data Sets for Use With PRIM-H (DRAFT), FTP stat library at Carnegue Mellon University.

Edge, O. P. and Friedberg, S. H. (1984). Factors affecting achievement in the first course in calculus. *Journal of Experimental Education*, 52, 136–40.

Fosback, N. G. (1987) *Stock Market Logic*, Fort Lauderdale, FL: Institute for Econometric Research.

Halio, M. P. (1990). Student writing: Can the machine maim the message? *Academic Computing*, January, 1990, 16–19, 45.

Juran, J. M. (1974), Editor. *Quality Control Handbook*. New York: McGraw-Hill.

Longley, J. W. (1967). An appraisal of least squares programs for the electronic computer from the point of view of the user. *Journal of the American Statistical Association*, 62, 819–31.

Neave, H. R. (1990). *The Deming Dimension*. Knoxville, TN: SPC Press.

Rosner, B. and Woods, C. (1988). Autoregressive modeling of baseball performance and salary data. *1988 Proceedings of the Statistical Graphics Section, American Statistical Association*, 132–37.

Shewhart, W. A. (1931). *Economic Control of Quality of Manufactured Product*. Princeton, NJ: D. Van Nostrand.

Tukey, J. W. (1987). *Exploratory Data Analysis*. Reading, MA: Addison-Wesley.

Weisberg, S. (1985). *Applied Linear Regression*, Second Edition. New York: Wiley.

Index

$ (dollar sign)
  cell references, 29–30
  currency style, 28
= (equals sign), 17–18
_ (underscore), 39

A

absolute reference, 29–30
absolute value (ABS), 19
acceptance region, 122
add-ins, 47–48. *See also* Analysis ToolPak Add-Ins
Advanced Filter, 32, 33–36
  and/or filters, 34
  calculated values, 35–36
  criteria formula requirements, 35
analysis of variance (ANOVA), 182
  assumptions of, 218, 233
  Bonferroni test, 225–228
  cells in, 231–232, 239–240
  computing, 221–222
  effects model, 228–230, 231
  formatting data, 236–237
  graphing data, 219–221, 234–236
  indicator variables, 228–230
  interaction plot, 232–233, 239–240
  interpreting, 182, 196–197, 222–224, 238–239
  means model, 218, 231
  models, 218, 228
  multiple comparisons, 218, 225–228
  one-way, 218–224
  outliers in, 221, 235
  replications, 231, 236
  robustness, 221
  standard deviation, 224
  statistics from, 224
  two-way, 231–240
Analysis ToolPak Add-Ins
  ANOVA commands, 221–222, 319–321
  commands, 319–332
  correlation matrix, 184, 321
  covariance command, 321–322
  descriptive statistics, 62–63, 322
  exponential smoothing, 322–323
  frequency table, 50
  f-test command, 323
  generating normal data, 100–102
  histograms, 48–50, 324
  moving average command, 324–325
  output options, 319
  and paired comparisons, 133–134
  random number generation, 325
  rank and percentile command, 327
  regression analysis, 178–183, 195–199, 327–328
  sampling command, 328–329
  t-test, 133–134, 329–331
  verifying availability of, 48, 319

z-test command, 331–332
applications, 3–4
arguments, 19
autocorrelation function (ACF), 251–255
  computing, 252–253, 254–255
  and constant variance, 252
  formulae, 251–252
  independent observations in, 254
  patterns of, 253–254
  random walk model, 255
  and seasonality, 272–273
AutoFill, 26
AutoFilter, 32–33
  adding second filter, 33
  gender filtering, 83–85
  and hidden data, 84
=AVERAGE, 106
average. *See* mean

B

bar charts, 151–152. *See also* histograms
bibliography, 352
bins
  changing intervals, 51–52
  values, 50, 110
Bold button, 15
Bonferroni approach, 185–186, 225–228
  vs. *t*-test, 226
  when to use, 226–227
boxplots, 53–62
  advantages of, 227–228, 235–236
  creating, 57–59
  display, 55–56
  outliers in, 55–56, 227–228
  template, 53–54
  and transformed data, 60–62
  uses of, 56, 60

C

calculated criteria, 32
categorical variables, 92, 146. *See also* qualitative variables
  charting, 146–155
  as factor, 231
  nominal, 146
  ordinal, 146
  relationships between, 152–155
causality, 175
C-chart, 297–299
cell reference box, 13
cells, 13
  active, 13–14, 74
  creating range name with reference box, 61
  formatting entries, 14–15
  Go To command, 17–18
  ranges of, 14

menu bar, 3, 10
    command shortcuts, 10
    opening, 10
Microsoft Office group icon, 9. *See also* Excel
minimize button, 3
minimum function, 52
mixed reference, 30
mode, 64
mouse
    pointer, 13, 21
    techniques, 4–5
moving averages, 255–257
    and exponential smoothing, 257–267
    and forecasting, 256–257
multiple comparisons, 218. *See also* analysis of variance
multiple correlation, 197
multiple regression, 192–216
    adding regression line, 213
    ANOVA table, 196–197
    assumptions, 193–194
        verifying, 200–205, 209–210
    coefficients in, 194, 198–199
        t test for, 198–199
    dependent vs. predicted values, 200–201
    descriptive statistics, 212
    determining fit, 196–197
    error estimate, 197
    examples
        predicting grades, 195–204
        sex discrimination, 205–214
    interpreting output, 196–197, 208, 213–214
    models, 192, 195
    and multiple correlation, 197
    normal probability plot, 204, 209–210
    performing, 196
    plotting results of, 199–204
    predicting with, 194–199
    predictor variables, selecting, 195–196
    residuals vs. predicted values, 201–202, 210–211
    residuals vs. predictor variables, 202–203, 209, 212
    SPLOM to determine relationships, 206–207

N

New Workbook button, 56
nominal variable, 25
nonparametric tests, 134–136, 177–178
normal distribution, 98–111
    characteristics of, 98–99
    determining fit, 99–113
    histogram of, 98–99
    and normal probability plot, 103–105
    and standard normal, 101–102
normal probability plot (PPlot), 103–105
    creating, 103–104
    interpreting, 104–105
    skewed data on, 105
normal scores, 104
null hypothesis, 120–121, 125. *See also* hypothesis testing

O

objects, 29
observations, 92
    independence of, 251
    multiple, 231

predicting with exponential smoothing, 257
Open button, 10–11
Open command, 10–11
Open dialog box, 12
ordinal variables, 25
    creating sort order for, 165–166
    determining relationships between, 163–164
    measures, 164
    as regression predictors, 206
    tables, 163–166
outliers, 55–56
    explaining, 79
    moderate, 56
    severe, 56
    and statistical validity, 55
overparametrization, 228, 231

P

paired comparisons, 129–136
    Analysis ToolPak Add-In, 133–134
    calculating t statistic, 131
    creating histogram of, 132
    creating matrix of, 225–226
    determining normality, 132–133
    planned vs. unplanned, 226–227
parametric tests, 134
Pareto chart, 302–305
    creating, 304–305
    interpreting, 305
    preparing data, 302–304
Paste Special, 84
P-chart, 299–301
Pearson Chi-Square Test, 156–159
    calculating p-value of, 158
    combining categories, 160–161
    degrees of freedom, 158
    formula, 158
    hiding categories, 160–161
    and small frequencies, 160–163
    validity of, 160–163
Pearson correlation, 175–177
Pearson, Karl, 156
percentages, calculating, 49–50
percentile function, 52
pie charts, 151
pivot tables
    creating, 147–149
    hiding blank rows, 149–150
    interaction plot, 232–233, 239–240
    missing values in, 149
    modifying, 150
    ordinal variables, 163–166
    percentages in, 150, 153–155
    restructuring for sparse data, 160–161
    selecting source data, 147
    two-way, 152–155
PivotTable Wizard, 147–149, 239–240
    choosing table structure, 148
points, 15
population, 96
    mean (μ), 97
    parameters, 97–98
    standard deviation (σ), 97
    standard normal, 101
predictor variables, 170
    correlations among, 186

F distribution, 192–193
    reducing number of, 198–199
Print button, 21
Print dialog box, 21–22
Print Preview button, 20–21
printer, selecting, 22
Printer Setup button, 22
probability density function, 93–98
    bivariate, 96
probability distributions, 91–115
    continuous random variables, 92–94
    and sample values, 94
    and stastical inference, 97
    uniform, 111–113
process, 285. *See also* statistical quality control
Program Manager, 3
p-value, 122
    adjusting with Bonferroni approach, 185–186

Q

qualitative variable, 25
    nominal, 25
    ordinal, 25
quality control, 284–307
quantitative variable, 25
querying. *See* filtering

R

random samples, 92
    calculating averages of, 106
    generating, 94–98, 100–102
random variables, 92
    continuous, 92
    extreme values, 122, 135
    discrete, 92
    probability distributions, 92–97
random walk model, 255
range, of data, 64
range chart, 288
    creating, 296–297
ranges, 14–15
    defining names, 28–29, 38
    deselecting, 15
    listing names, 29
    moving within, 26–27
    redefining, 30–31
    selecting, 14, 26–27
regression
    multiple, 191–216
    simple linear, 170–190
regression analysis, 178–183
    Adjusted R Square value, 180
    analysis of variance as, 228–230
    fitting effects model with, 229–230
    Multiple R value, 180
    Observations value, 180
    output, 179
    standard error, 180
    statistics, 180
regression equation, 170
rejection region, 122
relative reference, 29
Rename, 81
replicates, 231, 236

Reset button, 56
residuals, 170, 192
    vs. predicted values, 201–202
    vs. predictor variable, 181, 202–203
Restore button, 3
right-clicking, 4, 30
robustness, 129
rows
    headings, 13
    inserting new, 30–31
R Square value, 197. *See also* coefficient of determination (R^2)

S

sample
    mean (xbar), 97, 105–113
        confidence intervals, 117–120
        as estimator, 106–107
        precision of, 108
    from uniform distribution, 111–113
    size, 97, 106
        and hypothesis tests, 124
    standard deviation (*s*), 97
        as substitute for , 125–126
    two-sample comparisons, 137–141
sampling distribution, 106, 108
Save, 22–23
Save As, 22–23
saving work, 8, 22–23
scatterplots, 67–90
    adding linear regression line, 173–174
    activating, 73
    changing scale, 73–75
    creating, 68–72
    matrix, 85–89
        and correlation, 183–186
        creating, 86
        editing, 87–88
        grouping/ungrouping, 87–88
        individual charts in, 88
        in multiple regression, 206–207
        interpreting, 87
    moving averages, 255–257
    pasting in, 84
    sizing, 72
scroll bars, 15
scrolling, 12
    in worksheets, 15–16
seasonality, 245, 253–254. *See also* time series
    additive, 268
    adjusting for, 273–274
    autocorrelation function, 272–273
    creating boxplot, 270–271
    creating line plot, 271–272
    creating time series plot, 269–270
    multiplicative, 267–268
second quartile function, 52
Setup button, 21
sheets, 11, 13. *See also* Workbooks; Worksheets
    active, 13
    chart, 72, 79
    name tabs, 13
Shewhart, Walter A., 285
Shift Cells Down, 30
simulation, 101
skewness, 64
    determining, 60

on normal probability plot, 105
slope, 170, 171, 175
    and correlation, 176
smoothing constant, 257. *See also* exponential smoothing
    optimizing, 279–281
smoothing factor (*w*), 257. *See also* exponential smoothing
    choosing value for, 260
Solver, 279–281
Sort By, 31
sorting, 31
    creating custom sort orders, 165–166
spaghetti plots, 272
special causes, 285. *See also* statistical quality control
    identifying, 295–296
spin box, 70
SPLOM. *See* scatterplot matrix
spreadsheet, 8–9
standard deviation, 64–65
    definition, 46
standard error of the mean, 108, 118. *See also* sample, mean
standard normal population, 101
statistical inference, 92, 117–144
    and Central Limit Theorem, 112–113
    confidence intervals, 117–120, 125, 136, 141
    hypothesis testing, 120–125
    paired comparisons, 129–134
    t distribution, 125–137
    unknown σ, 125–129
statistical process control (SPC). See statistical quality control
statistical quality control (SQC), 285–307
    attributes, 288–289
    C-chart, 297–299
    control charts, 286–289
    definition, 285
    hypothesis testing, 288
    mean chart, 288–296
    Pareto chart, 302–305
    P-chart, 299–301
    range chart, 288, 296–297
    tampering, 285–286
    variables, 288–289
    variation types, 285–286
statistics, 46
    calculating, 52–53
    descriptive, table of, 62–63
    and context, 42
Stock Market Logic (Fosback), 256–257
Stuart tau-c, 164
Student Disk, 8
    saving work to, 22–23
Student Files
    accessing, 6–8
    ACCID.XLS, 297–299
    BASE.XLS, 98–109
    BEER.XLS, 268–278
    BIGTEN.XLS, 68–81, 171–178
    BOXPLOT.XLT, 54
    CALC.XLS, 184–188, 195–204
    COAT.XLS, 294–297
    COLA.XLS, 233–240
    copying, 6–7
    creating subdirectory for, 7–8
    DISCRIM.XLS, 205–214
    DOW.SLX, 245–268
    EXPSOLVE.XLS, 279–281
    4JRCOL.XLS, 82–89
    HOTEL.XLS, 218–230

    list of, 309–311
    PARK.XLS, 12–23
    PCINFO.XLS, 46–53
    PCPAIR.XLS, 129–136
    POWDER.XLS, 302–305
    SPACE.XLS, 137–141
    STEEL.XLS, 300–301
    TEACH.XLS, 290–292
    WBUS.XLS, 57–65
    WHEAT.TXT, 36–42
sum of squares, 222–224
    between-groups, 223
    error, 223
    within-groups, 223

T

tables, 145–166
    bar charts, 151–152
    nominal variables, 146–162
    ordinal variables, 163–166
    pie charts, 151
    pivot, 146–150, 152–155
    statistics, 159
    two-way, 152–155
tampering, 285–286
t distribution, 125–141
    degrees of freedom, 126
    one-sample, 126–127
    robustness, 129
    TINV function, 136
    two-sample, 127–129, 136
    vs. standard normal, 126
templates
    BOXPLOT.XLS, 53–54, 346
    CENTRAL.XLT, 109–113, 346–347
    CONF.XLT, 119–120, 347
    CPDF.XLT, 155–156, 347
    FPDF.XLT, 192–193, 348
    HYPOTH.XLT, 121–125, 348
    list of, 346–351
    PDF.XLT, 92–94, 349
    SMOOTH1.XLT, 258–260, 349
    SMOOTH2.XLT, 265–267, 350
    TARGET.XLT, 94–98, 350
    TPDF.XLT, 126–129, 351
text files, importing data from, 36–38
Text Import Wizard, 37–38
third quartile function, 52
time series, 244–283
    assumptions, 245
    autocorrelation function (ACF), 251–255
    calculating correlations, 263–264
    creating forecasts, 262–264
    cyclical, 253–254
    definition, 245
    descriptive statistics, 249
    exponential smoothing, 257–267
    forecasted vs. observed values, 262
    graphing data, 245–248
    lagged values, 249–251
    moving averages, 255–257
    oscillating, 253–254
    patterns of, 253–254
    random, 253–254
    random walk model, 255
    seasonality, 267–280

transformations in, 248
trend, 253–254
variance in, 247–248
toolbars
buttons in, 10–11
buttons vs. menu commands, 21
customizing, 11
Formatting, 10
Standard, 10–11
transformations, 60–62
logarithmic, 61, 245
percentage change, 248
trend parameter, 265
trends, avoiding, 233
t statistic
calculating, 127, 128
and confidence intervals, 136
for coefficients, 198–199
formulas, 128
determining significance, 127, 128–129, 134
for paired difference, 131
robustness, 129, 132
Tukey, John, 53
two-sample comparisons, 137–141
calculating confidence interval, 141
calculating standard error, 141
comparing distributions, 138
creating histograms for, 137–138
t statistic, 139–140
t-test, 138–140
variance in, 139–140
TXT file extension, 36

U

underscore (_), 39
uniform distribution, 111–113
sampling distribution of mean, 111–112
upper control limits (UCL), 286

V

variables
categorical, 146
continuous, 92
creating new, 39–40
descriptive statistics for multiple, 313
discrete, 92
eliminating nonsignificant, 199
functions for one, 312–313
graphing multiple, 68–90
graphing single, 45–66
grouping, 82–85
nominal, 146
ordinal, 146
random, 92
relationships between, 68
types, 24
variability, 64–65
in hypothesis testing, 124

and sample size, 97
variance, 65. *See also* analysis of variance
assumption of constant, 171, 180–181
with limited dependent variable, 201
variation
controlled, 285–286
uncontrolled, 285–286

W

Wilcoxon Sign Rank test, 134–136
Windows, Microsoft
elements of, 3–4
Help, 5–6
Program Manager, 3–6
starting, 2–3
tutorial, 5
windows, 3
resizing, 4–5, 10
Winter's method, 275. *See also* exponential smoothing
wizards, 37. *See also* Chart Wizard; Text Wizard
Workbooks, 11–13
opening, 11–12
scrolling, 16–17
sheets in, 13
Worksheets, 13–17
chart linkage, 50
entering data, 14, 25–44
formatting cell entries, 14–15
freeze panes, 41–42
headers and footers, 21
paging through, 17
previewing, 20–21
printing, 20–22
scrolling, 15–16
viewing parts simultaneously, 41
window, 13

X

χ^2 distribution, 163
x-axis
rescaling, 73–74
selecting data for, 70, 72
$\bar{x}$ xbar chart. See mean chart
XLS extension, 12, 23

Y

y-axis, 72
plotting multiple variables, 79–81
rescaling, 74–75

Z

Zoom Control, 73
z-scores, 102, 122